Solar Evaporation for Water-Energy-Resource Nexus

Guohua Liu • Ting Chen

Solar Evaporation for Water-Energy-Resource Nexus

Springer

Guohua Liu
Main Building
North China Electric Power University
Beijing, China

Ting Chen
School of Optoelectronic Materials and Technology
Jianghan University
Wuhan, China

ISBN 978-981-95-3814-0 ISBN 978-981-95-3815-7 (eBook)
https://doi.org/10.1007/978-981-95-3815-7

This Springer imprint is published by the registered company Springer Nature Singapore Pte Ltd.
The registered company address is: 152 Beach Road, #21-01/04 Gateway East, Singapore 189721, Singapore

Preface

In an era fraught with environmental challenges and growing demands on natural resources, pursuing sustainable solutions is imperative. This book delves into one of the most promising areas of sustainability: integrating solar interfacial evaporation within the water-energy-resource nexus. It offers insights into leveraging solar energy to meet critical needs for clean water, sustainable energy, and effective resource management.

Water bodies, covering over two-thirds of Earth's surface, offer immense opportunities for clean water sourcing, sustainable energy, and resource extraction. This book reviews recent advances in solar interfacial evaporation that stands at the confluence of these critical areas, and discusses its potential to drive economic growth and societal development. We begin with a primer on the fundamental principles that underpin solar interfacial evaporation, including photothermal conversion, heat and mass transport, and phase transition dynamics. A substantial portion of the discussion is dedicated to the exploration of innovative materials and structural designs that enhance the efficiency of energy conversion. We explore material selection strategies and structure design considerations focusing on energy management, anti-fouling, and steam handling.

We then explore a variety of simulation and modeling techniques in solar evaporation to capture both the macro and micro dynamics of the process, supporting the multifaceted design and optimization of the evaporator. The applications of solar interfacial evaporation are diverse and impactful. For clean water production, we discuss scalable and economical methods like solar desalination, pollutant degradation, and antibacterial disinfection to address water scarcity and contamination. In sustainable energy, the technology's role in energy storage, electricity generation, and fuel production contributes to a renewable energy framework. Additionally, the extraction of resources such as salt, lithium, and uranium, alongside enhanced food production through solar processes, are highlighted as innovative resource management and agricultural strategies. The book also tackles current challenges and future directions in the field, aiming to deepen understanding and spur ongoing innovation among researchers.

We hope this book serves as a valuable guide for those interested in the forefront of solar evaporation technology and its role in enhancing sustainability across various sectors. We encourage readers to engage with the ideas presented and consider how best to utilize this promising technology for a sustainable future.

Beijing, China Guohua Liu
Wuhan, China Ting Chen
2025-8-8

Contents

Chapter 1
Introduction

Water, energy, and resources play a crucial role in achieving sustainability, stimulating economic growth, and promoting societal development [1, 2]. Covering over two-thirds of Earth's surface, water bodies hold immense potential for producing clean water, sustainable energy, and metal resources [3]. Clean water, essential for human survival and development, is potable water sourced without health risks or contaminants. Despite the vast appearance of global water sources such as seas and rivers, usable clean water remains scarce [4, 5]. Only 3% of the world's water is fresh, with the rest trapped in oceans or locked in icecaps and glaciers, leaving less than 1% available for human use. This restricted supply satisfies almost 80% of the world's demand of clean water. Presently, 2.1 billion people encounter shortages of safe water, and by 2025, it is anticipated that more than half of the global population will experience significant water stress (Fig. 1.1) [6].

The scarcity of clean water is exacerbated by social development, climate change, and pollution. Rapid population growth, urbanization, and industrialization deplete local water resources, as seen in China, where per capita water availability is expected to fall below 500 m^3 by 2030 [7]. Climate change exacerbates water scarcity through alterations in the hydrological cycle, leading to more frequent and severe droughts and floods. For example, only 6.1% of the national runoff occurs in southern regions of Australia where most of urban cities and population are located [8]. Furthermore, water pollution, particularly from heavy metals, further strains clean water supplies, with countries like Bangladesh facing acute crises due to arsenic contamination. On the other hand, millions of people in the U.S. are at risk of consuming drinking water contaminated with heavy metals such as lead, arsenic, and mercury [9]. Therefore, it is of great urgency to implement cost-effective and sustainable water technologies to ensure access to clean water, green energy and valuable resources.

Solar energy emerges as a promising source due to its safety, cleanliness, renewability, affordability, and widespread availability on Earth [10, 11]. Figure 1.2 shows

G. Liu, T. Chen, *Solar Evaporation for Water-Energy-Resource Nexus*,
https://doi.org/10.1007/978-981-95-3815-7_1

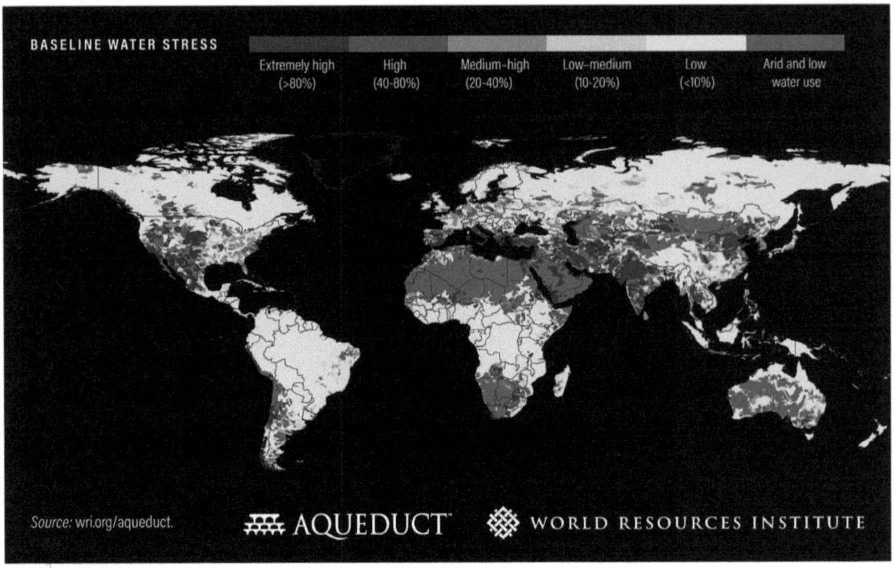

Fig. 1.1 Global water stress. (Source: wri.org/aqueduct)

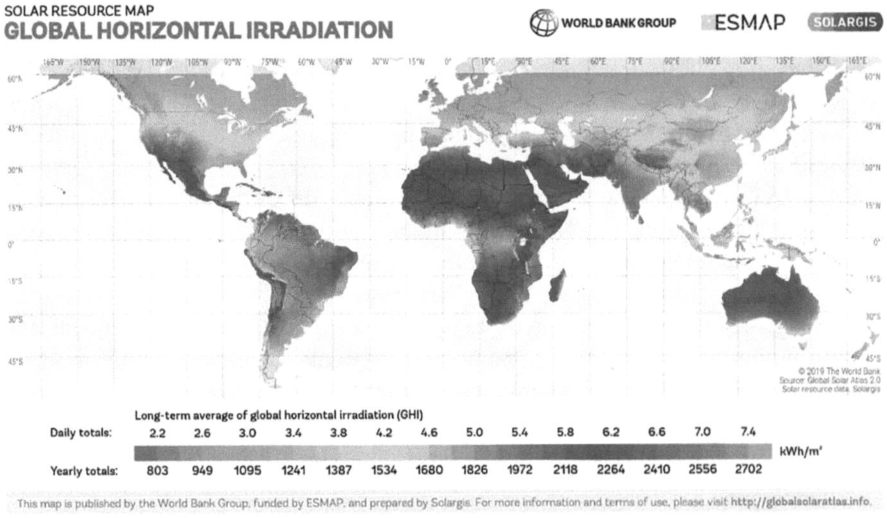

Fig. 1.2 Global annual average solar heat. (Source: global solar atlas)

the global annual average of direct solar irradiation. Utilizing solar energy for water desalination offers an optimal solution for producing both salt and fresh water without requiring external energy inputs [12, 13]. Solar evaporation is a well-established method for seawater desalination. Enhancing this process involves using a solar evaporator placed on the water surface to capture solar energy and localize it for the evaporation of water molecules [14, 15]. This method, known as solar interfacial evaporation, achieves high solar-to-vapor conversion efficiency, independent of water volume, and is adaptable for various applications [16–18]. After a decade of extensive research, this technology has reached its maximum thermodynamic efficiency of 92.5%, with fresh water production rates exceeding 80 liters per square meter per hour under concentrated sunlight, and it operates stably for over 600 h.

Solar evaporation offers versatile applications across the water-energy-resource nexus (Fig. 1.3) [2, 19]. This method can produce potable water from a variety of sources, including seawater, brackish water, and industrial wastewater, with the processed water being suitable for domestic use or for energy and resource production [20]. It can effectively treat wastewater, including domestic sewage and industrial effluents from sectors like textiles, thermal power plants, and pharmaceuticals, achieving zero liquid discharge by fully separating water from solutes [21]. Additionally, the hot vapor derived from solar evaporation can be used for disinfection and sterilization [22]. Solar evaporation can also be combined with other energy technologies to enhance evaporation rates, electricity production, and overall system efficiency, thereby improving practicality and viability [23]. For example, this technology can convert various forms of energy, including waste energy, supporting both water purification and electricity generation through hydrovoltaics [24]. Moreover, liquid water/vapor serves as a reactant for hydrogen fuel, enabling hydrogen production via photocatalysis or photovoltaic electrolysis [25, 26]. Effectively

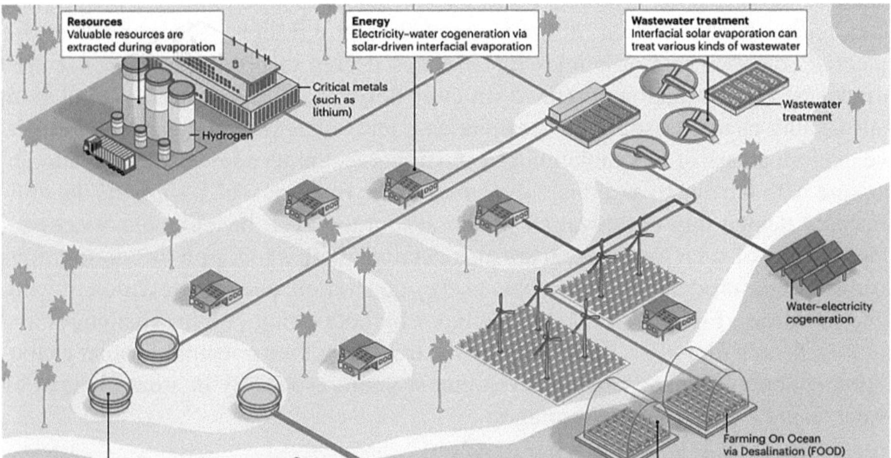

Fig. 1.3 Solar evaporation within the water-energy-resource nexus for producing clean water, sustainable energy and valuable resources. (Adapted with permission from [2] Copyright 2025 Nature)

harnessing these energy sources has the potential to usher in a new era in power generation, energy storage, and solar fuel production within the renewable energy sector.

Furthermore, seawater and wastewater are reservoirs of recyclable resources, such as organic compounds and valuable metals. Exploiting these resources in conjunction with water treatment offers substantial potential [17, 27]. Heavy metals including copper, nickel, and zinc can be recovered from wastewater using techniques like electrolysis, ion exchange, and chemical precipitation [28, 29]. However, these traditional methods are characterized by high energy consumption and large carbon footprints. The advancement of solar evaporation enhances resource recovery from wastewater, despite challenges like salt accumulation which impedes efficiency [30, 31]. By ongoing exploration and innovative efforts, solar evaporation has the potential to transform wastewater into valuable renewable resources, fostering a mutually benefits that promotes both sustainable development and environmental preservation [32, 33]. With precise scientific treatment and cutting-edge technologies, these resources can be efficiently repurposed, fulfilling the intertwined goals of resource recycling, environmental protection, and carbon neutrality [34, 35].

In this book, we explore recent advancements in solar evaporation within the water-energy-resource nexus, highlighting its effectiveness in generating clean water, sustainable energy, and valuable resources (Fig. 1.3). The discussion begins with an analysis of the fundamental physics, including photothermal conversion, heat and mass transfer dynamics, and phase change mechanisms. We then devote attention to the exploration of material and structural designs, offering criteria for material selection and principles for effective design to boost energy conversion efficiency. The evaluation of refined structure designs focuses on energy regulation, anti-fouling characteristics, and steam release and condensation dynamics. Further, we explore a variety of simulation and modeling techniques in solar evaporation to capture both the macro and micro dynamics of the process, which supports the multifaceted design and optimization of the evaporator. The benefits and successful applications of solar evaporation are also detailed, with a focus on its contributions to clean water production, sustainable energy generation, and the extraction of valuable resources. Finally, this book concludes by addressing current challenges and future research directions, underscoring the transformative potential of solar evaporation in supporting sustainable development goals, especially in areas plagued by water, energy, and resource shortages.

References

1. S. Chu, A. Majumdar, Opportunities and challenges for a sustainable energy future, ***Nature***, 488 (2012) 294-303.
2. Y. Song, S. Fang, N. Xu, J. Zhu, Solar-driven interfacial evaporation technologies for food, energy and water, ***Nature Reviews Clean Technology***, 1 (2025) 55-74.

3. P.E.C. Edited by Shiva Gorjian, Solar Energy Advancements in Agriculture and Food Production Systems, ***Academic Press***, (Copyright © 2022 Elsevier Inc.) ISBN: 978-970-323-89866-89869.
4. M.S. Irshad, N. Arshad, G. Maqsood, I. Ahmed, B. Shakoor, M.S. Asghar, U. Ghazanfar, L. Lin, M.A.K.Y. Shah, I. Ahmed, X. Chen, J. Wang, C. Yi, J. Li, J. Qian, W. Li, Z. Said, H. Li, N.X. Ho, H. Wang, X. Fan, N. Mushtaq, V.-D. Dao, T. Mei, X. Wang, Advancing water collection efficiency in hybrid solar evaporators: key factors, strategic innovations, and synergistic applications, ***Materials Science and Engineering: R: Reports***, 165 (2025) 101018.
5. A. Chu, S. Zhang, J. Jin, Recent Progress on Solar-Driven Interfacial Evaporation for Resource Recovery and Pollutant Removal, ***Advanced materials***, (2025) e2505656.
6. X. Wu, Y. Lu, X. Ren, P. Wu, D. Chu, X. Yang, H. Xu, Interfacial Solar Evaporation: From Fundamental Research to Applications, ***Advanced materials***, 36 (2024) e2313090.
7. S. Piao, P. Ciais, Y. Huang, Z. Shen, S. Peng, J. Li, L. Zhou, H. Liu, Y. Ma, Y. Ding, P. Friedlingstein, C. Liu, K. Tan, Y. Yu, T. Zhang, J. Fang, The impacts of climate change on water resources and agriculture in China, ***Nature***, 467 (2010) 43-51.
8. Z.K. B Bates, S Wu, Climate change and water, ***INTERGOVERNMENTAL PANEL ON CLIMATE CHANGE /IPCC***, (2008) 214.
9. S. Chowdhury, M.A.J. Mazumder, O. Al-Attas, T. Husain, Heavy metals in drinking water: Occurrences, implications, and future needs in developing countries, ***The Science of the total environment***, 569-570 (2016) 476-488.
10. P. Zhang, J. Li, S. Xing, Q. Jia, J. Wang, J. Bi, Z. Guo, L. Wang, Z.Y. Ji, L. Qu, Solar Evaporators for Saline Water: Sustainable Clean Water Harvesting and Critical Mineral Resources Extraction, ***ACS nano***, 19 (2025) 11625-11647.
11. X. Wang, J. Gao, Y. Wang, Y. Liu, X. Liu, L. Zhang, Over 12% efficiency solar-powered green hydrogen production from seawater, ***Energy & Environmental Science***, (2025).
12. P. Tao, G. Ni, C. Song, W. Shang, J. Wu, J. Zhu, G. Chen, T. Deng, Solar-driven interfacial evaporation, ***Nature Energy***, 3 (2018) 1031-1041.
13. M. Gao, L. Zhu, C.K. Peh, G.W. Ho, Solar absorber material and system designs for photothermal water vaporization towards clean water and energy production, ***Energy & Environmental Science***, 12 (2019) 841-864.
14. C. Dang, Y. Cao, H. Nie, W. Lang, J. Zhang, G. Xu, M. Zhu, Structure integration and architecture of solar-driven interfacial desalination from miniaturization designs to industrial applications, ***Nature Water***, 2 (2024) 115-126.
15. C. Chen, Y. Kuang, L. Hu, Challenges and Opportunities for Solar Evaporation, ***Joule***, 3 (2019) 683-718.
16. N. Xu, J. Li, C. Finnerty, Y. Song, L. Zhou, B. Zhu, P. Wang, B. Mi, J. Zhu, Going beyond efficiency for solar evaporation, ***Nature Water***, 1 (2023) 494-501.
17. P. Shi, J. Li, Y. Song, N. Xu, J. Zhu, Cogeneration of Clean Water and Valuable Energy/ Resources via Interfacial Solar Evaporation, ***Nano letters***, 24 (2024) 5673-5682.
18. T. Ding, Y. Zhou, W.L. Ong, G.W. Ho, Hybrid solar-driven interfacial evaporation systems: Beyond water production towards high solar energy utilization, ***Materials Today***, 42 (2021) 178-191.
19. L. Li, X. Sun, J. Miao, H. Wang, Y. Song, D. Tang, Gradient engineering in interfacial evaporation for water, energy, and mineral harvesting, ***Energy & Environmental Science***, (2025).
20. G.W. Ho, Y. Yamauchi, L. Hu, B. Mi, N. Xu, J. Zhu, P. Wang, Solar evaporation and clean water, ***Nature Water***, 3 (2025) 131-134.
21. L. Cui, P. Wang, H. Che, J. Chen, B. Liu, Y. Ao, Solar-driven interfacial water evaporation for wastewater purification: Recent advances and challenges, ***Chemical Engineering Journal***, 477 (2023) 147158.
22. L. Wang, Y. Feng, K. Wang, G. Liu, Solar water sterilization enabled by photothermal nanomaterials, ***Nano Energy***, 87 (2021) 106158.
23. B. Shao, Y. Song, Z. Song, Y. Wang, Y. Wang, R. Liu, B. Sun, Electricity Generation from Phase Transitions between Liquid and Gaseous Water, ***Advanced Energy Materials***, 13 (2023).

24. Q. Hu, X. Lin, G. Ren, J. Lü, W. Wang, D. Zhang, S. Zhou, Hydrovoltaic electricity generation induced by living leaf transpiration, ***Nature Water***, 2 (2024) 988-998.
25. M. Gao, P.K.N. Connor, G.W. Ho, Plasmonic photothermic directed broadband sunlight harnessing for seawater catalysis and desalination, ***Energy & Environmental Science***, 9 (2016) 3151-3160.
26. W.H. Lee, C.W. Lee, G.D. Cha, B.H. Lee, J.H. Jeong, H. Park, J. Heo, M.S. Bootharaju, S.H. Sunwoo, J.H. Kim, K.H. Ahn, D.H. Kim, T. Hyeon, Floatable photocatalytic hydrogel nanocomposites for large-scale solar hydrogen production, ***Nature nanotechnology***, 18 (2023) 754-762.
27. S.B. Darling, The brine of the times, ***Science***, 385 (2024) 1421-1422.
28. Z. Li, C. Li, X. Liu, L. Cao, P. Li, R. Wei, X. Li, D. Guo, K.-W. Huang, Z. Lai, Continuous electrical pumping membrane process for seawater lithium mining, ***Energy & Environmental Science***, 14 (2021) 3152-3159.
29. Z. Li, I.C. Chen, L. Cao, X. Liu, K.W. Huang, Z. Lai, Lithium extraction from brine through a decoupled and membrane-free electrochemical cell design, ***Science***, 385 (2024) 1438-1444.
30. Y. Song, S. Fang, N. Xu, M. Wang, S. Chen, J. Chen, B. Mi, J. Zhu, Solar transpiration-powered lithium extraction and storage, ***Science***, 385 (2024) 1444-1449.
31. J. Ren, T. Gu, S. Ma, X. Li, Z. Zhou, D. Hao, K. Fang, S. Tian, Key factor in continuous salt harvesting via solar interfacial evaporation: Water supply to evaporation ratio, ***Desalination***, 607 (2025) 118800.
32. K. Mao, Y. Zhang, S.C. Tan, Functionalizing solar-driven steam generation towards water and energy sustainability, ***Nature Water***, (2025).
33. Y. Wu, C. Ma, K. Zhu, L. Jin, L. Song, L. Li, Y. Lu, Y. Zheng, Y. Zhang, X. Zheng, S. Wu, Y. Pang, Z. Shen, S.C. Tan, H. Chen, Asymmetric evaporation for efficient and simultaneous extraction of freshwater, salt, and electrical energy from seawater, ***Energy & Environmental Science***, 17 (2024) 9303-9312.
34. J. Wu, G. Yin, J. Liu, Z.Z. Yu, X. Li, Multifunctional solar-driven interfacial evaporation system for simultaneous clean water production and high-value-added ion extraction, ***Materials horizons***, 12 (2025) 2878-2898.
35. W. Li, Y. Xu, K. Liu, L. Zhu, T. Xu, G. Wang, C. Si, Engineered Biomass-Based Solar Evaporators for Diversified and Sustainable Water Management, ***Advanced materials***, (2025) e2503658.

Chapter 2
Basic Physics

Solar evaporation is a process where solar energy is harnessed by photothermal nanomaterials to drive water evaporation at the liquid-vapor interface [1, 2]. The photothermal effect, a representative phenomenon of light-matter interaction, is characterized by a temperature increase in a material due to light absorption [3]. Different materials exhibit varying light-to-heat conversion efficiencies, depending on their electronic or bandgap structures' responses to electromagnetic radiation [4]. The phase change process relies on the absorption of solar radiation by the water surface, which increases the kinetic energy of water and solute molecules, causing their transition from a liquid to a vapor state or solid crystal [5–7]. This section will provide a concise overview of the basic physics underlying the solar evaporation process.

2.1 Light Transfer and Harvesting

In the scenario where photothermal materials are dispersed within homogeneously semitransparent media, the absorption of light occurs as a cumulative process characterized by exponential decay [8, 9]. This phenomenon adheres to Beer-Lambert's law, which is mathematically expressed as $I = I_0 e^{-\kappa cl}$ [10]. Here, I and I_0 represent the light intensities after and before absorption, respectively. κ is the extinction coefficient, c is the particle concentration, and l is the optical path length. To quantify the light absorbed by the material, absorbance a is introduced as the attenuation of the incident radiation power, which can be derived from the transmittance T as follows [4]:

$$a = -\ln T = \ln \frac{I_0}{I} = \kappa c l \tag{2.1}$$

G. Liu, T. Chen, *Solar Evaporation for Water-Energy-Resource Nexus*,
https://doi.org/10.1007/978-981-95-3815-7_2

Absorbance is proportional to both the extinction coefficient, an intrinsic property of the absorber, and the absorber's concentration per unit area. The extinction coefficient is influenced by factors such as the material, size, and shape of the absorber.

When solar radiation strikes a solar absorber's surface, the incoming energy can be reflected, transmitted, or absorbed. Absorption, representing the net inflow of energy, can be quantified by [11]:

$$q_{\text{absorption}} = q_{\text{solar}} - q_{\text{reflection}} - q_{\text{trasmission}} \tag{2.2}$$

The light-harvesting capability of a material is indicative of its efficiency in absorbing the energy from incident photons. The absorptivity A of a solar absorber is the fraction of solar radiation it absorbs, reflectivity R is the fraction it reflects, and transmissivity T is the fraction it transmits [12]. The energy absorbed by a material is contingent on its absorption spectrum across the range of incident light wavelengths and the intensity of absorbance at each wavelength. Solar radiation exhibits a spectrum similar to that of a 5778 K blackbody, with deviations primarily due to atmospheric absorption and scattering. The bulk of this radiation's energy is distributed across the ultraviolet (~3%), visible light (~45%), and infrared (~52%) bands, spanning wavelengths from 300 to 2500 nm [10, 13]. The assessment of a material's absorption, reflectivity, and transmissivity should be conducted within this spectral range. To compute absorptance, the absorption intensity is integrated over the spectral range. The overall absorptance at an incidence angle θ is then calculated as the ratio of the total absorbed light energy to the incident radiation [5, 14]:

$$A(\theta) = \frac{\int_{\lambda_{\min}}^{\lambda_{\max}} \left[1 - R(\theta,\lambda) - T(\theta,\lambda)\right] P(\theta,\lambda)\, d\lambda}{\int_{\lambda_{\min}}^{\lambda_{\max}} P(\theta,\lambda)\, d\lambda} \tag{2.3}$$

$P(\lambda)$ denotes the power of radiation that is dependent on the wavelength, while $\lambda_{\max}$ and $\lambda_{\min}$ indicate the maximum and minimum light wavelengths, respectively. $R(\theta, \lambda)$ and $T(\theta, \lambda)$ represent the reflectance and transmittance at the wavelength. As such the absorbed energy can be derived by:

$$q_{\text{absorption}} = A q_{\text{solar}} \tag{2.4}$$

Therefore, to enhance the absorptance, strategies should focus on minimizing both the reflectance and transmittance. Moreover, achieving broad-spectrum absorption is essential for maximizing the light-to-heat conversion [15, 16]. This is critical for the optimal performance of light absorbers in diverse applications.

2.2 Photothermal Conversion

The interaction of electromagnetic radiation with materials can be classified into three primary mechanisms: surface plasmon resonance (SPR) [17–19], non-radiative relaxation of electron-hole pairs [20, 21], and molecular thermal vibrations [22, 23]. This section offers a concise overview of these mechanisms that enable the conversion of light to heat. We then present fundamental equations essential for understanding the light-to-heat conversion process.

2.2.1 Basic Mechanism

2.2.1.1 Surface Plasmon Resonance

Photothermal conversion on metallic surfaces is predominantly driven by the SPR effect [2, 22]. This phenomenon occurs when the real part of a material's dielectric constant is negative and the imaginary part is small but positive, a condition commonly found in metals with a high density of free electrons on their surfaces. The SPR effect results from the coherent oscillation of surface electrons induced by resonant photons, which happens when the photon frequency aligns with the intrinsic frequency of the surface electrons (Fig. 2.1a) [5]. Two primary models describe the SPR effect: SPR and Localized Surface Plasmon Resonance (LSPR) [18, 24]. To quantitatively describe these phenomena, the concept of plasmon excitations is introduced. These excitations are quasi-particles, similar to magnetic vibrators and phonons, and they occur at any interface where there is a sign difference in the real parts of the complex dielectric constants. In SPR, when the frequency of light

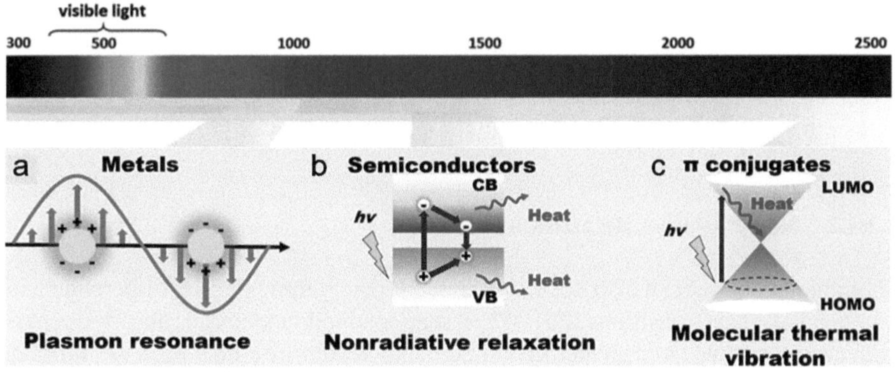

Fig. 2.1 Photothermal physics with different matters in the corresponding light-harvesting range. (**a**) plasmonic heating, (**b**) photoexcitation and relaxation of charge carriers, and (**c**) Thermal vibration of molecules. (Reproduced with permission from [5] Copyright 2024, Royal Society of Chemistry)

matches the resonance frequency of the particles, plasmonic excitons can propagate tens to hundreds of micrometers along the *x–y* plane at the metal-dielectric interface, while decaying exponentially in the *z*-axis direction within approximately 200 nm. In contrast, the LSPR model applies when the particle size is significantly smaller than the wavelength of the incident light. In this case, due to the small structure, plasmonic excitons on the particle surface are spatially confined and oscillate collectively around the nanoparticles at a specific frequency when excited by the incident light, leading to the spatially confined surface plasmonic exciton resonance effect known as LSPR [17].

In metal nanomaterials, when the photon frequency aligns with the intrinsic frequency of surface electrons, electromagnetic radiation excites conduction electrons, triggering coherent oscillations and activating the LSPR effect [17]. This effect leads to near-field enhancement, hot-electron generation, and photothermal conversion, resulting in significant light absorption and efficient transfer of photothermal energy to the surrounding medium through non-radiative dissipation [25, 26]. Initially, when a nanoparticle is irradiated at its resonant wavelength, it stimulates oscillations within the electron gas, elevating electrons from occupied to unoccupied states and producing 'hot electrons'. These hot electrons then oscillate in sync with the incident electromagnetic field, generating heat through the Joule heating mechanism [19]. The energy of these hot electrons is redistributed through electron-electron scattering decay, followed by collisions with the metal lattice, raising the lattice temperature via electron-phonon coupling. Heat is eventually transferred from the nanoparticle lattice to the external environment through phonon-phonon interactions. The energy of hot carriers is influenced by factors such as particle size, material morphology, the dielectric constant of the surrounding medium, and the energy of the Coulomb-charged plasma resonance between nanoparticles [26]. Thus, localized heating in metallic nanostructures through the plasmonic effect involves the excitation of LSPRs with enhanced absorption, the generation and relaxation of hot electrons, and the subsequent heat transfer to the surrounding medium. The energy transfer from absorbed photons to the conduction band electrons of metallic nanostructures involves several key processes: (1) resonant excitation of surface plasmons; (2) a thermal process involving electron-electron collisions; (3) rapid lattice thermalization via electron-phonon scattering; and (4) slower thermal dissipation through phonon-phonon collisions [27].

2.2.1.2 Non-radiative Relaxation

Photothermal effect can also occur in semiconductors through direct interband and intraband electron transitions [22]. When semiconducting materials absorb photons with energies equal to or greater than their bandgap, electron-hole pairs are formed. The energy of these excited electrons can be emitted as photons or transferred to the material's lattice through nonradiative relaxation mechanisms [23, 25]. In semiconductors, the recombination of charge carriers often results in the release of phonons instead of photons, enhancing heat generation and consequently elevating the local lattice temperature. The thermal distribution within semiconductor is heavily influenced by the specific characteristics of light absorption and the recombination occurring at the surface and within the bulk of the material. Two nonradiative relaxation mechanisms including Auger and Shockley-Read-Hall recombination play critical roles in facilitating the heat production (Fig. 2.1b) [4].

Auger recombination, involving three carriers, is an intrinsic process determined by the material's properties. This recombination becomes more significant as bandgaps decrease. In this process, when an electron-hole pair recombines without emitting photons, their energy is transferred to another electron, raising it higher in the conduction band, or to another hole, pushing it deeper into the valence band [4, 20]. The third carrier, now energetically excited, typically returns to the band edge through lattice vibrations in a process known as thermalization. Shockley-Read-Hall recombination is another nonradiative relaxation mechanism influenced by the material's quality [4, 28]. Defects or impurities in a semiconductor typically introduce midgap energy states within the bandgap, altering the electronic structure. These states, known as trap levels, capture charge carriers and facilitate trap-assisted recombination. This process occurs in two stages: electrons transition from the conduction band to the trap level, then recombine with a hole in the valence band, exchanging thermal energy with the lattice. When defect traps are located at or near the semiconductor surface, this recombination process is specifically termed surface recombination [25]. This process heavily depends on the density of surface defects and significantly affects the semiconductor's thermal behavior.

2.2.1.3 Thermal Vibration of Molecules

Photothermal conversion in carbon-based materials and their modified polymers primarily arises from molecular thermal vibrations [29]. These black materials utilize these vibrations for efficient photothermal transformation. The fundamental mechanism involves the photothermal material absorbing light energy when the photon energy matches the energy needed for electron transitions within the molecule, resulting in electron excitation [22, 30]. These excited electrons then transfer their energy throughout the atomic lattice, which is subsequently converted into thermal energy via lattice vibrations (Fig. 2.1c) [22]. However, many single-carbon bonds, such as C–O, C–H, C–C, and O–H, have significant energy gaps between the σ and σ^* orbitals, corresponding to wavelengths in the solar spectrum below 350 nm, making σ-to-σ^* transitions unfeasible under solar irradiation [2]. In contrast, π(p) bonds are generally weaker than s bonds, allowing electrons in π bonds to transition from π orbitals to π^* orbitals at lower energy levels [31, 32]. Additionally, the presence of conjugated π bonds can cause a red shift in the absorption spectrum, enhancing the material's ability to absorb longer wavelengths of light.

As the number of π-bonds increases within a molecule, the energy gap between the highest occupied molecular orbital (HOMO) and the lowest unoccupied molecular orbital (LUMO) narrows, enhancing their susceptibility to excitation. The specific wavelengths at which these transitions occur depend on the spacing between these electronic energy levels [4]. In structures like graphene isomers, which feature a high density of conjugated π-bonds, the material is capable of absorbing a broad range of the solar spectrum. This extensive absorption induces various π–π* transitions, contributing to the material's characteristic black color [33, 34]. When the photon energy of incident sunlight matches the energy needed for these electronic transitions, electrons are excited from the ground state (HOMO) to a higher energy state (LUMO) [4, 23]. These electrons then relax through electron-phonon coupling, transferring the absorbed light energy to the vibrational modes of the atomic lattice. This process effectively converts the absorbed light energy into thermal energy, raising the macroscopic temperature.

2.2.2 The Conversion Efficiency

To determine the photothermal conversion efficiency, one can monitor the temperature increase and calculate the heat generated by incident light. The photothermal conversion efficiency η can be expressed by the equation [4]:

$$\eta = \frac{Q}{E} = \frac{cm\nabla T}{pst} \tag{2.5}$$

Here, Q represents the thermal energy of absorber, and E is the incident light. The variables c and m are the specific heat and mass of the material, respectively, while ∇T indicates the temperature increase of the material due to light exposure. p denotes the power density of the light source, with s and t representing the radiation area and duration, respectively. This method considers all incident light, including reflected, scattered, absorbed, and transmitted photons, as the input energy. A key advantage of this approach is that only the absorbed photons contribute to the generation of thermal energy within the photothermal material [35]. However, this method's dependence on the quantity of photothermal material used makes it difficult to directly compare efficiency across different materials. Additionally, this calculation does not account for the heat transfer from the photothermal material to its surroundings.

To estimate heat loss to the environment, the temperature decay after the cessation of light exposure can be monitored [36, 37]. Consequently, the photothermal conversion efficiency can be expressed as [4]:

$$\eta = \frac{hs\left(T_{\text{max}} - T_{\text{sur}}\right) - Q_{\text{dis}}}{I\left(1 - 10^{a_\lambda}\right)} \tag{2.6}$$

where h represents the heat transfer coefficient, s is the surface area involved in heat transfer, T_{max} is the equilibrium temperature, and T_{sur} is the temperature of the surrounding environment. Q_{dis} is the heat dissipated into the environment, I is the intensity of the incident light, and a is the absorbance of the photothermal material at wavelength λ. This formulation addresses the influence of heat transfer and material concentration on the light-to-heat conversion efficiency, thereby allowing for a more straightforward comparison of the photothermal capabilities across different materials [22, 38, 39]. However, accurately determining this efficiency requires the experimental measurement of several parameters, some of which are complex and can significantly influence the final efficiency value.

2.3 Heat and Mass Transport

The evaporation efficiency is influenced by heat delivery and mass transport dynamics. Solar energy enhances evaporation in localized heated areas, with heat losses occurring through reflection, radiation, convection, and conduction [40, 41]. Materials with low thermal conductivity and high thermal insulation are typically used as water carriers at the interface. The properties of these materials govern the substantial heat losses due to radiation and convection while also controlling the movement of water molecules [14, 42]. Consequently, an integrated discussion of thermal management and water transport offers a novel viewpoint for enhancing the performance of evaporation processes.

2.3.1 Heat Dissipation

When absorbed by a photothermal material, incident light is converted into thermal energy through the photothermal process. This heat is subsequently transferred to nearby materials at lower temperatures or dissipated into the surrounding environment [2, 43]. Thus, heat transfer constitutes a critical process within photothermal systems. The movement of thermal energy between materials is driven by thermal gradients and can occur through three primary mechanisms: conduction, convection, and radiation (Fig. 2.2) [4, 18].

Thermal conduction takes place within a material or between adjacent objects, enabling spontaneous heat transfer from a higher-temperature region to a lower-temperature region. The energy transferred through conduction can be quantified as [12]:

$$Q_{\text{cond}} = \frac{ks}{L}\left(T - T_{\text{obj}}\right) \tag{2.7}$$

Here, k denotes the material's thermal conductivity, s is the surface area, L represents the length of conduction path, and T and T_{obj} are the temperatures of the absorber and the heat transfer object, respectively.

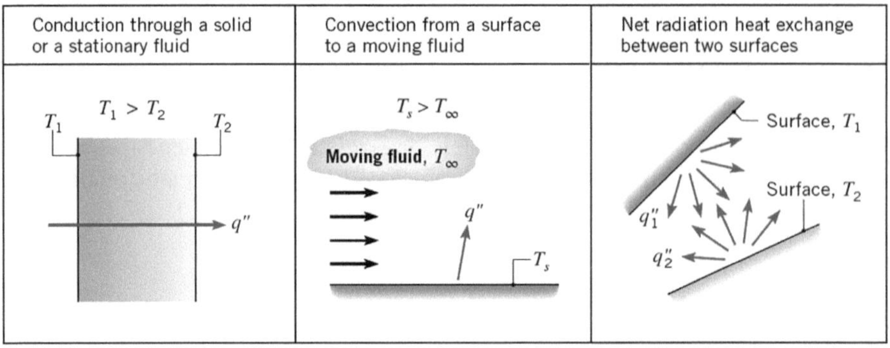

Fig. 2.2 Conduction, convection and radiation heat transfer. (Reproduced with permission from [18] Copyright 2022, Elsevier)

Thermal convection involves the transfer of heat energy facilitated by fluid movement. The energy transferred through convection from the heating source to the fluid is described by [12]:

$$Q_{\text{conv}} = hs\left(T - T_{\text{fluid}}\right) \tag{2.8}$$

where h is the heat transfer coefficient, which is determined by the fluid's physical properties and the thermohydraulic conditions, and inversely related to thermal insulance, and T_{fluid} is fluid temperature.

Thermal radiation occurs across all objects via electromagnetic waves without necessitating a medium. According to the Stefan-Boltzmann law, the emitted energy is proportional to the fourth power of the object's temperature. The net thermal radiation energy, representing the balance between emission and absorption of radiation from surrounding objects, can be expressed as [10]:

$$Q_{\text{conv}} = \varepsilon\sigma s\left(T^4 - T_{\text{env}}^4\right) \tag{2.9}$$

Here, ε represents the emissivity, σ is the Stefan-Boltzmann constant, and T_{env} denotes the environmental temperature. This formulation underscores the dynamic exchange of energy between the heated object and its cooler surroundings.

Heat transfer is influenced by both the physical properties of the materials and the conditions of the surrounding environment. In solar evaporation, thermal conduction is crucial, requiring the minimization of heat loss to the environment to concentrate heat on the surface, thereby improving vapor generation efficiency [18]. Conversely, in volumetric water heating, the convection process is critical, requiring enhanced heat transfer for rapid temperature increases in the surrounding medium [8, 44]. Given the empirical nature of heat flux calculations, the estimated heat losses derived from the equations are approximate. Nonetheless, they offer crucial insights for developing strategies to minimize energy loss. For example, if the temperature of absorber is 41 °C and the surrounding temperature is 27 °C, the estimated conductive, convective, and radiative heat losses are approximately 2.3%, 3.0%, and 4.0% of the solar energy, respectively [3]. This indicates that the three mechanisms of heat loss contribute similarly to the overall energy efficiency. Therefore, selecting appropriate photothermal materials and their surrounding environments is crucial for optimizing the performance of photothermal systems for specific applications.

2.3.2 Mass Transport

Mass transport is crucial for continuously supplying liquid water to the evaporation interface and efficiently removing the generated vapor, both of which are essential for maintaining high-performance evaporation [14]. When solar energy is absorbed by photothermal materials, it is converted into heat localized at the liquid-air interface. This localized heating induces evaporation, creating a concentration gradient that drives liquid water from the bulk reservoir to the evaporation front. Simultaneously, vapor diffuses away from the interface into the surrounding environment. The mass transport process involves multiple mechanisms [43, 45] (1) Diffusion: Water molecules migrate through porous media via concentration gradients, while vapor molecules diffuse into the air governed by Fick's law. (2) Capillary action: in porous photothermal structures (e.g., hydrogels, aerogels), capillary forces dominate liquid replenishment, overcoming gravitational and viscous resistances. (3) Convection: Buoyancy-driven flows or external airflow may enhance vapor dissipation, reducing saturation near the interface. Inefficient mass transport leads to interfacial dry-out (insufficient water supply) or vapor condensation (reduced effective evaporation), both degrading system efficiency.

Vapor generation depends on the interplay between heat and mass transport. According to the energy balance, the total incoming solar energy, denoted as (units: W), is partitioned into thermal energy driving evaporation and heat losses [46]:

$$Q_{\text{sun}} = Q_{\text{evap}} + Q_{\text{loss}} = S_{\text{evap}} D_{\text{a}} h_{\text{fg}} \left(c_{\text{sat}} \left(T_{\text{s}} \right) - \phi_{\infty} c_{\text{sat}} \left(T_{\infty} \right) \right) + A h_{\text{loss}} \Delta T \tag{2.10}$$

Key variables include the evaporative shape factor S_{evap}, the vapor mass diffusivity in air D_{a}, and the surface area of evaporator A. The vapor saturation concentrations at the surface temperature T_{s} and the near-field temperature T_{∞} are represented as $c_{\text{sat}}(T_{\text{s}})$ and $c_{\text{sat}}(T_{\infty})$, respectively, while ϕ_{∞} represents the far-field relative humidity. In the diffusion scenario, the heat flow due to evaporation, Q_{evap}, is determined by the evaporative shape factor and the vapor concentration gradient between the evaporator and the far-field ambient air, $\Delta c = c_{\text{sat}}(T_{\text{s}}) - \phi_{\infty} c_{\text{sat}}(T_{\infty})$ as described by Fick's law [12]. The evaporative shape factor depends solely on the geometry of the evaporator and is related to the effective boundary layer thickness for vapor transport, calculated as $\delta = A/S_{\text{evap}}$. A smaller δ value indicates enhanced vapor transport, which results in a reduced temperature difference ΔT under steady-state conditions. Equation 2.10 highlights that vapor transport is independent of material design, relying only on the evaporator's geometry and the thermal properties of vapor and air. The intricate coupling of heat and mass transport poses a significant challenge in achieving solar-to-vapor efficiencies above 90%. This issue becomes critically important when scaling the technology from laboratory dimensions (e.g., diameter ~ 1 cm) to practical applications (e.g., diameter ~ 1 m), particularly under the diffusion limit. Future research should meticulously explore this aspect. Additionally, there is significant potential to enhance solar-vapor conversion efficiency by optimizing the evaporator geometry or modifying external airflow to reduce the effective boundary layer thickness for vapor transport [47]. This approach opens up promising avenues for optimizing performance in practical applications.

Efficient water transport is critical in solar evaporation to sustain continuous evaporation, prevent dryout and ensure system stability [48]. Capillary forces dominate seawater diffusion within porous evaporators, enabling passive water replenishment [49, 50]. In such systems, water transport is quantified by Darcy's law, with capillary pressure defined via the Young-Laplace equation. Under solar irradiation, thermal localization at the evaporation interface increases temperatures and salt concentrations, dynamically coupling thermal convection (driven by temperature gradients) and chemical potential (from concentration gradients) to enhance water transport. Gravitational effects further modulate flow dynamics when seawater inlets are elevated relative to brine outlets. The structural design and material properties of photothermal systems critically determine water supply efficiency [16]. A balanced water influx must match the evaporation rate to avoid interfacial dryout. Transient salt-ion transport, governed by the equation [14]:

$$\frac{\partial c}{\partial t} + \nabla \bullet (\boldsymbol{u}c) = D_{\text{eff}} \nabla^2 c \tag{2.11}$$

underscoring the need for rapid brine dilution to maintain ion concentration gradients. Here t is the diffusion time, $\boldsymbol{u}$ is the flow field and D_{eff} is the mass diffusivity of salt ions and c represents the ion concentration. The critical solar flux inducing dryout can be estimated by integrating these principles, revealing limitations under high solar intensities (>10 suns) [46, 51]:

$$q_{\text{dry}} = \frac{2h_{\text{fg}}\rho_l K_{\text{s}}}{\mu_1} \frac{\gamma \cos\theta_{\text{rec}}}{t_{\text{s}} r_{\text{wick}}} \tag{2.12}$$

here ρ_l, K_{s}, μ_1, γ, θ_{rec}, t_{s}, and r_{wick} are the liquid density, effective permeability of the wick, dynamic viscosity, liquid-vapor surface tension, receding contact angle, thickness of the capillary wick and pore radius of the wick, respectively. While capillary wicks typically suffice for moderate conditions, material innovations such as hierarchical-pore structures or asymmetric wettability are essential to enhance transport capacity in high-flux or back-side supply configurations.

2.4 Phase Transitions

2.4.1 Vapor Generation

Liquid water enters the control volume at a temperature and an enthalpy, undergoes evaporation at the absorber surface, and exits the control volume as vapor at a temperature and an enthalpy. The enthalpy change is calculated by summing the sensible heat and phase change enthalpies, expressed as [6, 11]:

$$\Delta h = h_{\text{vapor}} - h_{\text{liquid}} = c_{\text{p}} \left(T_{\text{vapor}} - T_{\text{liquid}} \right) + h_{\text{fg}} \tag{2.13}$$

Here, c_{p} represents the specific heat capacity and h_{fg} denotes the latent heat of vaporization at T_{vapor}. This calculation assumes that the variation of c_{p} with temperature is minimal. For enhanced accuracy, it is advisable to compute the enthalpy change using experimentally determined thermodynamic properties from reliable thermodynamic databases whenever possible. The mass flow rate is typically assessed by continuously monitoring the change in water weight with an analytical balance [52, 53]. It is important to recognize that in the absence of light, water can absorb heat from the surroundings and evaporate. Therefore, conducting a control experiment to determine the baseline mass flow rate under dark conditions is essential, and this baseline should be subtracted from the total mass flow rate for accurate measurements.

The solar-to-steam conversion efficiency is defined as the ratio of the energy used for water vaporization to the total energy supplied. It can be calculated as follows [54, 55]:

$$\eta = \frac{\dot{m}}{q_{\text{solar}}} \Delta h \tag{2.14}$$

Here, $\dot{m}$ represents the mass flow rate of the vapor, and q_{solar} is the total power input. The evaporation rate, equivalent to the mass flow rate serves as a measure of the productivity of the vapor generation. As explored in subsequent sections, enhancing both the photothermal efficiency and the evaporation rate are primary objectives. To date, advancements have enabled researchers to achieve solar-to-steam efficiencies as high as 94%. In an ideal scenario, water evaporates isothermally with minimal change in enthalpy, allowing for the complete conversion of solar energy into the latent heat of water vaporization. Considering standard conditions of one sun illumination (1000 W/m^2) and an initial water temperature of 25 °C, where Δh is approximately 2400 kJ/kg, the estimated mass flow rate of the generated vapor is about 1.50 kg/m^2/h [11]. This estimation provides a useful benchmark for evaluating the performance of solar evaporation systems.

2.4.2 Vapor Condensation

Desalinated water is obtained by condensing the generated vapor, which is achieved by reducing the vapor temperature below its saturation temperature, also known as the dew point [15, 56]. Vapor condensation can occur through three primary mechanisms: homogeneous condensation, where water droplets form within the vapor creating fog; direct contact condensation, where vapor directly interacts with liquid water and integrates into it; and surface condensation, where vapor contacts a cooler surface and condenses on it [11]. Due to its straightforward implementation, surface condensation is the predominant method used in solar stills for water collection.

Surface condensation begins when vapor encounters a surface with a temperature lower than saturation temperature, causing liquid to nucleate on the surface. The condensation process is then driven by free convection of the vapor phase [57, 58]. As a result, the geometry of the condensing surface and the flow conditions of the vapor phase significantly impact the efficiency of the condensation process. During this process, latent heat is released and transferred to the surface. The total heat transfer associated with this phase change can be quantitatively assessed to optimize the design and performance of condensation systems [6]:

$$Q_{\text{condens}} = h_{\text{condens}} A_{\text{s}} \left(T_{\text{sat}} - T_{\text{surf}}\right) \tag{2.15}$$

h_{condens} is the heat transfer coefficient with condensation, and A_{s} denotes the condensing area. The coefficient h_{condens} reflects the efficiency of heat transfer and can be determined using empirical formulas. Once it is derived, the mass flow rate of the condensate can be calculated as:

$$\dot{m} = \frac{h_{\text{condens}} A_{\text{s}} \left(T_{\text{sat}} - T_{\text{surf}}\right)}{\Delta h} \tag{2.16}$$

here, h_{condens} the enthalpy difference between the saturated vapor and liquid water. To optimize water generation, it is crucial to maximize the condensation heat transfer. Eq. 2.16 suggests four primary strategies to enhance the mass flow rate: reducing Δh, increasing h_{condens} or A_{s}, and increasing the temperature differential between T_{sat} and T_{surf}. These strategies collectively aim to improve the efficiency and effectiveness of the condensation process, thereby enhancing water yield.

Upon condensation, the condensate can form either a continuous film or discrete droplets, known as film condensation and dropwise condensation, respectively, depending on the surface properties [59, 60]. Generally, hydrophilic surfaces like glass promote film condensation, while hydrophobic surfaces such as plastics favor dropwise condensation. Regardless of the type, condensate acts as a thermal barrier, impeding heat transfer. In dropwise condensation, the limited surface coverage allows for a higher heat transfer coefficient compared to film condensation. However, in solar stills, where solar irradiation must penetrate the condensation surface to reach photothermal materials, droplets can significantly reflect light, reducing solar absorption efficiency. Consequently, hydrophilic surfaces are recommended for condensation in solar stills [61]. The surface inclination also influences condensate removal and accumulation, impacting the heat transfer coefficient [62, 63]. Research has explored how varying the inclination angle affects vapor condensation. Techniques like wiping or vibrating the surface have been explored to actively remove condensate and enhance heat transfer [64, 65]. Increasing the area of the condensing surface is a straightforward method to boost condensate production. Methods to increase the temperature differential between T_{sat} and T_{surf} include enhancing solar absorption and water evaporation, which elevate chamber humidity and T_{sat}. Since condensation releases latent heat, effective heat dissipation from the surface to the surroundings is crucial to prevent an increase in surface temperature. Strategies such as surface cooling and using thinner materials for faster heat conduction have been tested to achieve this.

2.4.3 Salt Crystallization

Crystallization is a separation technique based on the limited solubility of a salt in a solvent under specific temperature and pressure conditions [66–68]. Nucleation occurs spontaneously when ion concentrations exceed those of a saturated solution, forming a cluster of crystalline atoms [7, 69]. The free energy gain from this cluster must exceed the energy required to create an interface between the crystalline and liquid phases. Nucleation thermodynamics are governed by the balance between interfacial free energy (γ_s) and the free energy difference between the liquid and crystal ($\Delta\mu_V$) [70]. The overall free energy change for a spherical crystalline nucleus of radius r results from the opposing contributions of a surface term and a volume term [7]:

$$\Delta G_N = 4\pi r^2 \gamma_s - \frac{4}{3}\pi r^3 \Delta\mu_V \tag{2.17}$$

This function reaches its maximum when the crystal attains its critical size (Fig. 2.1a):

$$n^* = \frac{32\pi\rho_C}{3} \frac{\gamma_s^{\ 3}}{\Delta\mu_V^{\ 3}} \tag{2.18}$$

Here, ρ_C represents the crystalline density. The critical nucleus size is the minimum number of atoms needed in a cluster to initiate spontaneous free energy reduction. Clusters formed through stochastic fluctuations allow the system to overcome the homogeneous nucleation barrier (represented by the purple curve in Fig. 2.3a) [70]:

$$\Delta G_N^{\ *} = \frac{16\pi}{3} \frac{\gamma_s^{\ 3}}{\Delta\mu_V^{\ 2}} \tag{2.19}$$

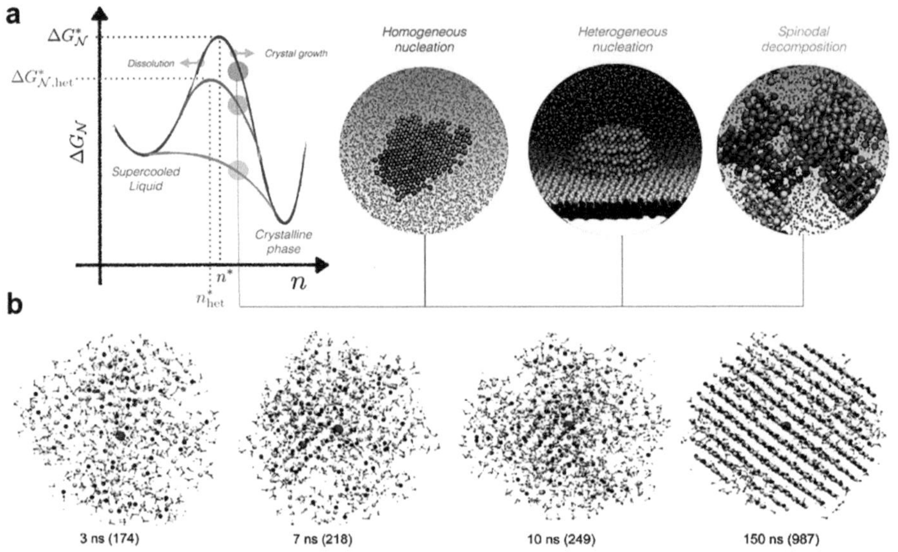

Fig. 2.3 Crystal Nucleation (**a**) The free energy versus the nucleus size n. (**b**) Ion snapshots at different times within a radius of 2 nm, the reference Na^+ ion (larger and blue), Cl^- ion (larger and purple), oxygen atoms in red, and hydrogen atoms in white. (Reprinted with permission from [7, 70]. Copyright 2016 American Chemical Society)

Heterogeneous nucleation (green) features a lower free energy barrier and smaller critical size due to the presence of a foreign impurity, while multi-crystalline cluster symbiosis in metastable liquid is observed during spinodal decomposition (orange) [71, 72]. In solar desalination, nucleation free energy depends on supersaturation. Molecular dynamic simulations illustrate the nucleation process of salt crystals (Fig. 2.3b) [73]. Initially, a disordered structure forms near a reference Na-ion (blue), with the total ion count within 2 nm reaching 174, exceeding the expected 156 for an unstable solution based on the radial distribution function. The ion count then increases to 218 by 7 ns as ordered crystalline structures begin to appear. From 10 to 150 ns, the growth of these crystalline regions is clearly observable, confirming that crystal nucleation in solution arises from solute density fluctuations.

Crystallization in saline environments is affected by both the solution composition and the substrate structures and transport properties [68]. When saline solutions evaporate on hydrophilic surfaces, cubic crystals form, reducing the liquid-vapor interfacial tension and causing the droplet to spread, leading to a ring-like crystalline deposit at the edge, known as the coffee-ring effect [74, 75]. This phenomenon is due to capillary flow from the drop's center to its edge, driven by evaporation and anchored by surface irregularities. On hydrophobic surfaces, crystallization typically occurs at the liquid/solid interface, forming a cauliflower-like structure over cubic crystals [76]. In practical scenarios, crystallization often occurs in confined spaces, influencing dynamics such as melting or freezing points, nucleation rates, and crystal polymorphism and orientation. Subflorescence forms within porous media, starting at the evaporative front and controlled by evaporative flux. Saline migrates to the edges of forming crystals, promoting self-amplification and new crystal initiation in the dry area, leading to exponential mass growth. This process demonstrates that subflorescence can extend centimeters from the liquid surface. Therefore, understanding interfacial properties and transport mechanisms is crucial for predicting and controlling whether salts will effloresce or subfloresce during evaporation.

References

1. G.W. Ho, Y. Yamauchi, L. Hu, B. Mi, N. Xu, J. Zhu, P. Wang, Solar evaporation and clean water, ***Nature Water***, 3 (2025) 131-134.
2. M. Gao, L. Zhu, C.K. Peh, G.W. Ho, Solar absorber material and system designs for photothermal water vaporization towards clean water and energy production, ***Energy & Environmental Science***, 12 (2019) 841-864.
3. G. Ni, G. Li, Svetlana V. Boriskina, H. Li, W. Yang, T. Zhang, G. Chen, Steam generation under one sun enabled by a floating structure with thermal concentration, ***Nature Energy***, 1 (2016).
4. X. Cui, Q. Ruan, X. Zhuo, X. Xia, J. Hu, R. Fu, Y. Li, J. Wang, H. Xu, Photothermal Nanomaterials: A Powerful Light-to-Heat Converter, ***Chemical reviews***, 123 (2023) 6891-6952.
5. B. Wang, X. Huang, Z. Liu, J. Zhang, C. Wei, B. Cai, P. Xie, A. Zhou, Solar interfacial evaporation systems with multi-field synergies boost water purification and blue energy harvesting technologies, ***Energy & Environmental Science***, 17 (2024) 7600-7626.
6. V.P. Carey, Liquid-Vapor Phase-Change Phenomena An Introduction to the Thermophysics of Vaporization and Condensation Processes in Heat... , ***CRC Press***, 3rd Edition (2020) ISBN 9780429082221.
7. G. Liu, T. Chen, J. Xu, G. Yao, J. Xie, Y. Cheng, Z. Miao, K. Wang, Salt-Rejecting Solar Interfacial Evaporation, ***Cell Reports Physical Science***, 2 (2021) 100310.
8. X. Liu, X. Wang, J. Huang, G. Cheng, Y. He, Volumetric solar steam generation enhanced by reduced graphene oxide nanofluid, ***Applied Energy***, 220 (2018) 302-312.
9. J. Zhang, L. Shu, C. Chang, X. Li, R. Lin, B. Fu, C. Song, W. Shang, P. Tao, T. Deng, Crumpled particles of ethanol-wetted graphene oxide for medium-temperature nanofluidic solar-thermal energy harvesting, ***Carbon***, 186 (2022) 492-500.
10. J.R. Howell, M.P. Mengüç, K. Daun, R. Siegel, Thermal radiation heat transfer, ***CRC Press***, 7th Edition (2020).
11. Z. Xu, Z. Li, Y. Jiang, G. Xu, M. Zhu, W.-C. Law, K.-T. Yong, Y. Wang, C. Yang, B. Dong, F. Xing, Recent advances in solar-driven evaporation systems, ***Journal of Materials Chemistry A***, 8 (2020) 25571-25600.
12. T.L. Bergman, Lavine, Adrienne S., Fundamentals of Heat and Mass Transfer, ***John Wiley & Sons Inc***, ISBN 9781119330103 (2017).
13. P.E.C. Edited by Shiva Gorjian, Solar Energy Advancements in Agriculture and Food Production Systems, ***Academic Press***, (Copyright © 2022 Elsevier Inc.) ISBN: 978-970-323-89866-89869.
14. C. Dang, Y. Cao, H. Nie, W. Lang, J. Zhang, G. Xu, M. Zhu, Structure integration and architecture of solar-driven interfacial desalination from miniaturization designs to industrial applications, ***Nature Water***, 2 (2024) 115-126.
15. Y. Shen, R. Zhang, G. Hou, Y. Chen, P. Yang, X. Qian, S. Tang, Sunflower-inspired smart evaporator with omnidirectional solar tracking for highly enhanced energy-capturing and sustainable freshwater harvesting, ***Chemical Engineering Journal***, 503 (2025) 157917.

16. B. Wang, C. Wang, Y. Li, J. Jin, X. Lin, C. Shi, Bionic design: insights from nature for solar interfacial evaporators, ***Energy & Environmental Science***, 18 (2025) 3432-3461.
17. G. Liu, Thermoplasmonics, ***Springer***, ISBN 978-981-97-8331-1 (2024).
18. G. Liu, J. Xu, T. Chen, K. Wang, Progress in thermoplasmonics for solar energy applications, ***Physics Reports***, 981 (2022) 1-50.
19. P. Cheng, M. Klingenhof, H. Honig, L. Zhang, P. Strasser, P. Schaaf, D. Lei, D. Wang, Synergizing Plasmonic Local Heating and 3D Nanostructures to Boost the Solar-to-Vapor Efficiency Beyond 100, ***Advanced materials***, (2024) e2415655.
20. K.W. Tan, C.M. Yap, Z. Zheng, C.Y. Haw, P.S. Khiew, W.S. Chiu, State-of-the-Art Advances, Development, and Challenges of Metal Oxide Semiconductor Nanomaterials for Photothermal Solar Steam Generation, ***Advanced Sustainable Systems***, 6 (2022).
21. H. Liu, C. Chen, H. Wen, R. Guo, N.A. Williams, B. Wang, F. Chen, L. Hu, Narrow bandgap semiconductor decorated wood membrane for high-efficiency solar-assisted water purification, ***Journal of Materials Chemistry A***, 6 (2018) 18839-18846.
22. J. Hu, Y. Sun, Z. Liu, B. Zhu, L. Zhang, N. Xu, M. Zhu, J. Zhu, Z. Chen, Photothermal fabrics for solar-driven seawater desalination, ***Progress in Materials Science***, 150 (2025) 101407.
23. G. Liu, J. Xu, K. Wang, Solar water evaporation by black photothermal sheets, ***Nano Energy***, 41 (2017) 269-284.
24. N. Jiang, X. Zhuo, J. Wang, Active Plasmonics: Principles, Structures, and Applications, ***Chemical reviews***, 118 (2018) 3054-3099.
25. B. Yang, C. Li, Z. Wang, Q. Dai, Thermoplasmonics in Solar Energy Conversion: Materials, Nanostructured Designs, and Applications, ***Advanced materials***, 34 (2022) e2107351.
26. S. Santoro, A.H. Avci, A. Politano, E. Curcio, The advent of thermoplasmonic membrane distillation, ***Chemical Society reviews***, 51 (2022) 6087-6125.
27. M.L. Brongersma, N.J. Halas, P. Nordlander, Plasmon-induced hot carrier science and technology, ***Nature nanotechnology***, 10 (2015) 25-34.
28. D. Shin, G. Kang, P. Gupta, S. Behera, H. Lee, A.M. Urbas, W. Park, K. Kim, Thermoplasmonic and Photothermal Metamaterials for Solar Energy Applications, ***Advanced Optical Materials***, 6 (2018).
29. C. Chen, Y. Kuang, L. Hu, Challenges and Opportunities for Solar Evaporation, ***Joule***, 3 (2019) 683-718.
30. T. Ding, Y. Zhou, W.L. Ong, G.W. Ho, Hybrid solar-driven interfacial evaporation systems: Beyond water production towards high solar energy utilization, ***Materials Today***, 42 (2021) 178-191.
31. W. Li, Y. Xu, K. Liu, L. Zhu, T. Xu, G. Wang, C. Si, Engineered Biomass-Based Solar Evaporators for Diversified and Sustainable Water Management, ***Advanced materials***, (2025) e2503658.
32. A.M. Saleque, N. Nowshin, M.N.A.S. Ivan, S. Ahmed, Y.H. Tsang, Natural Porous Materials for Interfacial Solar Steam Generation toward Clean Water Production, ***Solar RRL***, 6 (2022).
33. Y. Kong, Y. Gao, Y.K. Sun, Y.F. Qi, W.Y. Yin, S.Q. Wang, F.J. Yin, Z.G. Dai, B.Y. Gao, Q.Y. Yue, Manipulating a vertical temperature-gradient of Fe@/graphene aerogel to enhanced solar evaporation and sterilization, ***Journal of Materials Chemistry A***, 10 (2022) 3750-3759.
34. Y. Peng, W. Zhao, F. Ni, W. Yu, X. Liu, Forest-like Laser-Induced Graphene Film with Ultrahigh Solar Energy Utilization Efficiency, ***ACS nano***, 15 (2021) 19490-19502.
35. M. Yu, X. Li, S. Jia, H. Wang, C. Li, D. Bao, W. Tian, P. Wang, J. Guo, L. Huang, R. Li, Z. Zhang, Solar-driven interfacial evaporators from traditional microstructures to smart microstructures, ***Separation and Purification Technology***, 369 (2025) 133099.
36. H. Cao, D. Li, J. Fu, Y. Yang, Y. Zheng, Y. Que, X. He, H. Wang, J. Liu, H. Zhou, Multifunctional biomass-based solar evaporator with enhanced sterilization performance for solar-driven clean water evaporation, ***Desalination***, 591 (2024) 118052.
37. L. Cui, P. Wang, H. Che, J. Chen, B. Liu, Y. Ao, Solar-driven interfacial water evaporation for wastewater purification: Recent advances and challenges, ***Chemical Engineering Journal***, 477 (2023) 147158.

38. N. Ding, J. Zhu, Y. Qiao, D. Yao, X. Gao, J. Chen, C. Lu, X. Pang, Janus-structured self-floating hydrogel via thermal localization and arched evaporation interface design toward all-weather efficient solar desalination, ***Chemical Engineering Journal***, 513 (2025) 162851.
39. M. Ebrahimian Mashhadi, M.M. Hassan, R. Yang, Q. Lu, All-in-One Hybrid Solar-Driven Interfacial Evaporators for Cogeneration of Clean Water and Electricity, ***Advanced Functional Materials***, (2024).
40. J. Gao, L. Zhang, J. You, Z. Ye, Y. Zhong, R. Wang, E.N. Wang, Z. Xu, Extreme salt-resisting multistage solar distillation with thermohaline convection, ***Joule***, 7 (2023) 2274-2290.
41. A. LaPotin, Y. Zhong, L. Zhang, L. Zhao, A. Leroy, H. Kim, S.R. Rao, E.N. Wang, Dual-Stage Atmospheric Water Harvesting Device for Scalable Solar-Driven Water Production, ***Joule***, 5 (2021) 166-182.
42. X. Mu, Y. Gu, P. Wang, A. Wei, Y. Tian, J. Zhou, Y. Chen, J. Zhang, Z. Sun, J. Liu, L. Sun, S. Tanemura, L. Miao, Strategies for breaking theoretical evaporation limitation in direct solar steam generation, ***Solar Energy Materials and Solar Cells***, 220 (2021) 110842.
43. X. Sun, J. Miao, H. Wang, Y. Song, D. Tang, L. Li, Multiscale Heat and Mass Transport Optimization for Solar-Driven Interfacial Evaporation: Opportunities and Challenges for Industrial Applications, ***Small***, (2025) e2502639.
44. G. Ni, N. Miljkovic, H. Ghasemi, X. Huang, S.V. Boriskina, C.-T. Lin, J. Wang, Y. Xu, M.M. Rahman, T. Zhang, G. Chen, Volumetric solar heating of nanofluids for direct vapor generation, ***Nano Energy***, 17 (2015) 290-301.
45. X. Wu, Y. Lu, X. Ren, P. Wu, D. Chu, X. Yang, H. Xu, Interfacial Solar Evaporation: From Fundamental Research to Applications, ***Advanced materials***, 36 (2024) e2313090.
46. L. Zhang, Z. Xu, L. Zhao, B. Bhatia, Y. Zhong, S. Gong, E.N. Wang, Passive, high-efficiency thermally-localized solar desalination, ***Energy & Environmental Science***, 14 (2021) 1771-1793.
47. L. Wang, Z. Liu, J. Xu, K. Wang, Q. Wang, G. Liu, Conical solar-thermo-radiative evaporator for sustainable desalination and salt recovery, ***Desalination***, 567 (2023) 116993.
48. Z. Wang, X. Wu, F. He, S. Peng, Y. Li, Confinement Capillarity of Thin Coating for Boosting Solar-Driven Water Evaporation, ***Advanced Functional Materials***, 31 (2021).
49. C. Sheng, N. Yang, Y. Yan, X. Shen, C. Jin, Z. Wang, Q. Sun, Bamboo decorated with plasmonic nanoparticles for efficient solar steam generation, ***Applied Thermal Engineering***, 167 (2020) 114712.
50. Y. Tian, H. Yang, S. Wu, J. Yan, K. Cen, T. Luo, G. Xiong, Y. Hou, Z. Bo, K. Ostrikov, Beyond lotus: Plasma nanostructuring enables efficient energy and water conversion and use, ***Nano Energy***, 66 (2019) 104125.
51. X. Zhao, H. Zhang, K.Y. Chan, X. Huang, Y. Yang, X. Shen, Tree-Inspired Structurally Graded Aerogel with Synergistic Water, Salt, and Thermal Transport for High-Salinity Solar-Powered Evaporation, ***Nano-micro letters***, 16 (2024) 222.
52. L. Cheng, L. Zuo, C. Li, J. Shen, L. Wang, Y. Li, D. Dukashin, P. Senin, S. Yan, T. Bian, Sandwiched solar interfacial evaporation system: Integrating phase change material/hydrogel/NiCo2O4 nanoflower for superior desalination, ***Separation and Purification Technology***, 348 (2024) 127816.
53. W. Geng, H. Zhang, W. Lei, X. Zhao, C. Chen, Welding Pollen-Based Solar Evaporator for Clean Water Production, ***Small***, 21 (2025) e2408576.
54. C. Chang, P. Tao, J. Xu, B. Fu, C. Song, J. Wu, W. Shang, T. Deng, High-Efficiency Superheated Steam Generation for Portable Sterilization under Ambient Pressure and Low Solar Flux, ***ACS applied materials & interfaces***, 11 (2019) 18466-18474.
55. K. Mao, Y. Zhang, S.C. Tan, Functionalizing solar-driven steam generation towards water and energy sustainability, ***Nature Water***, (2025).
56. H. Liu, Z. Huang, K. Liu, X. Hu, J. Zhou, Interfacial Solar-to-Heat Conversion for Desalination, ***Advanced Energy Materials***, 9 (2019).

57. C. Shi, Y. Li, Z. Wu, A. Chen, Q. Wang, X. Li, H. Jin, C. Yan, Y. Shi, Y. Shi, Z. Lin, B. Su, A Green and Energy-Supply–Free Artificial Plant for Efficient and Non-Selective Enrichment of Heavy Metal Ions Out of Soil, ***Advanced Functional Materials***, 34 (2024).
58. T. Wang, F. Bai, D. Lei, L. Wang, X. Zhang, G. Yuan, Z. Wang, Experimental investigation of an asymmetric CPC concentrated solar interfacial evaporation device (CSIED), ***Desalination***, 604 (2025) 118739.
59. S. Cao, Q. Jiang, X. Wu, D. Ghim, H. Gholami Derami, P.-I. Chou, Y.-S. Jun, S. Singamaneni, Advances in solar evaporator materials for freshwater generation, ***Journal of Materials Chemistry A***, 7 (2019) 24092-24123.
60. M.S. Irshad, N. Arshad, G. Maqsood, I. Ahmed, B. Shakoor, M.S. Asghar, U. Ghazanfar, L. Lin, M.A.K.Y. Shah, I. Ahmed, X. Chen, J. Wang, C. Yi, J. Li, J. Qian, W. Li, Z. Said, H. Li, N.X. Ho, H. Wang, X. Fan, N. Mushtaq, V.-D. Dao, T. Mei, X. Wang, Advancing water collection efficiency in hybrid solar evaporators: key factors, strategic innovations, and synergistic applications, ***Materials Science and Engineering: R: Reports***, 165 (2025) 101018.
61. X. Geng, P. Yang, Y. Wan, How to reduce enthalpy in the interfacial solar water generation system for enhancing efficiency?, ***Nano Energy***, 123 (2024) 109434.
62. X. Li, P. Bista, A.Z. Stetten, H. Bonart, M.T. Schür, S. Hardt, F. Bodziony, H. Marschall, A. Saal, X. Deng, R. Berger, S.A.L. Weber, H.-J. Butt, Spontaneous charging affects the motion of sliding drops, ***Nature Physics***, 18 (2022) 713-719.
63. M. Mohsenzadeh, L. Aye, P. Christopher, Development and experimental analysis of an innovative self-cleaning low vacuum hemispherical floating solar still for low-cost desalination, ***Energy Conversion and Management***, 251 (2022) 114902.
64. W. Deng, T. Fan, Y. Li, Water wave vibration-promoted solar evaporation with super high productivity, ***Nano Energy***, 92 (2022) 106745.
65. Z. Zheng, W. Li, H. Liu, X. Wang, Sustainable Interfacial Evaporation System Based on Hierarchical MXene/Polydopamine/Magnetic Phase-Change Microcapsule Composites for Solar-Driven Seawater Desalination, ***ACS applied materials & interfaces***, 14 (2022) 50966-50981.
66. Y. Song, S. Fang, N. Xu, M. Wang, S. Chen, J. Chen, B. Mi, J. Zhu, Solar transpiration-powered lithium extraction and storage, ***Science***, 385 (2024) 1444-1449.
67. Y. Wu, C. Ma, K. Zhu, L. Jin, L. Song, L. Li, Y. Lu, Y. Zheng, Y. Zhang, X. Zheng, S. Wu, Y. Pang, Z. Shen, S.C. Tan, H. Chen, Asymmetric evaporation for efficient and simultaneous extraction of freshwater, salt, and electrical energy from seawater, ***Energy & Environmental Science***, 17 (2024) 9303-9312.
68. X. Chen, M. Yang, S. Zheng, F. Temprano-Coleto, Q. Dong, G. Cheng, N. Yao, H.A. Stone, L. Hu, Z.J. Ren, Spatially separated crystallization for selective lithium extraction from saline water, ***Nature Water***, 1 (2023) 808-817.
69. S. Sun, C. Shi, Y. Kuang, M. Li, S. Li, H. Chan, S. Zhang, G. Chen, A. Nilghaz, R. Cao, J. Tian, 3D-printed solar evaporator with seashell ornamentation-inspired structure for zero liquid discharge desalination, ***Water research***, 226 (2022) 119279.
70. G.C. Sosso, J. Chen, S.J. Cox, M. Fitzner, P. Pedevilla, A. Zen, A. Michaelides, Crystal Nucleation in Liquids: Open Questions and Future Challenges in Molecular Dynamics Simulations, ***Chemical reviews***, 116 (2016) 7078-7116.
71. M. Schiro, E. Ruiz-Agudo, C. Rodriguez-Navarro, Damage mechanisms of porous materials due to in-pore salt crystallization, ***Physical review letters***, 109 (2012) 265503.
72. P. Shi, J. Li, Y. Song, N. Xu, J. Zhu, Cogeneration of Clean Water and Valuable Energy/Resources via Interfacial Solar Evaporation, ***Nano letters***, 24 (2024) 5673-5682.
73. D. Chakraborty, G.N. Patey, How Crystals Nucleate and Grow in Aqueous NaCl Solution, ***The journal of physical chemistry letters***, 4 (2013) 573-578.
74. J. Li, X. Zhou, J. Zhang, C. Liu, F. Wang, Y. Zhao, H. Sun, Z. Zhu, W. Liang, A. Li, Migration Crystallization Device Based on Biomass Photothermal Materials for Efficient Salt-Rejection Solar Steam Generation, ***ACS Applied Energy Materials***, 3 (2020) 3024-3032.

75. J. Yao, Q. Zhong, J. Zhang, J. Zhao, Z. Wang, Interfacial solar evaporation for zero liquid discharge desalination, ***Communications Materials***, 5 (2024).
76. N. Shahidzadeh-Bonn, S. Rafai, D. Bonn, G. Wegdam, Salt crystallization during evaporation: impact of interfacial properties, ***Langmuir : the ACS journal of surfaces and colloids***, 24 (2008) 8599-8605.

Chapter 3
Photothermal Nanomaterials

Photothermal materials that convert sunlight into heat are crucial for solar evaporation. The field advanced significantly, leading to the development of various materials with improved light absorption capabilities, including metallic nanostructures, semiconductors, carbon-based materials, and conjugated polymers [1, 2]. Although many reviews covered photothermal materials [3–9], this section highlights the design principles, material types heterojunctions, and the nanoengineering and fabrication processes of solar absorbing materials.

3.1 Design Principles

Photothermal conversion, crucial for solar energy harvesting, relies on the intrinsic properties of materials, such as composition, chemical structure, and other relevant characteristics. These attributes significantly impact the photothermal efficiency, determining their ability to harness solar energy [3, 10]. To optimize solar energy utilization, photothermal materials are designed to maximize the solar absorption while minimizing transmission and reflection losses. These materials are also engineered to efficiently convert the photon energy into thermal energy [11–13]. The development of advanced photothermal nanomaterials is guided by several key criteria: (i) exceptional light absorption across solar spectrum; (ii) highly efficient photothermal conversion; and (iii) high performance and stability in real applications.

Selecting suitable photothermal materials involves a sophisticated process of fine-tuning their composition and structural properties, influenced by factors like the choice of heat carrier. The field is currently enhanced by a range of emerging technologies and design methodologies, driving continuous advancements [14–16]. Building complex and extensive photothermal conversion systems fundamentally depends on the careful selection of materials and strict adherence to preparation

G. Liu, T. Chen, *Solar Evaporation for Water-Energy-Resource Nexus*,
https://doi.org/10.1007/978-981-95-3815-7_3

protocols. Photothermal materials, known for their resistance to high temperatures, corrosion, and UV radiation, form a unique class with exceptional properties [1, 3]. The development of photothermal materials is at the front of innovation in solar energy utilization, making it one of the most dynamic evolving research fields.

3.2 Materials Selection

Selecting the proper material for specific application is critical for optimizing the photothermal systems. Carbon-based materials like graphene [17, 18], carbon nanotubes (CNT) [19, 20], and MXene [21, 22] are exceptionally effective due to their superior optical absorption across ultraviolet, visible, and near-infrared light spectrums. Their unique electronic properties facilitate the efficient conversion of light into heat, while their high thermal conductivity allows for rapid heat dissipation, improving thermal management. These properties make them suitable for applications in solar energy harvesting, seawater desalination, environmental remediation and photothermal therapy. Plasmonic metals such as gold (Au), platinum (Pt), and silver (Ag) are also advantageous for photothermal applications due to their ability to absorb light effectively through plasmonic resonance, a process that involves the excitation of surface electrons under specific light wavelengths [23, 24]. This results in enhanced light absorption and localized heating. The photothermal properties of these metals can be optimized by adjusting their size, shape, and the surrounding environment, making them particularly useful in photothermal therapy where gold and silica nanoparticles are favored for their low toxicity and functional adaptability. Metal sulfides like copper sulfide (CuS) [25] and molybdenum disulfide (MoS_2) [26, 27], along with metal oxides such as iron oxides (Fe_2O_3, Fe_3O_4) [28], manganese oxides (MnO_2, Mn_2O_3, Mn_3O_4) [29], and cobalt oxide (Co_3O_4) [30, 31] are valued for their strong light absorption. These materials can convert a wide spectrum of solar radiation into heat, making them ideal for photothermal catalysis and power generation.

3.2.1 Metallic Nanostructures

Metallic nanostructures, particularly gold and silver, are highly effective for solar steam generation due to their quick light-to-heat conversion and enhanced photothermal effects driven by localized surface plasmon resonance [32, 33]. In a study by Deng's group, a self-assembled porous Au film was engineered at the air-water interface, acting as an effective photothermal structure for solar evaporation, rapidly generating localized plasmonic heat that enhances water evaporation [34, 35] (Fig. 3.1). This contrasts with Au nanofluid systems where a broader hot zone is formed through bulk liquid heating, whereas in the floating system, heat is more confined to the evaporation zone, thus boosting efficiency.

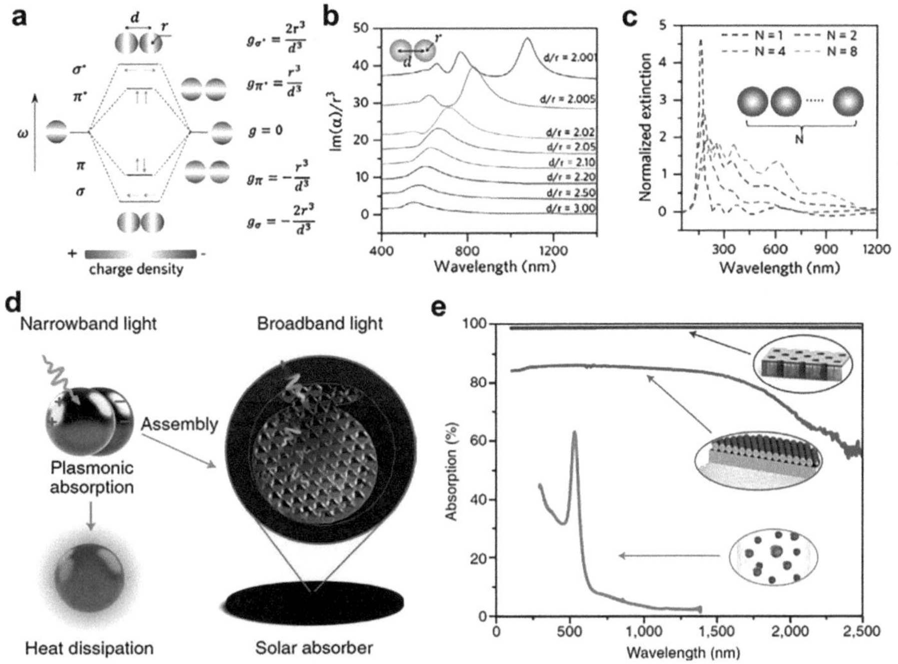

Fig. 3.1 Particle coupling effect (**a**) Energy hybridization within a plasmonic dimer. (**b**) Absorption spectra correlating to different interparticle distances. (**c**) Normalized extinction cross section ratio (C_{ext}/C_{geo}) as a function of particle count. (Adapted with permission from [46] Copyright 2021 Wiley-VCH). (**d**) Individual particles exhibit distinct wavelength-dependent light absorption. (**e**) Representative absorption spectra for dispersed plasmonic particles (depicted with a blue line), self-assembled films (illustrated with a red line), and three-dimensional template films (shown with a purple line). (Reproduced with permission from [33] Copyright 2018 Nature Publishing Group)

To enhance the practicality of these systems, Liu and colleagues developed a paper-based Au film, which demonstrated durability over more than 30 cycles, with air-laid paper's intrinsic roughness and microscale pores improving light absorption (~87% across 400 to 800 nm) and thereby evaporation efficiency [36–38]. Further advancements were made by Zhu's group, who employed porous alumina (Al_2O_3) templates with finely controlled pore sizes as scaffolds for Au nanoparticle deposition [39]. A physical vapor deposition (PVD) process enabled scalable production, achieving an impressive <99% light absorption across a broad spectral range (400 nm to 10 μm) and an energy conversion efficiency exceeding 90% at 4 kW/m^2 for solar evaporation.

Kim et al. explored the tunability of light absorption in an Au nanowire-based membrane by varying the taper angle of anodic aluminum oxide (AAO) nanowires, achieving an average light absorption of 91% in the visible to near-infrared region (400 to 2500 nm), which resulted in a 57% evaporation efficiency under intense irradiation (20 kW/m^2) [40]. Deng's team also highlighted the critical role of substrate surface hydrophilicity/wettability on the evaporation performance. Their findings showed that systems with hydrophilic substrates facilitate rapid water replenishment for continuous evaporation, whereas systems with hydrophobic substrates encounter vapor bubble accumulation, forming an insulating layer that impedes heat and mass transfer [41]. Explorations have also extended to less expensive plasmonic materials like copper (Cu) and aluminum (Al) as cost-effective alternatives [42, 43]. Although substrates such as cellulose paper and porous ceramics enhance mechanical strength, their inherent fragility poses challenges for practical applications. To overcome these, robust substrates like wood and hydrogels have been developed [44, 45], offering enhanced mechanical properties and durability.

3.2.2 Semiconductors

Semiconductor nanomaterials, including metal oxides and chalcogenides, are distinguished candidates for photothermal conversion due to their cost-effectiveness, simple synthesis, and low toxicity [1, 2, 47]. Compared to organic photothermal materials, semiconductors offer greater resistance to photodegradation and photobleaching. They feature tunable absorption wavelengths and high extinction coefficients in the near-infrared region. The optical properties of semiconductors, such as light absorption, are mainly determined by their bandgap energies, which enable the generation of free charge carriers [48]. Figure 3.2 illustrates the bandgap energies and band edge positions for various commonly studied semiconductors, including oxides, chalcogenides, nitrides, and silicon [7].

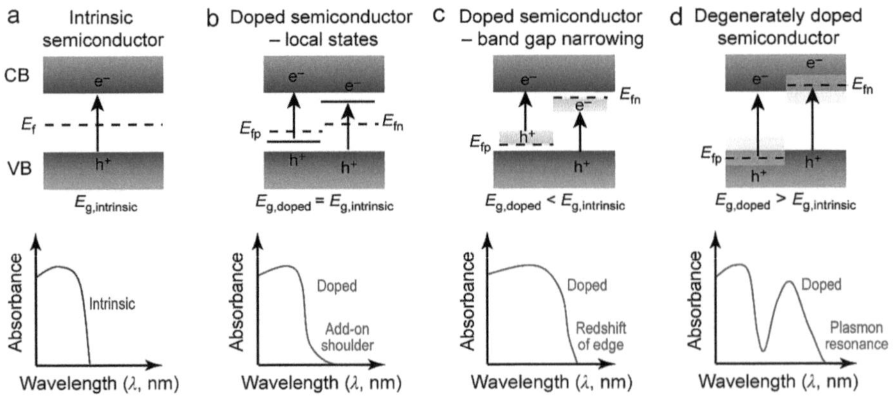

Fig. 3.2 Band Structure Modifications. (**a**) An intrinsic semiconductor. (**b**) Formation of shallow-level energy states. (**c**) Formation of deep-level energy states. (**d**) Localized surface plasmon resonance induced by degenerate doping. Each panel presents the electronic band structures at the top and the corresponding optical absorption spectra at the bottom. (Reprinted with permission from [7], Copyright 2019, Royal Society of Chemistry)

When incident light photons with energies matching the semiconductor's bandgap interact with the material, electron-hole pairs are generated, as shown in Fig. 3.2a. The nonradiative recombination of these electron-hole pairs produces thermal energy [32, 49]. Typically, semiconductors have much lower free charge carrier concentrations than metals, making them generally translucent in the IR spectrum but opaque in the visible spectrum [1]. Doping is a typical method used to increase the free charge carrier concentration in semiconductors, thereby extending their absorption range and enhancing photothermal conversion efficiency through increased nonradiative recombination probabilities.

Introducing defects or impurities not only shifts the band energy but also creates new energy states within the bandgap. This doping-induced bandgap engineering leads to several changes: the emergence of intraband energy states, bandgap narrowing, and the formation of impurity bands in heavily doped semiconductors [7]. In the first case, doping introduces defect and trap-level states that serve as charge recombination centers, extending the absorption spectrum to longer wavelengths and often appearing as an additional shoulder on the absorption curve.Since the initial synthesis of N-doped TiO_2 in 2001 [50], doping has been extensively applied to TiO_2, incorporating various non-metal anions and transition metal cations [1, 51].

In the second doping scenario, the shift in the position of the valence or conduction band, without the introduction of new energy levels, results in a narrower bandgap [52]. This narrowing effect broadens and redshifts the absorption edge, allowing photons with energies higher than the narrow bandgap to generate above-bandgap electron-hole pairs. The subsequent relaxation of these pairs to the band edges converts excess energy into heat through thermalization [53, 54]. In contrast, photons absorbed by broad bandgap semiconductors are predominantly released radiatively, resulting in lower photothermal conversion efficiency. Various narrow bandgap semiconductors have been developed for efficient light-to-heat conversion, with black TiO_2 gaining significant attention due to its broadband light absorption and enhanced photothermal properties facilitated by surface disorder and oxygen vacancies [55].

Recently, degenerately doped semiconductors have shown strong localized surface plasmon resonances (LSPRs) in the NIR region due to high free charge carrier concentrations [7, 56, 57]. These LSPRs, similar to those seen in noble metals, can be tuned from the visible to the far-infrared region by adjusting the free charge carrier concentrations [32, 58]. However, high-concentration doping can disrupt the band structure, requiring careful consideration of factors such as the active dopant fraction, solid solubility limits, and doping compensation effects [59].

3.2.3 Carbon-Based Nanomaterials

Carbon, one of the most abundant elements in the universe, is an essential component of all known life forms. Carbon atoms can form bonds in various hybridization states (sp, sp^2, sp^3), resulting in the formation of multiple carbon allotropes (Fig. 3.3)

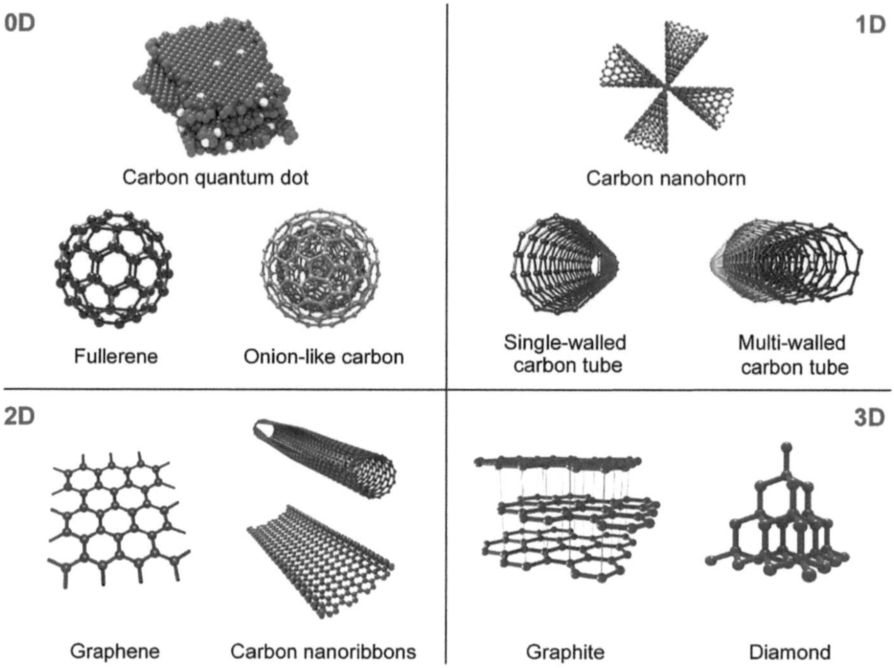

Fig. 3.3 Carbon-based materials. (Reprinted with permission from [60]. Copyright 2015 American Chemical Society)

[1, 60]. Notable carbon-based materials encompass carbon dots [61], nanodiamonds [62], fullerenes [63], graphite [64], carbon nanotubes [19, 65], graphene [66, 67], and their derivatives [60]. These allotropes are distinctly advantageous as photothermal agents owing to their robust chemical stability, broad spectrum light absorption, lightweight and cost-effectiveness nature. The light-to-heat conversion in the materials primarily involves the excitation of delocalized π electrons followed by their relaxation back to the ground state [56].

Efforts have focused on enhancing absorption intensity and reducing surface light reflection, rather than broadening the absorption spectrum. The optical properties of carbon-based nanomaterials are influenced by structural parameters like size, shape, doping levels, and the number of layers [68, 69]. These parameters are closely linked to the fabrication methods, including both top-down and bottom-up approaches [1, 70]. Top-down methods generally involve reducing the size of bulk materials, such as through mechanical exfoliation, while bottom-up methods, like chemical vapor deposition and epitaxial growth, construct materials from the atomic level.

When engineered into porous nanostructures, these carbon-based materials exhibit a significant reduction in light reflection by minimizing the impact of the angle of incidence and lowering the effective refractive index [71–73]. For instance, a dual-layer structure combining exfoliated graphite with a carbon foam layer has demonstrated a reflectivity as low as 3% across a solar spectrum range of 250 to 2250 nm [74]. Similarly, CNTs and porous graphene have been engineered to enhance solar absorption [75]. The formation of microcavities between the two sides of a spacer layer in these porous structures effectively confines light, significantly enhancing material-light interactions. A network of porous graphene has achieved an impressive 97% light absorption across the 200–2500 nm range due to multiscattering effects, while vertically aligned graphene sheets have demonstrated 98% absorption of visible light and nearly 100% absorption of near-infrared light [76, 77]. Moreover, hollow CNTs structured into a hierarchically nanoporous network have showed wide solar absorption of 99% and a photothermal-conversion efficiency of 86.8% [6, 78].

3.2.4 Organic Polymer Nanomaterials

Conjugated organic polymers become a leading class of photothermal materials, characterized by their adjustable molecular structures, excellent near-infrared absorption, high light-to-heat conversion efficiencies, and good biocompatibility [1, 79, 80]. Similar to carbon nanomaterials, the optical absorption properties of these conjugated polymers in the visible and NIR spectra arise from the nonradiative relaxation of their delocalized π-electron systems [7, 81]. Strategic development of polymer-based photothermal nanomaterials has been primarily focused on fundamental conjugated structures, including (PT) [82], polypyrrole [83], polyaniline [84], and polydopamine (Fig. 3.4) [85, 86].

The photothermal conversion mechanism in these materials is facilitated by the dense packing of monomeric units, which enhances intermolecular collisions and promotes nonradiative relaxation processes [1, 79, 87]. This structural arrangement effectively quenches molecular fluorescence while optimizing photothermal conversion efficiency. Recent advancements in molecular design have introduced donor–acceptor (D–A) strategies to engineer novel conjugated polymers with extended light absorption spectra and enhanced conversion efficiencies [88, 89]. A representative class of D–A structured polymer photothermal materials, illustrated in Fig. 3.4b, incorporates diketopyrrolopyrrole (DPP) as the electron-accepting moiety paired with various electron-donating polymers [79]. In this configuration, the conjugated polymer donor absorbs incident photons, while the acceptor polymer facilitates nonradiative decay of excited electrons, requiring a Fermi level lower than the donor's excited state [90]. The electronic bandgap of these D–A systems can be precisely modulated by adjusting the relative proportions and electronic strengths of the donor and acceptor components [91].

Upon photoexcitation, these D–A structures exhibit intramolecular charge transfer along the polymer backbone, effectively suppressing radiative electron-hole recombination and promoting heat generation. This design principle has been successfully implemented in the synthesis of conjugated polymer nanoparticles for NIR light harvesting, achieving remarkable photothermal conversion efficiencies.

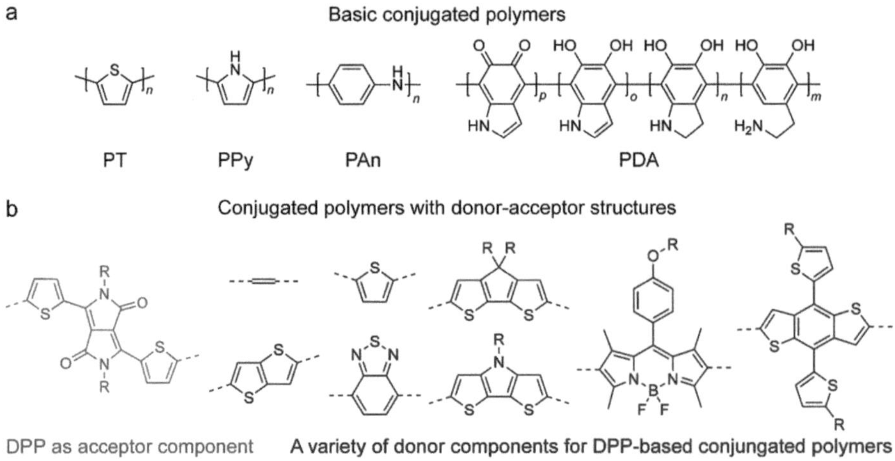

Fig. 3.4 Polymer photothermal materials (**a**) Fundamental conjugated polymers. The monomeric units serve as essential building blocks for constructing donor-acceptor structured conjugated polymers. (**b**) Advanced photothermal materials based on conjugated polymers with donor-acceptor architectures. (Reprinted with permission from [79]. Copyright 2020 American Chemical Society)

Notably, the incorporation of a porphyrin-pyrene pendant as an additional light-harvesting unit has yielded a conversion efficiency of 62.3% [88]. Further innovation in molecular architecture has led to the development of biodegradable conjugated polymers featuring an acceptor-donor-acceptor (A–D–A) configuration. This structural motif achieves full fluorescence quenching and establishes a narrow energy gap, leading to a high photothermal conversion efficiency of 82% [91]. To address the dual challenges of enhancing conversion efficiency and improving material stability, researchers have increasingly focused on supramolecular assembly strategies for fabricating advanced polymer-based photothermal materials.

Supramolecular assembly marks a paradigm shift from traditional molecular design, utilizing flexible noncovalent interactions such as van der Waals forces, hydrogen bonding, and electrostatic interactions [92]. Analogous to the aggregation-caused quenching phenomenon, supramolecular systems can significantly attenuate intrinsic molecular fluorescence, thereby enhancing thermal energy generation [93]. Furthermore, supramolecular architectures confer remarkable material stability by encapsulating monomeric species within their assemblies, where only peripheral molecules are susceptible to environmental influences [94]. This spatial confinement effectively mitigates molecular decomposition and oxidation, preserving the photo-physicochemical properties of the photothermal components and substantially improving their conversion efficiency.

The unique advantages of supramolecular assembly have spurred extensive research into assembly methodologies and the optimization of photothermal performance [1, 95]. A coordination self-assembly approach has been developed for synthesizing supramolecular nanodrugs, producing particles with well-defined spherical morphology, uniform size distribution, and exceptional colloidal stability [96, 97]. Recent advancements enabled the fabrication of dual-peak absorbing photothermal nanoagents via supramolecular assembly strategies [98]. These systems incorporate two distinct donor-acceptor structured conjugated polymers, allowing efficient light absorption in both NIR-I and NIR-II spectral windows. The resulting photothermal conversion efficiencies reach 44.9% at 808 nm and 43.4% at 1064 nm, highlighting the potential of supramolecular engineering in developing advanced photothermal materials.

3.2.5 Two-Dimensional Nanomaterials

The single-layer graphene marked a pivotal milestone, catalyzing rapid advancements in the exploration of 2D nanomaterials with exceptional physicochemical properties [99, 100]. Over the past decades, a diverse array of 2D materials such as transition metal dichalcogenides (TMDCs) [101], MXenes [102], black phosphorus (BP) [103], tellurene [104], metal-organic frameworks (MOFs) [105, 106], and layered double hydroxides [3], has been engineered with tailored photothermal conversion capabilities. Unlike elemental 2D materials such as graphene or BP, TMDCs and MXenes have emerged as two rapidly expanding families, distinguished by their compositional versatility and tunable optoelectronic properties [102, 107]. The layered architecture of 2D nanomaterials enables precise modulation of their electronic bandgaps through layer-number control, thereby broadening their optical absorption spectra. For instance, BP exhibits a tunable bandgap spanning 0.3–2.0 eV across visible to far-infrared (FIR) regions via layer-thickness engineering [108]. As the thinnest known materials, 2D systems surpass zero-dimensional, one-dimensional, and bulk counterparts in specific surface area. Their ultrathin geometry facilitates ultrafast photoresponsivity and high in-plane carrier mobility, which synergistically enhance photothermal conversion efficiency [109]. Notably, MXenes such as Ti_3C_2 demonstrate near-perfect light-to-heat conversion efficiencies approaching 100%, positioning them as front runners for photothermal applications [27, 110].

As a rising star in 2D systems [102], MXenes have garnered significant attention as a novel class of transition metal carbides/nitrides with the general formula $M_{n+1}X_nT_x$ ($n = 1$–4), where M denotes early transition metals (e.g., Sc, Ti, V, Mo), X represents carbon/nitrogen, and T_x corresponds to surface terminations (e.g., –OH, –O, –F) [111, 112]. These materials are derived from selective etching of MAX-phase precursors, where A typically denotes group IIIA/IVA elements (Fig. 3.5) [2, 113]. Computational studies predict a vast compositional landscape for MXenes, with numerous variants already experimentally realized. MXenes exhibit extraordinary metallic conductivity (up to 20,000 S cm^{-1}) and free charge carrier densities ($\sim 10^{22}$ cm^{-3}), properties rooted in their transition metal constituents [114]. These attributes underpin their exceptional electromagnetic interference shielding performance, where incident waves undergo surface reflection and

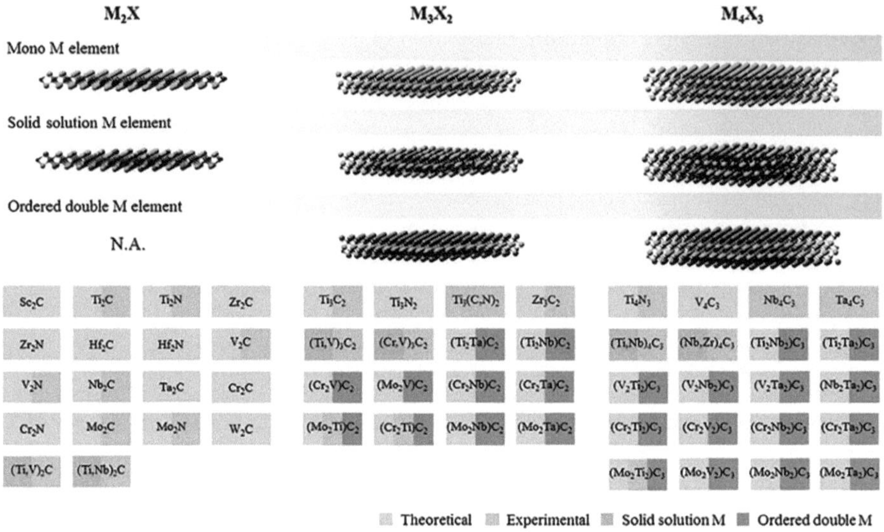

Fig. 3.5 Different types of MXene. (Reproduced with permission [2, 116] Copyright 2024, Wiley-VCH GmbH)

multi-scale internal scattering, amplifying energy absorption. Furthermore, MXenes support tunable localized surface plasmon resonances across visible to near-infrared regimes, modulated by structural parameters such as M/X site configuration and surface termination stoichiometry [115]. Their interband transitions also induce strong UV absorption, collectively enabling broadband light harvesting and efficient thermal energy generation.

Due to the integration of carbon and nitrogen atoms into its metal lattice, MXene shows characteristics of a semimetal, merging attributes from both ceramic and metal precursors [116]. For example, MXenes exhibit metallic properties or possess a minimal bandgap, alongside a robust capacity for sunlight absorption [22, 117]. The light absorption can be enhanced through the effects of electromagnetic shielding and localized surface plasmon resonance associated with MXene, achieving nearly 100% photothermal conversion efficiency [114]. Furthermore, the abundant surface groups on MXenes render them superhydrophilic, enhancing their suitability for interfacial solar steam generation [118]. As detailed by Wang and colleagues, exfoliated Ti_3C_2 MXene flakes were deposited onto a hydrophilic polyvinylidene fluoride substrate to create a self-standing photothermal membrane. This membrane was subsequently modified on its surface with poly(dimethylsiloxane) to enable the evaporator to remain afloat on water surfaces. Under standard solar irradiation, an evaporation efficiency of 84% was observed [119]. In a novel approach inspired by the crest-ridge scales of West African Gaboon vipers, Chen engineered $Ti_3C_2T_x$ MXene evaporators with isotropically crumpled structures that enhance light absorption through multiple scattering and reflection of incident light across a broad spectrum [120]. With a minimal MXene loading of 0.32 mg per square centimeter, a light absorption rate of 93.2% and a corresponding steam generation rate of 1.33 kg per square meter per hour were recorded.

3.2.6 Hybrid Materials

Hybridization acts as a versatile method to combine complementary and synergistic properties from different components. Constructing hybrid nanostructures has become a leading strategy to enhance light absorption and optimize photothermal efficiency [1, 21, 121]. For instance, dual-plasmonic systems engineered through the integration of plasmonic semiconductors with metallic nanostructures exhibit multiple plasmon resonance modes, effectively broadening the near-infrared absorption spectrum [57, 98, 122]. Strategic modifications, such as decorating semiconductors with palladium nanoparticles, enhance photothermal catalytic activity by introducing metallic active sites [123], while silver nanoparticle incorporation synergistically elevates antibacterial and photothermal functionalities [124]. To overcome limitations in photostability and excessive photoluminescence, conjugated polymers are often combined with carbon nanomaterials or plasmonic metals [97, 125]. with polymeric coatings concurrently mitigating long-term toxicity concerns in inorganic counterparts. These hybrid architectures demonstrate the capacity to tailor optical properties for application-specific photothermal optimization.

The integration of complementary photothermal materials with synergistic optical characteristics represents a strategic advance toward full-spectrum solar absorption, surpassing the limitations of single-component systems [9, 126]. For example, Zhou et al. demonstrated enhanced solar evaporation using Au-GO hybrid structures, where plasmonic coupling between Au nanorods and GO nanosheets extended light absorption into the NIR region, achieving about 20% performance improvement over pristine GO [127]. Liang et al. engineered a hierarchical Co-CNS/M membrane comprising vertically aligned $Ti_3C_2T_x$ nanosheets decorated with cobalt-embedded carbon nanoplates (Fig. 3.6). This multifunctional integration delivered exceptional broadband absorption (>97% across 200–2000 nm), efficient photothermal conversion, and robust corrosion/oxidation resistance [128]. Nevertheless, despite optimized absorber designs, the theoretical absorption ceiling of 100% imposes fundamental constraints on evaporation performance enhancement.

Researchers pioneered innovative strategies to transcend inherent limitations in solar evaporation. Recent advancements in solar steam generation technology highlight transformative progress in evaporator architecture and energy management [129, 130]. Breakthroughs in efficiency and scalability have been achieved through structural innovation and systematic thermal optimization, as elaborated in subsequent discussions.

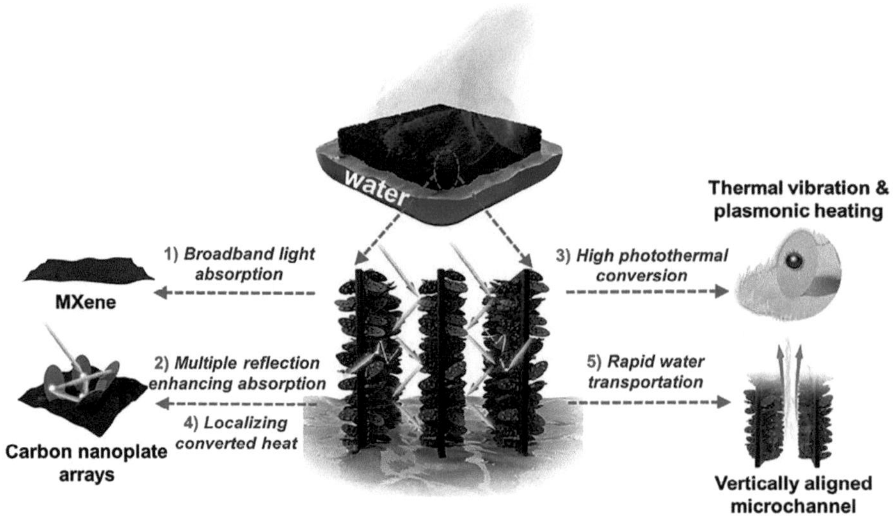

Fig. 3.6 Solar steam generation from Co-CNS/M foam. (Reproduced with permission from [128] Copyright 2020, Wiley-VCH)

3.3 Material Nanoengineering

Nanostructure engineering serves as a critical factor in the improvement of optical properties [47, 131, 132]. The photothermal properties of materials, characterized by their ability to convert absorbed electromagnetic radiation into heat, are inmpacted by their structures, including zero-dimensional, one-dimensional, two-dimensional, and three-dimensional forms [133]. These structural dimensions are essential in determining the solar absorption, the light-to-heat conversion, and the distribution and dissipation of the generated heat [1, 48, 122]. The variations dependent on these dimensions stem from the unique surface-to-volume ratios, quantum confinement effects, and the specific interactions between light and matter that are characteristic of nanomaterials. The development of multidimensional nanomaterials tailored for photothermal applications requires a dimension-specific approach to optimize their photothermal performance [134]. Each category exhibits distinct physicochemical properties and poses unique engineering challenges that necessitate customized strategies to harness their full potentials. A thorough comparative analysis, as depicted in Fig. 3.7, outlines the design paradigms specific to each dimension for photothermal nanomaterials [3, 135]. This approach offers vital insights into the comparative performance of nanomaterials across different dimensions in terms of photothermal conversion, thus providing a systematic basis for the optimization of photothermal nanomaterials.

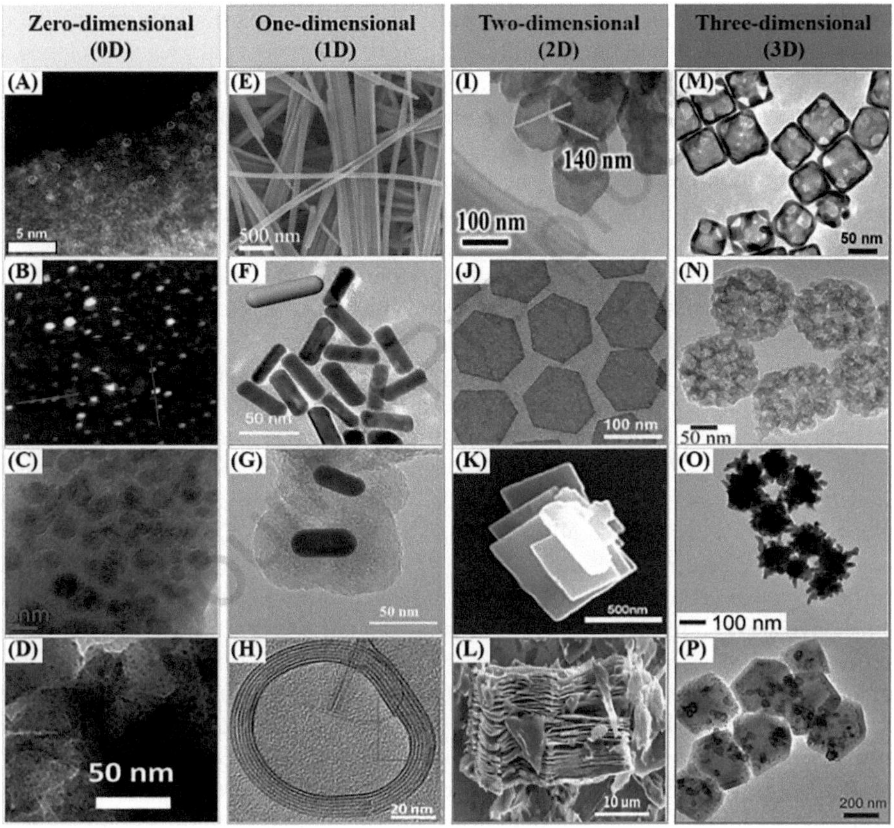

Fig. 3.7 Multi-dimensional nanostructures. (Reproduced with permission from [3] Copyright 2024 Elsevier)

3.3.1 Zero-Dimensional Nanomaterials

Zero-dimensional (0D) materials, featured by their spatial confinement within the nanoscale regime (≤100 nm) across all three dimensions, exhibit distinct quantum mechanical phenomena that fundamentally alter their physical and electronic properties compared to bulk materials [136]. This quantum confinement effect manifests most prominently in discrete electronic states, leading to size-dependent optical and electronic properties. The 0D nanomaterial family encompasses a diverse range of structures, such as single atoms [137], quantum dots [138], and various particles [91, 139, 140], each demonstrating unique quantum confinement characteristics. The quantum size effect in these nanostructures enables precise tuning of light absorption spectra, particularly through size-dependent bandgap engineering, which significantly enhances photothermal conversion efficiency at specific wavelengths [141, 142]. Furthermore, the nanoscale dimensions of these materials facilitate localized thermal energy generation with unprecedented spatial precision, a feature that has revolutionized targeted therapeutic applications, particularly in precision oncology [62, 91]. A notable example is the work by Chang et al., who developed a polyporous Cu single-atom nanozyme functionalized with licogliflozin for heat shock protein silencing-mediated mild photothermal therapy (Fig. 3.7a) [143].

The photothermal conversion of QDs exhibits substantial variation between metallic and non-metallic systems, primarily due to fundamental differences in their electronic structure and light-matter interaction mechanisms [59, 144]. Metallic quantum dots, often made of metals like gold or silver, exhibit strong plasmonic resonance effects. These effects stem from the collective oscillation of conduction electrons at the metal-dielectric interface, leading to efficient localized surface plasmon resonance-mediated photothermal conversion [145]. The photothermal performance of metallic quantum dots has been further improved through innovative nanostructure engineering. This is exemplified by Song et al.'s creation of a carbon nanotube ring (CTR) architecture adorned with gold nanoparticles (CTR@Au NPs) [146]. This system showed enhanced photothermal conversion by synergistic coupling between the CTR's electronic properties and the plasmonic modes of the Au NPs. In contrast, non-metallic QDs, primarily semiconductor systems, rely on interband electronic transitions for photon absorption and energy conversion [57, 59]. Recent advancements in non-metallic QD systems have showed remarkable photothermal capabilities, particularly in the near-infrared region. For example, black phosphorus quantum dots, with lateral dimensions of 2.6 nm and a thickness of approximately 1.5 nm (Fig. 3.7b), have demonstrated remarkable photothermal performance. They exhibit an extinction coefficient of 14.8 L g^{-1} cm^{-1} at 808 nm and a photothermal conversion efficiency of 28.4% [147]. $CdSe/Bi_2Se_3$ QDs was reported with a photothermal conversion efficiency of 27.1% (Fig. 3.7c), demonstrating excellent NIR photostability [148]. These findings underscore the significant potential of quantum dots in photothermal applications, particularly in biomedical contexts where precise thermal control is paramount.

3.3.2 One-Dimensional Nanomaterials

One-dimensional (1D) materials are a unique class of nanostructures characterized by their high aspect ratios, where their length extends beyond the nanoscale in one dimension while remaining confined to the nanoscale in the others [48]. This unique morphological characteristic endows them with directional properties that are particularly advantageous in applications where length-dependent phenomena are critical. Typical examples of 1D nanostructures include nanowires (Fig. 3.7e) [149, 150], nanorods (Fig. 3.7f) [151, 152], nano dumbbells (Fig. 3.7g) [153], nanobelts (Fig. 3.7h) [154], and nanotubes [155, 156], which exhibit anisotropic heat conduction-a phenomenon where thermal conductivity is significantly higher along their longitudinal axis compared to other directions. This anisotropic property can be strategically harnessed in the design of advanced thermal management systems, enabling efficient heat dissipation or targeted thermal delivery in various applications [157, 158].

Moreover, the elongated morphology of 1D nanostructures enhances their interaction with electromagnetic radiation, facilitating broadband light absorption across a wide range of wavelengths [159, 160]. This enhanced light-matter interaction significantly amplifies their photothermal conversion efficiency, making them highly suitable for photothermal applications. A notable demonstration of this principle was reported by Miao et al., who engineered a graphene oxide (GO) by gold nanorods, forming a plasmonic nanofluid with strong optical absorption and enhanced solar steam generation performance [127]. The Au-GO dispersions showed increased absorption across a broad spectrum, extending into the near-infrared region. Due to the coupling effects between nanorods and GO nanostructures for localized heating, the steam generation efficiency reached 84.1% under one sun, representing a 20% enhancement compared to GO dispersion alone. Guo et al. further advanced this concept by constructing a donor-acceptor system composed of a superhydrophilic covalent triazine framework and carbon nanotubes [161]. These structures enhance water extraction and supply, suppressing energy loss through controllable charge migration. The donor-acceptor charge migration not only extends light harvesting but also increases energy conversion efficiency by reducing undesired radiation relaxation, achieving high performance in solar vapor generation with a rate of 1.59 kg m^{-2} h^{-1}. These advancements underscore the significant potential of 1D nanomaterials in photothermal applications and beyond, highlighting their versatility in addressing complex challenges in nanotechnology and energy conversion.

3.3.3 Two-Dimensional Nanomaterials

Two-dimensional nanomaterials, extending beyond the nanoscale in two dimensions, manifest as structures with atomic or molecular-level thickness, forming flat, expansive surfaces [100, 162]. These materials including layered double hydroxides (Fig. 3.7i) [163], graphene [164], nanosheets (Fig. 3.7j) [165], 2D transition metal compounds (Fig. 3.7k) [166], and nitrides (Fig. 3.7l), are distinguished by their high surface areas. This intrinsic property allows extensive interaction with incident light, significantly enhancing photothermal effects. Characterized by their ultra-thin yet laterally expansive nature, these materials can span several micrometers in width while maintaining a thickness of just a few atomic layers [77, 102]. Surface functionalization of 2D materials, through chemical modification or integration of various molecules or nanoparticles, further extends their applicability, enabling precise tuning of photothermal properties for diverse applications.

Innovative research into these materials has led to significant advancements. For example, Yu et al. developed a nanocomposite catalyst, Pt-Mn_2O_3 on a carbon nitride nanosheet, utilizing the nanosheet's ability to effectively disperse Pt nanoparticles, with Mn_2O_3 enhancing photothermal conversion [167]. Gitashree Darabdhara and colleagues demonstrated that the photothermal conversion efficiency of these nanomaterials can be substantially increased by integrating different components into a single material, such as embedding bimetallic Au-Pd on reduced graphene oxide [168]. Further illustrating the potential of 2D materials in photothermal applications, Song et al. reported that Cu nanoparticle-decorated two-dimensional carbon nanosheets exhibit an impressive photothermal conversion efficiency of 65%, highlighting the exceptional capabilities of 2D nanomaterials in harnessing and converting light into thermal energy [169]. These advancements underscore the vast potential of 2D nanomaterials as platforms for innovative photothermal technologies.

3.3.4 Three-Dimensional Nanomaterials

Three-dimensional nanomaterials, which extend beyond the nanoscale in all spatial dimensions, exemplify the integration of nanostructures into bulk forms [45, 170]. These materials combine various nanoscale architectures, such as nanocages (Fig. 3.7m) [171], nanospheres (Fig. 3.7n) [172], nanostars (Fig. 3.7o) [173], nanocubes [174], nanoflowers [175, 176], and core-shell configurations [177], into cohesive three-dimensional frameworks. Notably, zeolitic imidazolate frameworks (ZIFs) represent a distinct class of 3D nanostructures, maintaining the intrinsic properties of nanoscale materials while enabling the creation of larger, functionally versatile constructs [178]. A key feature of 3D nanostructures is their ability to uniformly absorb light and convert it into heat across their entire volume, facilitating large-scale heat generation while utilizing the unique attributes of nanoscale materials.

The intricate 3D nanostructures significantly enhance light trapping, improving light absorption and providing thermal insulation to reduce heat dissipation. This property is advantageous for applications requiring efficient light-to-heat conversion over extended volumes. For instance, Zhang et al. encapsulated Pt nanocubes and Au nanocages within a zeolitic imidazolate framework (ZIF-8), creating a hybrid structure (Au&Pt@ZIF) [179]. This study highlighted the role of metal-organic frameworks as effective thermal insulators, which trap heat within the MOFs and create localized high-temperature zones, boosting the catalytic efficiency of Pt nanocubes. Similarly, polyethylene glycol was used as a phase change material to capture and release heat generated by hollow CuS nanofillers at or near the material's phase transition temperature [180]. This method showed strong photothermal activity and efficient heat retention, highlighting the potential of 3D nmaterials to integrate nanoscale properties for industrial applications.

3.4 Heterojunction Constructing

Designing heterogeneous interfaces is crucial for enhancing photocatalytic and photothermal process [3, 126]. These interfaces categorized by dimensionality into point, line, and plane types, optimize charge transfer, light absorption, and thermal management. Point interfaces introduce localized active sites, line interfaces enable directional energy transfer, and plane interfaces provide large contact areas for efficient interactions. Each interface type uniquely modifies material properties, such as electronic structure, optical absorption, and thermal conductivity. By strategically designing these interfaces, researchers can tailor materials for applications like photocatalysis (e.g., water splitting) [51, 181] and photothermal processes (e.g., solar-driven evaporation or chemical reactions) [182], addressing specific functional challenges.

3.4.1 Point Heterogeneous Interfaces

Point heterogeneous interfaces are crafted by integrating structures of different dimensions, such as zero-dimensional entities with one-dimensional or two-dimensional nanomaterials [136, 183]. This approach is exemplified by layered configurations where 1D structures are strategically placed on 2D substrates, creating point interfaces that primarily promote carrier migration along the 1D axis. For instance, Xu et al. positioned 0D CdS nanoparticles on 2D Bi_2Se_3 nanosheets, forming a point heterojunction interface that significantly enhances charge separation under light exposure [184]. Similarly, Li et al. engineered point interfaces by incorporating 0D Pt and $Cu_{1.94}S$ nanoparticles along the sides of 1D $Zn_xCd_{1-x}S$ nanorods, resulting in enhanced light absorption and improved charge tra nsfer efficiency [185].

Moreover, quantum dots, due to their nanoscopic scale, serve as localized heat sources upon irradiation, significantly enhancing photothermal conversion efficiency [144, 147]. This enhancement is attributed to quantum confinement effects, which optimize their electronic and optical properties for photothermal applications. For example, 0D Ni_2P quantum dots were used as a cocatalyst on the surface of ultra-thin TiO_2 nanosheets, creating a point heterogeneous interface that enhances charge transfer, inhibits charge recombination, and accelerates the rate of photothermal catalytic hydrogen production [186]. This assembly of heterogeneous interfaces demonstrates a complex approach to tuning material properties for advanced photothermal catalytic applications, highlighting the critical importance of dimensional integration in optimizing functional performance.

3.4.2 Linear Heterogeneous Interfaces

The design and fabrication of linear interfaces are pivotal in photoinduced and photothermal catalysis [48]. These interfaces are fabricated by either promoting the growth of one-dimensional structures along existing 1D frameworks or by integrating 1D materials onto two-dimensional substrates [149, 187]. For instance, a one-step solvothermal method was introduced to facilitate the in-situ formation of 1D CdS nanorods on 2D Nb_2CT_X matrices [188]. This technique ensures a close interfacial contact, creating Schottky heterojunctions at the CdS/Nb_2CT_X interface, which enhances charge transfer and electron-hole separation across the catalyst, thereby boosting its photocatalytic performance. Furthermore, Avijit Pramanik et al. introduced a linear heterostructure composed of 1D WO_3 nanowires and 2D reduced graphene oxide, exhibiting exceptional anti-superbug properties under near-infrared light [189]. This activity is driven by a synergistic photothermal and photocatalytic mechanism activated by NIR illumination.

In addition, linear interfaces can be established by vertically stacking two-dimensional layers or by epitaxially growing a 2D structure on the edge of another 2D or three-dimensional entity. For example, Pan et al. constructed a linear heterojunction interface with snowflake-like CuS embedded within graphite carbon nitride nanosheets [190]. This arrangement boosts charge transfer and separation, minimizes the distance over which charge transfer occurs, and optimally harnesses the photothermal effects induced by near-infrared light. In a similar vein, a linear heterojunction was created through the in situ growth of 2D CdS nanosheets on 2D $Ti_3C_2T_x$ MXene nanosheets, thereby broadening the NIR absorption spectrum [191]. This heterojunction enhances photochemical reaction rates and supports the swift separation of photocarriers, leading to improved thermal conductivity and enhanced efficiency in heat distribution. These deliberate designs of linear heterogeneous interfaces contribute to advancements in the field by fostering enhanced thermal management and photothermal efficiency in catalysis, underscoring the pivotal role of structural integration in augmenting material properties and functionalities.

3.4.3 Plane Heterogeneous Interfaces

The engineering of heterogeneous interfaces within two-dimensional materials, including graphene layers, nanosheets, layered double hydroxides, and thin film coatings on substrates, represents a pioneering effort in materials science to enhance photothermal properties [192–194]. A example of this is the synthesis of photocatalysts by integrating $MoSe_2$ nanoplates with CuSe nanosheets [195]. This approach optimizes the interface between CuSe and $MoSe_2$, reducing internal resistance, improving separation efficiency, and promoting strong binding. The photocatalysts have a large specific surface area, offering superior full-spectrum absorption compared to their unmodified counterparts. Additionally, Xu et al. have precisely engineered a face-to-face contact 2D/2D Bi_2WO_6/Ti_3C_2 heterostructure [196]. This innovative configuration enhances light absorption, accelerates charge carrier transfer, and improves photothermal conversion efficiency. The planar interfaces offer large surfaces for light absorption and act as effective barriers to heat dissipation, thereby localizing thermal effects. The interaction between different layers can induce novel electronic states, which are crucial for optimizing photothermal processes.

The strategic design of heterogeneous interfaces within materials is paramount for the manipulation of their photothermal properties [197]. These interfaces can precipitate a series of beneficial outcomes [144, 191], including: (1) Enhanced light absorption and heat generation: By significantly improving light absorption through maximizing the interaction surface between different materials, these interfaces subsequently enhance heat generation, a critical factor for efficient photothermal conversion. (2) Improved heat flow and dissipation control: The capacity to modulate heat flow and enhance dissipation through these interfaces allows for more precise thermal management, essential for maintaining optimal operating temperatures in photothermal applications. (3) Tailored electronic and optical properties: Customizing electronic and optical properties through the design of these interfaces enables the material to be finely tuned for specific applications, leveraging the unique benefits of photothermal processes.

3.5 Fabrication Techniques

Utilizing advanced fabrication techniques such as layer-by-layer assembly [198], sol-gel processes [199], and aerosol jet printing [200], facilitates an unprecedented level of precision in the design and control of material properties. These methodologies not only enhance the performance and efficiency of materials but also allow for the adjustment of their microstructures and chemical compositions. Layer-by-Layer Assembly: Layer-by-layer (LbL) assembly is a versatile technique that allows for the construction of nanoscale multi-layered structures with precisely tailored properties [201]. This method involves the sequential deposition of oppositely charged layers to build up films or coatings of nanometer thickness. The primary advantage of LbL assembly is its ability to create composite materials with customized optical, electrical, or mechanical properties that are finely tuned for specific applications. For instance, by alternating layers of conductive polymers and insulating materials, one can fabricate advanced electronic devices with enhanced performance characteristics. Additionally, LbL assembly can be used to encapsulate bioactive molecules, thus providing controlled release mechanisms for pharmaceutical applications.

Sol–gel process: The sol–gel process is another fundamental fabrication technique, renowned for its low-temperature synthesis of materials [202]. This method involves the transition from a colloidal solution (sol) to a gel phase, which can then be dried and heat-treated to form ceramic or glassy materials. The sol-gel process offers fine control over the porosity and surface area of the resultant materials, which is crucial for applications such as catalysts and sensors. By adjusting the precursors and reaction conditions, researchers can tailor the chemical composition and microstructure of the sol–gel-derived materials, thereby optimizing their functional properties. Moreover, the sol–gel technique is particularly beneficial for producing coatings on substrates, enabling the creation of protective or functional layers with unique characteristics. Aerosol Jet Printing: Aerosol jet printing provides a high-resolution method for depositing materials onto substrates, including electronic, structural, and biological materials. This technique uses an aerosol jet to deliver a fine mist of material, which is then directed onto a target substrate using a focused gas stream [203]. The ability to deposit materials with such precision makes aerosol jet printing ideal for applications requiring complex, patterned coatings, such as electronic circuitries and sensor interfaces. Furthermore, this method is compatible with a wide range of materials, from metals and ceramics to polymers and biological agents, thereby broadening its applicability in various fields of material science and engineering.

Integrating Fabrication Techniques: By integrating these advanced fabrication techniques, scientists and engineers can create materials with unprecedented functionality and performance. For example, combining the layer-by-layer assembly with the sol-gel process can result in hybrid materials with enhanced mechanical strength and thermal stability, while maintaining desired surface characteristics [204]. In a similar vein, combining aerosol jet printing with other techniques facilitates the creation of multifunctional devices that integrate both electronic and structural elements. Extensive research has been dedicated to developing inks with properties such as conductivity, semiconductivity, magnetism, piezoresistivity, and piezoelectricity [205]. The methods of layer-by-layer assembly, sol-gel processes, and aerosol jet printing form the foundation for advancing high-performance materials across diverse applications. These techniques extend the possibilities in materials science and engineering by providing advanced tools for the precise control of material properties. As research progresses, these fabrication methods are expected to play increasingly significant roles in addressing the challenges of modern technology and industry, thereby contributing to the advancement of numerous scientific and engineering disciplines.

References

1. X. Cui, Q. Ruan, X. Zhuo, X. Xia, J. Hu, R. Fu, Y. Li, J. Wang, H. Xu, Photothermal Nanomaterials: A Powerful Light-to-Heat Converter, ***Chemical reviews***, 123 (2023) 6891-6952.
2. X. Wu, Y. Lu, X. Ren, P. Wu, D. Chu, X. Yang, H. Xu, Interfacial Solar Evaporation: From Fundamental Research to Applications, ***Advanced materials***, 36 (2024) e2313090.
3. X. Yu, S. Fan, B. Zhu, S.I. El-Hout, J. Zhang, C. Chen, Recent progress on photothermal nanomaterials: Design, mechanism, and applications, ***Green Energy & Environment***, (2024).
4. Y. Wang, W. Zhao, Y. Lee, Y. Li, Z. Wang, K.C. Tam, Thermo-adaptive interfacial solar evaporation enhanced by dynamic water gating, ***Nature communications***, 15 (2024) 6157.

5. A. Chu, S. Zhang, J. Jin, Recent Progress on Solar-Driven Interfacial Evaporation for Resource Recovery and Pollutant Removal, ***Advanced materials***, (2025) e2505656.
6. L. Hou, S. Li, Y. Qi, J. Liu, Z. Cui, X. Liu, Y. Zhang, N. Wang, Y. Zhao, Advancing Efficiency in Solar-Driven Interfacial Evaporation: Strategies and Applications, ***ACS nano***, 19 (2025) 9636-9683.
7. M. Gao, L. Zhu, C.K. Peh, G.W. Ho, Solar absorber material and system designs for photothermal water vaporization towards clean water and energy production, ***Energy & Environmental Science***, 12 (2019) 841-864.
8. C. Chen, Y. Kuang, L. Hu, Challenges and Opportunities for Solar Evaporation, ***Joule***, 3 (2019) 683-718.
9. M.S. Irshad, N. Arshad, G. Maqsood, I. Ahmed, B. Shakoor, M.S. Asghar, U. Ghazanfar, L. Lin, M.A.K.Y. Shah, I. Ahmed, X. Chen, J. Wang, C. Yi, J. Li, J. Qian, W. Li, Z. Said, H. Li, N.X. Ho, H. Wang, X. Fan, N. Mushtaq, V.-D. Dao, T. Mei, X. Wang, Advancing water collection efficiency in hybrid solar evaporators: key factors, strategic innovations, and synergistic applications, ***Materials Science and Engineering: R: Reports***, 165 (2025) 101018.
10. L. Zhang, Z. Yuan, X. Fu, S. Shi, X. Chen, P. Chen, D. Ye, Biomass-Derived Gradient and Aligned Structured Aerogel for Sustainable Agricultural Irrigation, ***Nano letters***, 25 (2025) 5383-5390.
11. C. Zhang, S. Xie, G. Chen, J. Liu, S. Li, X. Zheng, Z. Guo, Callistemon Viminalis-Inspired Rotational Structure for Omnidirectional Fog Collection and Power Generation, ***Advanced Functional Materials***, (2025).
12. J.H. Zhang, R. Mittapally, A. Oluwade, G. Chen, Mechanisms and scale-up potential of 3D solar interfacial-evaporators, ***Energy & Environmental Science***, (2025).
13. H. Zheng, J. Fan, A. Chen, X. Li, X. Xie, Y. Liu, Z. Ding, Enhancing Solar-Driven Water Purification by Multiscale Biomimetic Evaporators Featuring Lamellar MoS(2)/GO Heterojunctions, ***ACS nano***, 18 (2024) 3115-3124.
14. S. Cao, P. Rathi, X. Wu, D. Ghim, Y.S. Jun, S. Singamaneni, Cellulose Nanomaterials in Interfacial Evaporators for Desalination: A "Natural" Choice, ***Advanced materials***, 33 (2021) e2000922.
15. T. Gao, Y. Li, C. Chen, Z. Yang, Y. Kuang, C. Jia, J. Song, E.M. Hitz, B. Liu, H. Huang, J. Yu, B. Yang, L. Hu, Architecting a Floatable, Durable, and Scalable Steam Generator: Hydrophobic/Hydrophilic Bifunctional Structure for Solar Evaporation Enhancement, ***Small Methods***, 3 (2018).
16. Q. Huang, X. Liang, C. Yan, Y. Liu, Review of interface solar-driven steam generation systems: High-efficiency strategies, applications and challenges, ***Applied Energy***, 283 (2021) 116361.
17. Y. Dong, Y. Tan, K. Wang, Y. Cai, J. Li, C. Sonne, C. Li, Reviewing wood-based solar-driven interfacial evaporators for desalination, ***Water research***, 223 (2022) 119011.
18. C. Finnerty, L. Zhang, D.L. Sedlak, K.L. Nelson, B. Mi, Synthetic Graphene Oxide Leaf for Solar Desalination with Zero Liquid Discharge, ***Environmental science & technology***, 51 (2017) 11701-11709.
19. M.S. Asghar, M.S. Irshad, N. Arshad, M.A. Huwayz, M. Alomar, G. Maqsood, M.A. Ali, U. Ghanzanfar, M. Sabir, J. Li, V.D. Dao, N.X. Ho, X. Wang, Z. Guo, Functionalized carbon nanotubes interconnected with metal-organic frameworks for in-situ solar-driven evaporation and salt recovery from seawater, ***Water research***, 282 (2025) 123643.
20. P. Cao, L. Zhao, Z. Yang, P. Yuan, Y. Zhang, Q. Li, Carbon Nanotube Network-Based Solar-Thermal Water Evaporator and Thermoelectric Module for Electricity Generation, ***ACS Applied Nano Materials***, 4 (2021) 8906-8912.
21. D. Fan, Y. Lu, H. Zhang, H. Xu, C. Lu, Y. Tang, X. Yang, Synergy of photocatalysis and photothermal effect in integrated 0D perovskite oxide/2D MXene heterostructures for simultaneous water purification and solar steam generation, ***Applied Catalysis B: Environmental***, 295 (2021) 120285.

22. P. Zhou, K. Yang, L. Liu, Q. Liu, N. Zhang, J. Xu, Janus-structured MXene-PA/MS with an ultrathin intermediate layer for high-salinity water desalination and wastewater purification, ***Journal of colloid and interface science***, 682 (2024) 795-803.
23. M. Gao, P.K.N. Connor, G.W. Ho, Plasmonic photothermic directed broadband sunlight harnessing for seawater catalysis and desalination, ***Energy & Environmental Science***, 9 (2016) 3151-3160.
24. C. Chen, L. Zhou, J. Yu, Y. Wang, S. Nie, S. Zhu, J. Zhu, Dual functional asymmetric plasmonic structures for solar water purification and pollution detection, ***Nano Energy***, 51 (2018) 451-456.
25. X. Wu, M.E. Robson, J.L. Phelps, J.S. Tan, B. Shao, G. Owens, H. Xu, A flexible photothermal cotton-CuS nanocage-agarose aerogel towards portable solar steam generation, ***Nano Energy***, 56 (2019) 708-715.
26. T. Wu, B. Liu, C. Liu, J. Wan, A. Yang, K. Liu, F. Shi, J. Zhao, Z. Lu, G. Chen, A. Pei, H.Y. Hwang, Y. Cui, Solar-driven efficient heterogeneous subminute water disinfection nanosystem assembled with fingerprint MoS2, ***Nature Water***, 1 (2023) 462-470.
27. Z. Guo, W. Zhou, N. Arshad, Z. Zhang, D. Yan, M.S. Irshad, L. Yu, X. Wang, Excellent energy capture of hierarchical MoS2 nanosheets coupled with MXene for efficient solar evaporators and thermal packs, ***Carbon***, 186 (2022) 19-27.
28. L. Wang, Z. Liu, J. Xu, K. Wang, Q. Wang, G. Liu, Conical solar-thermo-radiative evaporator for sustainable desalination and salt recovery, ***Desalination***, 567 (2023) 116993.
29. Y. Shi, R. Li, Y. Jin, S. Zhuo, L. Shi, J. Chang, S. Hong, K.-C. Ng, P. Wang, A 3D Photothermal Structure toward Improved Energy Efficiency in Solar Steam Generation, ***Joule***, 2 (2018) 1171-1186.
30. L. Cheng, L. Zuo, C. Li, J. Shen, L. Wang, Y. Li, D. Dukashin, P. Senin, S. Yan, T. Bian, Sandwiched solar interfacial evaporation system: Integrating phase change material/hydrogel/NiCo2O4 nanoflower for superior desalination, ***Separation and Purification Technology***, 348 (2024) 127816.
31. L. Zhu, L. Zhu, T. Shi, K. Zhao, W. He, L. Han, Z. Jin, J. Kang, S. Sun, N. Cao, Z. Yu, Solar-Powered Hierarchical Microenvironments with Authigenic Multi-Field Synergies for Simultaneous Extraction of Freshwater and Cesium, ***Advanced science***, (2025) e2505997.
32. G. Liu, Thermoplasmonics, ***Springer***, ISBN 978-981-97-8331-1 (2024).
33. P. Tao, G. Ni, C. Song, W. Shang, J. Wu, J. Zhu, G. Chen, T. Deng, Solar-driven interfacial evaporation, ***Nature Energy***, 3 (2018) 1031-1041.
34. X. Wu, G.Y. Chen, G. Owens, D. Chu, H. Xu, Photothermal materials: A key platform enabling highly efficient water evaporation driven by solar energy, ***Materials Today Energy***, 12 (2019) 277-296.
35. H. Yu, H. Jin, Y. Liang, D. Wang, Y. Lu, X. Yang, H. Xu, A New Era of Passive Continuous Freshwater Production: When Interfacial Solar Evaporation Marries Moisture Harvest, ***ACS Energy Letters***, 10 (2025) 1192-1215.
36. Y. Liu, S. Yu, R. Feng, A. Bernard, Y. Liu, Y. Zhang, H. Duan, W. Shang, P. Tao, C. Song, T. Deng, A bioinspired, reusable, paper-based system for high-performance large-scale evaporation, ***Advanced materials***, 27 (2015) 2768-2774.
37. Q. Liao, K. Zhu, X. Hao, C. Wu, J. Li, H. Cheng, J. Yan, L. Jiang, L. Qu, Bio-Inspired Ultrathin Perfect Absorber for High-Performance Photothermal Conversion, ***Advanced materials***, 36 (2024) e2313366.
38. Y. Zhang, Y. Wang, B. Yu, K. Yin, Z. Zhang, Hierarchically Structured Black Gold Film with Ultrahigh Porosity for Solar Steam Generation, ***Advanced materials***, 34 (2022) e2200108.
39. L. Zhou, Y. Tan, D. Ji, B. Zhu, P. Zhang, J. Xu, Q. Gan, Z. Yu, J. Zhu, Self-assembly of highly efficient, broadband plasmonic absorbers for solar steam generation, ***Science advances***, 2 (2016) e1501227.
40. K. Bae, G. Kang, S.K. Cho, W. Park, K. Kim, W.J. Padilla, Flexible thin-film black gold membranes with ultrabroadband plasmonic nanofocusing for efficient solar vapour generation, ***Nature communications***, 6 (2015) 10103.

41. F. Meng, W. Hao, S. Yu, R. Feng, Y. Liu, F. Yu, P. Tao, W. Shang, J. Wu, C. Song, T. Deng, Vapor-Enabled Propulsion for Plasmonic Photothermal Motor at the Liquid/Air Interface, ***Journal of the American Chemical Society***, 139 (2017) 12362-12365.
42. L. Zhou, Y. Tan, J. Wang, W. Xu, Y. Yuan, W. Cai, S. Zhu, J. Zhu, 3D self-assembly of aluminium nanoparticles for plasmon-enhanced solar desalination, ***Nature Photonics***, 10 (2016) 393-398.
43. C. Sheng, N. Yang, Y. Yan, X. Shen, C. Jin, Z. Wang, Q. Sun, Bamboo decorated with plasmonic nanoparticles for efficient solar steam generation, ***Applied Thermal Engineering***, 167 (2020) 114712.
44. S. Chen, Z. Sun, W. Xiang, C. Shen, Z. Wang, X. Jia, J. Sun, C.-J. Liu, Plasmonic wooden flower for highly efficient solar vapor generation, ***Nano Energy***, 76 (2020) 104998.
45. P. Cheng, M. Klingenhof, H. Honig, L. Zhang, P. Strasser, P. Schaaf, D. Lei, D. Wang, Synergizing Plasmonic Local Heating and 3D Nanostructures to Boost the Solar-to-Vapor Efficiency Beyond 100, ***Advanced materials***, (2024) e2415655.
46. J. Chen, Z. Ye, F. Yang, Y. Yin, Plasmonic Nanostructures for Photothermal Conversion, ***Small science***, 1 (2021) 2000055.
47. K.W. Tan, C.M. Yap, Z. Zheng, C.Y. Haw, P.S. Khiew, W.S. Chiu, State-of-the-Art Advances, Development, and Challenges of Metal Oxide Semiconductor Nanomaterials for Photothermal Solar Steam Generation, ***Advanced Sustainable Systems***, 6 (2022).
48. J. Hu, Y. Sun, Z. Liu, B. Zhu, L. Zhang, N. Xu, M. Zhu, J. Zhu, Z. Chen, Photothermal fabrics for solar-driven seawater desalination, ***Progress in Materials Science***, 150 (2025) 101407.
49. L. Zhu, L. Tian, S. Jiang, L. Han, Y. Liang, Q. Li, S. Chen, Advances in photothermal regulation strategies: from efficient solar heating to daytime passive cooling, ***Chemical Society reviews***, 52 (2023) 7389-7460.
50. R. Asahi, T. Morikawa, T. Ohwaki, K. Aoki, Y. Taga, Visible-light photocatalysis in nitrogen-doped titanium oxides, ***Science***, 293 (2001) 269-271.
51. H. Huang, R. Shi, X. Zhang, J. Zhao, C. Su, T. Zhang, Photothermal-Assisted Triphase Photocatalysis Over a Multifunctional Bilayer Paper, ***Angewandte Chemie***, 60 (2021) 22963-22969.
52. G. Liu, J. Xu, K. Wang, Solar water evaporation by black photothermal sheets, ***Nano Energy***, 41 (2017) 269-284.
53. W. Li, H. Jian, W. Wang, D. Yu, Highly efficient solar vapour generation via self-floating three-dimensional Ti2O3-based aerogels, ***Colloids and Surfaces A: Physicochemical and Engineering Aspects***, 634 (2022) 128031.
54. J. Wang, Y. Li, L. Deng, N. Wei, Y. Weng, S. Dong, D. Qi, J. Qiu, X. Chen, T. Wu, High-Performance Photothermal Conversion of Narrow-Bandgap Ti2O3 Nanoparticles, ***Advanced materials***, 29 (2017).
55. X. Chen, L. Liu, P.Y. Yu, S.S. Mao, Increasing solar absorption for photocatalysis with black hydrogenated titanium dioxide nanocrystals, ***Science***, 331 (2011) 746-750.
56. G. Liu, T. Chen, J. Xu, G. Li, K. Wang, Solar evaporation for simultaneous steam and power generation, ***Journal of Materials Chemistry A***, 8 (2020) 513-531.
57. A. Agrawal, S.H. Cho, O. Zandi, S. Ghosh, R.W. Johns, D.J. Milliron, Localized Surface Plasmon Resonance in Semiconductor Nanocrystals, ***Chemical reviews***, 118 (2018) 3121-3207.
58. S. Santoro, A.H. Avci, A. Politano, E. Curcio, The advent of thermoplasmonic membrane distillation, ***Chemical Society reviews***, 51 (2022) 6087-6125.
59. A. Agrawal, R.W. Johns, D.J. Milliron, Control of Localized Surface Plasmon Resonances in Metal Oxide Nanocrystals, ***Annual Review of Materials Research***, 47 (2017) 1-31.
60. V. Georgakilas, J.A. Perman, J. Tucek, R. Zboril, Broad family of carbon nanoallotropes: classification, chemistry, and applications of fullerenes, carbon dots, nanotubes, graphene, nanodiamonds, and combined superstructures, ***Chemical reviews***, 115 (2015) 4744-4822.
61. B. Wang, S. Lu, The light of carbon dots: From mechanism to applications, ***Matter***, 5 (2022) 110-149.

62. V.N. Mochalin, O. Shenderova, D. Ho, Y. Gogotsi, The properties and applications of nanodiamonds, ***Nature nanotechnology***, 7 (2011) 11-23.
63. X. Chang, Y. Xu, M. von Delius, Recent advances in supramolecular fullerene chemistry, ***Chemical Society reviews***, 53 (2024) 47-83.
64. Y. Liu, H. Shi, Z.-S. Wu, Recent status, key strategies and challenging perspectives of fast-charging graphite anodes for lithium-ion batteries, ***Energy & Environmental Science***, 16 (2023) 4834-4871.
65. S. Sun, H. Li, M. Zhang, B. Sun, Y. Xie, W. Zhou, P. Yang, H.Y. Mi, Z. Guo, C. Liu, C. Shen, A Multifunctional Asymmetric Fabric for Sustained Electricity Generation from Multiple Sources and Simultaneous Solar Steam Generation, ***Small***, 19 (2023) e2303716.
66. C.T.K. Finnerty, A.K. Menon, K.M. Conway, D. Lee, M. Nelson, J.J. Urban, D. Sedlak, B. Mi, Interfacial Solar Evaporation by a 3D Graphene Oxide Stalk for Highly Concentrated Brine Treatment, ***Environmental science & technology***, 55 (2021) 15435-15445.
67. T.D. Le, D. Yang, H.K. Nam, Y. Lee, C.T. Lim, B.J. Lee, S.W. Kim, Y.J. Kim, Low-Cost, Eco-Friendly, and High-Performance 3D Laser-Induced Graphene Evaporator for Continuous Solar-Powered Water Desalination, ***ACS nano***, 18 (2024) 33220-33231.
68. Y.S. Jun, X. Wu, D. Ghim, Q. Jiang, S. Cao, S. Singamaneni, Photothermal Membrane Water Treatment for Two Worlds, ***Accounts of chemical research***, 52 (2019) 1215-1225.
69. Y. Zeng, K. Wang, J. Yao, H. Wang, Hollow carbon beads for significant water evaporation enhancement, ***Chemical Engineering Science***, 116 (2014) 704-709.
70. W. Li, X. Tian, X. Li, S. Han, C. Li, X.-Z. Zhai, Y. Kang, Z.-Z. Yu, Ultrahigh solar steam generation rate of a vertically aligned reduced graphene oxide foam realized by dynamic compression, ***Journal of Materials Chemistry A***, 9 (2021) 14859-14867.
71. C.W. Abney, R.T. Mayes, T. Saito, S. Dai, Materials for the Recovery of Uranium from Seawater, ***Chemical reviews***, 117 (2017) 13935-14013.
72. H. Liu, Z. Huang, K. Liu, X. Hu, J. Zhou, Interfacial Solar-to-Heat Conversion for Desalination, ***Advanced Energy Materials***, 9 (2019).
73. T. Meng, Z. Li, Z. Wan, J. Zhang, L. Wang, K. Shi, X. Bu, S.M. Alshehri, Y. Bando, Y. Yamauchi, D. Li, X. Xu, MOF-Derived nanoarchitectured carbons in wood sponge enable solar-driven pumping for high-efficiency soil water extraction, ***Chemical Engineering Journal***, 452 (2023) 139193.
74. G. Ni, G. Li, Svetlana V. Boriskina, H. Li, W. Yang, T. Zhang, G. Chen, Steam generation under one sun enabled by a floating structure with thermal concentration, ***Nature Energy***, 1 (2016).
75. Y. Luo, Z. Chen, Q. Li, X. Chen, Laser-Induced Porous Graphene on a Polyimide Membrane with a Melamine Sponge Framework (PI@MS) for Long-Term Stable Steam Generation, ***ACS Applied Energy Materials***, 4 (2021) 9766-9774.
76. X. Li, B. Zhu, J. Zhu, Graphene oxide based materials for desalination, ***Carbon***, 146 (2019) 320-328.
77. X. Li, W. Xu, M. Tang, L. Zhou, B. Zhu, S. Zhu, J. Zhu, Graphene oxide-based efficient and scalable solar desalination under one sun with a confined 2D water path, ***Proceedings of the National Academy of Sciences of the United States of America***, 113 (2016) 13953-13958.
78. B. Wang, C. Wang, Y. Li, J. Jin, X. Lin, C. Shi, Bionic design: insights from nature for solar interfacial evaporators, ***Energy & Environmental Science***, 18 (2025) 3432-3461.
79. L. Xiao, X. Chen, X. Yang, J. Sun, J. Geng, Recent Advances in Polymer-Based Photothermal Materials for Biological Applications, ***ACS Applied Polymer Materials***, 2 (2020) 4273-4288.
80. M.A. Baird, R.S. Kingsbury, Y. Wang, A. Gang, C. Zhu, M.L. Whittaker, B.A. Helms, Microporous Polymer Sorbents for Direct Lithium Extraction, ***ACS Energy Letters***, 9 (2024) 4361-4368.
81. C. Jiang, Z. Yan, Y. Bai, R. Li, M. Wu, W.-H. Yu, H.-M. Chen, P. Hu, K.-Q. Zhao, K. Niu, Y. Liu, S.-H. Ho, W. Wang, Y. Wei, Scalable, high-efficiency porous monolithic polymer foam for solar-driven interfacial water evaporation and lithium extraction, ***npj Clean Water***, 8 (2025).

82. J. Sun, T. Wu, H. Wu, W. Li, L. Li, S. Liu, J. Wang, W.J. Malfait, S. Zhao, Aerogel-based solar-powered water production from atmosphere and ocean: A review, ***Materials Science and Engineering: R: Reports***, 154 (2023) 100735.
83. H.W. Lim, S.H. Park, S.J. Lee, 3D thermoresponsive hydrogel with enhanced water uptake and active evaporation for effective interfacial solar steam generation, ***Desalination***, 550 (2023) 116368.
84. Z. Lin, T. Wu, J. Shi, B. Zhou, C. Zhu, Y. Wang, R. Liang, M. Mizuno, Poly(N-phenylglycine)-Based Bioinspired System for Stably and Efficiently Enhancing Solar Evaporation, ***ACS Sustainable Chemistry & Engineering***, 9 (2020) 448-457.
85. Z. Zheng, W. Li, H. Liu, X. Wang, Sustainable Interfacial Evaporation System Based on Hierarchical MXene/Polydopamine/Magnetic Phase-Change Microcapsule Composites for Solar-Driven Seawater Desalination, ***ACS applied materials & interfaces***, 14 (2022) 50966-50981.
86. D. Huang, J. Zhang, G. Wu, S.-C. Chen, Y.-Z. Wang, A solar evaporator based on hollow polydopamine nanotubes with all-in-one synergic design for highly-efficient water purification, ***Journal of Materials Chemistry A***, 9 (2021) 15776-15786.
87. S. Cao, A. Thomas, C. Li, Emerging Materials for Interfacial Solar-Driven Water Purification, ***Angewandte Chemie***, 62 (2023) e202214391.
88. J. Zhang, C. Yang, R. Zhang, R. Chen, Z. Zhang, W. Zhang, S.H. Peng, X. Chen, G. Liu, C.S. Hsu, C.S. Lee, Biocompatible D-A Semiconducting Polymer Nanoparticle with Light-Harvesting Unit for Highly Effective Photoacoustic Imaging Guided Photothermal Therapy, ***Adv Funct Mater***, 27 (2017).
89. Q. Zou, M. Abbas, L. Zhao, S. Li, G. Shen, X. Yan, Biological Photothermal Nanodots Based on Self-Assembly of Peptide-Porphyrin Conjugates for Antitumor Therapy, ***Journal of the American Chemical Society***, 139 (2017) 1921-1927.
90. B. Guo, Z. Sheng, D. Hu, A. Li, S. Xu, P.N. Manghnani, C. Liu, L. Guo, H. Zheng, B. Liu, Molecular Engineering of Conjugated Polymers for Biocompatible Organic Nanoparticles with Highly Efficient Photoacoustic and Photothermal Performance in Cancer Theranostics, ***ACS nano***, 11 (2017) 10124-10134.
91. X. Li, L. Liu, S. Li, Y. Wan, J.X. Chen, S. Tian, Z. Huang, Y.F. Xiao, X. Cui, C. Xiang, Q. Tan, X.H. Zhang, W. Guo, X.J. Liang, C.S. Lee, Biodegradable pi-Conjugated Oligomer Nanoparticles with High Photothermal Conversion Efficiency for Cancer Theranostics, ***ACS nano***, 13 (2019) 12901-12911.
92. L. Zhao, Y. Liu, R. Chang, R. Xing, X. Yan, Supramolecular Photothermal Nanomaterials as an Emerging Paradigm toward Precision Cancer Therapy, ***Advanced Functional Materials***, 29 (2018).
93. X. Li, S. Lee, J. Yoon, Supramolecular photosensitizers rejuvenate photodynamic therapy, ***Chemical Society reviews***, 47 (2018) 1174-1188.
94. L. Zhao, Y. Liu, R. Xing, X. Yan, Supramolecular Photothermal Effects: A Promising Mechanism for Efficient Thermal Conversion, ***Angewandte Chemie***, 59 (2020) 3793-3801.
95. O. Goor, S.I.S. Hendrikse, P.Y.W. Dankers, E.W. Meijer, From supramolecular polymers to multi-component biomaterials, ***Chemical Society reviews***, 46 (2017) 6621-6637.
96. S. Li, Q. Zou, Y. Li, C. Yuan, R. Xing, X. Yan, Smart Peptide-Based Supramolecular Photodynamic Metallo-Nanodrugs Designed by Multicomponent Coordination Self-Assembly, ***Journal of the American Chemical Society***, 140 (2018) 10794-10802.
97. P. Chen, Y. Ma, Z. Zheng, C. Wu, Y. Wang, G. Liang, Facile syntheses of conjugated polymers for photothermal tumour therapy, ***Nature communications***, 10 (2019) 1192.
98. Y. Jiang, J. Li, X. Zhen, C. Xie, K. Pu, Dual-Peak Absorbing Semiconducting Copolymer Nanoparticles for First and Second Near-Infrared Window Photothermal Therapy: A Comparative Study, ***Advanced materials***, 30 (2018) e1705980.
99. F. Liu, Z. Fan, Defect engineering of two-dimensional materials for advanced energy conversion and storage, ***Chemical Society reviews***, 52 (2023) 1723-1772.

100. C. Tan, X. Cao, X.J. Wu, Q. He, J. Yang, X. Zhang, J. Chen, W. Zhao, S. Han, G.H. Nam, M. Sindoro, H. Zhang, Recent Advances in Ultrathin Two-Dimensional Nanomaterials, ***Chemical reviews***, 117 (2017) 6225-6331.
101. B. Zeng, T. Kumar, H. Wu, S. Stark, H. Hamza, H. Zhao, H. Xu, X. Zhang, Evolution of Morphology and Distribution of Salt Crystals on a Photothermal Layer during Solar Interfacial Evaporation, ***Langmuir : the ACS journal of surfaces and colloids***, 39 (2023) 14737-14747.
102. A. VahidMohammadi, J. Rosen, Y. Gogotsi, The world of two-dimensional carbides and nitrides (MXenes), ***Science***, 372 (2021).
103. Y. Yuan, B. Niu, Q. Yu, X. Guo, Z. Guo, J. Wen, T. Liu, H. Zhang, N. Wang, Photoinduced Multiple Effects to Enhance Uranium Extraction from Natural Seawater by Black Phosphorus Nanosheets, ***Angewandte Chemie***, 59 (2020) 1220-1227.
104. A.M. Saleque, S. Ahmed, M.N.A.S. Ivan, M.I. Hossain, W. Qarony, P.K. Cheng, J. Qiao, Z.L. Guo, L. Zeng, Y.H. Tsang, High-temperature solar steam generation by MWCNT-HfTe2 van der Waals heterostructure for low-cost sterilization, ***Nano Energy***, 94 (2022) 106916.
105. S. Ge, J. Song, J. Zhang, X.-L. Yang, M.-H. Xie, Flexible, scalable, and bifunctional MOF-based nanofibrous membranes with gems-on-string structure for simultaneous efficient water evaporation and pollution remediation, ***Chemical Engineering Journal***, 514 (2025) 163143.
106. Q. Ma, P. Yin, M. Zhao, Z. Luo, Y. Huang, Q. He, Y. Yu, Z. Liu, Z. Hu, B. Chen, H. Zhang, MOF-Based Hierarchical Structures for Solar-Thermal Clean Water Production, ***Advanced materials***, 31 (2019) e1808249.
107. M. Chhowalla, H.S. Shin, G. Eda, L.J. Li, K.P. Loh, H. Zhang, The chemistry of two-dimensional layered transition metal dichalcogenide nanosheets, ***Nature chemistry***, 5 (2013) 263-275.
108. J. Kim, S.S. Baik, S.H. Ryu, Y. Sohn, S. Park, B.G. Park, J. Denlinger, Y. Yi, H.J. Choi, K.S. Kim, 2D MATERIALS. Observation of tunable band gap and anisotropic Dirac semi-metal state in black phosphorus, ***Science***, 349 (2015) 723-726.
109. H. Liu, Y. Du, Y. Deng, P.D. Ye, Semiconducting black phosphorus: synthesis, transport properties and electronic applications, ***Chemical Society reviews***, 44 (2015) 2732-2743.
110. Z. Lei, X. Sun, S. Zhu, K. Dong, X. Liu, L. Wang, X. Zhang, L. Qu, X. Zhang, Nature Inspired MXene-Decorated 3D Honeycomb-Fabric Architectures Toward Efficient Water Desalination and Salt Harvesting, ***Nano-micro letters***, 14 (2021) 10.
111. D. Wang, C.K. Zhou, A.S. Filatov, W.J. Cho, F. Lagunas, M.Z. Wang, S. Vaikuntanathan, C. Liu, R.F. Klie, D.V. Talapin, Direct synthesis and chemical vapor deposition of 2D carbide and nitride MXenes, ***Science***, 379 (2023) 1242-1247.
112. K.R.G. Lim, M. Shekhirev, B.C. Wyatt, B. Anasori, Y. Gogotsi, Z.W. Seh, Fundamentals of MXene synthesis, ***Nature Synthesis***, 1 (2022) 601-614.
113. Z. Wang, K. Yu, S. Gong, H. Mao, R. Huang, Z. Zhu, Cu(3)BiS(3)/MXenes with Excellent Solar-Thermal Conversion for Continuous and Efficient Seawater Desalination, ***ACS applied materials & interfaces***, 13 (2021) 16246-16258.
114. F. Shahzad, M. Alhabeb, C.B. Hatter, B. Anasori, S. Man Hong, C.M. Koo, Y. Gogotsi, Electromagnetic interference shielding with 2D transition metal carbides (MXenes), ***Science***, 353 (2016) 1137-1140.
115. X. Wu, J. Wang, Z. Wang, F. Sun, Y. Liu, K. Wu, X. Meng, J. Qiu, Boosting the Electrocatalysis of MXenes by Plasmon-Induced Thermalization and Hot-Electron Injection, ***Angewandte Chemie***, 60 (2021) 9416-9420.
116. P. Zhou, Q. Zhu, X. Sun, L. Liu, Z. Cai, J. Xu, Recent advances in MXene-based membrane for solar-driven interfacial evaporation desalination, ***Chemical Engineering Journal***, 464 (2023) 142508.
117. C. Niu, L. Yang, H. Sun, Z. Zhu, W. Liang, J. Li, A. Li, Three-dimensional macroporous foamy MXene membrane with high efficiency on interfacial evaporation and wastewater purification, ***Chemical Engineering Journal***, 476 (2023) 146522.

118. W. Xiao, J. Yan, S. Gao, X. Huang, J. Luo, L. Wang, S. Zhang, Z. Wu, X. Lai, J. Gao, Superhydrophobic MXene based fabric composite for high efficiency solar desalination, ***Desalination***, 524 (2022) 115475.
119. D. Xu, Z. Li, L. Li, J. Wang, Insights into the Photothermal Conversion of 2D MXene Nanomaterials: Synthesis, Mechanism, and Applications, ***Advanced Functional Materials***, 30 (2020).
120. K. Li, T.H. Chang, Z. Li, H. Yang, F. Fu, T. Li, J.S. Ho, P.Y. Chen, Biomimetic MXene Textures with Enhanced Light-to-Heat Conversion for Solar Steam Generation and Wearable Thermal Management, ***Advanced Energy Materials***, 9 (2019).
121. M. He, H. Dai, H. Liu, Q. Cai, Y. Liu, L. Wang, X. Qin, J. Yu, High-Performance Solar Steam Generator Based on Polypyrrole-Coated Fabric via 3D Macro- and Microstructure Design, ***ACS applied materials & interfaces***, 13 (2021) 40664-40672.
122. G. Liu, J. Xu, T. Chen, K. Wang, Progress in thermoplasmonics for solar energy applications, ***Physics Reports***, 981 (2022) 1-50.
123. Z. Yang, Z.Y. Wu, Z. Lin, T. Liu, L. Ding, W. Zhai, Z. Chen, Y. Jiang, J. Li, S. Ren, Z. Lin, W. Liu, J. Feng, X. Zhang, W. Li, Y. Yu, B. Zhu, F. Ding, Z. Li, J. Zhu, Optically selective catalyst design with minimized thermal emission for facilitating photothermal catalysis, ***Nature communications***, 15 (2024) 7599.
124. Y. Xu, J. Ma, Y. Han, H. Xu, Y. Wang, D. Qi, W. Wang, A simple and universal strategy to deposit Ag/polypyrrole on various substrates for enhanced interfacial solar evaporation and antibacterial activity, ***Chemical Engineering Journal***, 384 (2020) 123379.
125. T. Weng, H. Du, H. Luo, M. Jiang, Z. Chen, J. Wang, F. Jiang, H. Chen, Synergistic photothermal conversion on Mn3CeOx/Coal-based carbon material membrane for simultaneous interface evaporation and organic pollutant removal, ***Separation and Purification Technology***, 355 (2025) 129573.
126. W. Li, Y. Xu, K. Liu, L. Zhu, T. Xu, G. Wang, C. Si, Engineered Biomass-Based Solar Evaporators for Diversified and Sustainable Water Management, ***Advanced materials***, (2025) e2503658.
127. J. Zhou, Y. Gu, Z. Deng, L. Miao, H. Su, P. Wang, J. Shi, The dispersion of Au nanorods decorated on graphene oxide nanosheets for solar steam generation, ***Sustainable Materials and Technologies***, 19 (2019) e00090.
128. X. Fan, Y. Yang, X. Shi, Y. Liu, H. Li, J. Liang, Y. Chen, A MXene-Based Hierarchical Design Enabling Highly Efficient and Stable Solar-Water Desalination with Good Salt Resistance, ***Advanced Functional Materials***, 30 (2020).
129. X. Mu, Y. Gu, P. Wang, A. Wei, Y. Tian, J. Zhou, Y. Chen, J. Zhang, Z. Sun, J. Liu, L. Sun, S. Tanemura, L. Miao, Strategies for breaking theoretical evaporation limitation in direct solar steam generation, ***Solar Energy Materials and Solar Cells***, 220 (2021) 110842.
130. J. Sun, S. Zhao, X. Wang, W. Kong, W. Li, S. Wang, S. Liu, S. Nie, Minimizing enthalpy of evaporation in solar steam generation: An emerging strategy beyond theoretical evaporation limitation, ***Materials Today***, 80 (2024) 619-647.
131. B. Yang, C. Li, Z. Wang, Q. Dai, Thermoplasmonics in Solar Energy Conversion: Materials, Nanostructured Designs, and Applications, ***Advanced materials***, 34 (2022) e2107351.
132. J.U. Kim, S. Lee, S.J. Kang, T.I. Kim, Materials and design of nanostructured broadband light absorbers for advanced light-to-heat conversion, ***Nanoscale***, 10 (2018) 21555-21574.
133. F. Tao, M. Green, A.V. Garcia, T. Xiao, A.T. Van Tran, Y. Zhang, Y. Yin, X. Chen, Recent progress of nanostructured interfacial solar vapor generators, ***Applied Materials Today***, 17 (2019) 45-84.
134. X. Zhao, T. Wang, R. Wang, Y. Lan, Y. Jiang, J. Pan, Superwetting photothermal membranes enabled by polyphenol-mediated nanostructured coating with raspberry-like architectures for solar-driven interfacial evaporation, ***Desalination***, 542 (2022) 116046.
135. X. Lu, C. Mu, Y. Liu, L. Wu, Z. Tong, K. Huang, Recent advances in solar-driven interfacial evaporation coupling systems: Energy conversion, water purification, and seawater resource extraction, ***Nano Energy***, 120 (2024) 109180.

136. Z. Xiang, Y. Shi, X. Zhu, L. Cai, W. Lu, Flexible and Waterproof 2D/1D/0D Construction of MXene-Based Nanocomposites for Electromagnetic Wave Absorption, EMI Shielding, and Photothermal Conversion, ***Nano-micro letters***, 13 (2021) 150.
137. X. Li, X.I. Pereira-Hernandez, Y. Chen, J. Xu, J. Zhao, C.W. Pao, C.Y. Fang, J. Zeng, Y. Wang, B.C. Gates, J. Liu, Functional CeO(x) nanoglues for robust atomically dispersed catalysts, ***Nature***, 611 (2022) 284-288.
138. D.A. Hanifi, N.D. Bronstein, B.A. Koscher, Z. Nett, J.K. Swabeck, K. Takano, A.M. Schwartzberg, L. Maserati, K. Vandewal, Y. van de Burgt, A. Salleo, A.P. Alivisatos, Redefining near-unity luminescence in quantum dots with photothermal threshold quantum yield, ***Science***, 363 (2019) 1199-1202.
139. Y. Lin, Z. Chen, L. Fang, M. Meng, Z. Liu, Y. Di, W. Cai, S. Huang, Z. Gan, Copper nanoparticles with near-unity, omnidirectional, and broadband optical absorption for highly efficient solar steam generation, ***Nanotechnology***, 30 (2019) 015402.
140. C. Ma, J. Yan, Y. Huang, C. Wang, G. Yang, The optical duality of tellurium nanoparticles for broadband solar energy harvesting and efficient photothermal conversion, ***Science advances***, 4 (2018) eaas9894.
141. N.J. Hogan, A.S. Urban, C. Ayala-Orozco, A. Pimpinelli, P. Nordlander, N.J. Halas, Nanoparticles heat through light localization, ***Nano letters***, 14 (2014) 4640-4645.
142. S. Ishii, R.P. Sugavaneshwar, K. Chen, T.D. Dao, T. Nagao, Solar water heating and vaporization with silicon nanoparticles at mie resonances, ***Optical Materials Express***, 6 (2016) 640.
143. M. Chang, Z. Hou, M. Wang, D. Wen, C. Li, Y. Liu, Y. Zhao, J. Lin, Cu Single Atom Nanozyme Based High-Efficiency Mild Photothermal Therapy through Cellular Metabolic Regulation, ***Angewandte Chemie***, 134 (2022).
144. C. Cheng, Q. Liang, M. Yan, Z. Liu, Q. He, T. Wu, S. Luo, Y. Pan, C. Zhao, Y. Liu, Advances in preparation, mechanism and applications of graphene quantum dots/semiconductor composite photocatalysts: A review, ***Journal of hazardous materials***, 424 (2022) 127721.
145. N. Jiang, X. Zhuo, J. Wang, Active Plasmonics: Principles, Structures, and Applications, ***Chemical reviews***, 118 (2018) 3054-3099.
146. J. Song, F. Wang, X. Yang, B. Ning, M.G. Harp, S.H. Culp, S. Hu, P. Huang, L. Nie, J. Chen, X. Chen, Gold Nanoparticle Coated Carbon Nanotube Ring with Enhanced Raman Scattering and Photothermal Conversion Property for Theranostic Applications, ***Journal of the American Chemical Society***, 138 (2016) 7005-7015.
147. Z. Sun, H. Xie, S. Tang, X.F. Yu, Z. Guo, J. Shao, H. Zhang, H. Huang, H. Wang, P.K. Chu, Ultrasmall Black Phosphorus Quantum Dots: Synthesis and Use as Photothermal Agents, ***Angewandte Chemie***, 54 (2015) 11526-11530.
148. G.Z. Jia, W.K. Lou, F. Cheng, X.L. Wang, J.H. Yao, N. Dai, H.Q. Lin, K. Chang, Excellent photothermal conversion of core/shell CdSe/Bi2Se3 quantum dots, ***Nano Research***, 8 (2015) 1443-1453.
149. C. Chen, H. Liu, H. Wang, Y. Zhao, M. Li, A scalable broadband plasmonic cuprous telluride nanowire-based hybrid photothermal membrane for efficient solar vapor generation, ***Nano Energy***, 84 (2021) 105868.
150. W. Duan, B. Shao, Z. Wang, K. Ni, S. Liu, X. Yuan, Y. Wang, B. Sun, X. Zhang, R. Liu, Silicon nanowire/ionic hydrogel-based hybrid moist-electric generators with enhanced voltage output and operational stability, ***Energy & Environmental Science***, 17 (2024) 3788-3796.
151. Y. Feng, X. Ning, J. Wang, Z. Wen, F. Cao, Q. You, J. Zou, X. Zhou, T. Sun, J. Cao, X. Chen, Mace-Like Plasmonic Au-Pd Heterostructures Boost Near-Infrared Photoimmunotherapy, ***Advanced science***, 10 (2023) e2204842.
152. X. Ma, X. Wan, Z. Fang, Z. Li, X. Wang, Y. Hu, M. Dong, Z. Ye, X. Peng, Orientational seawater transportation through Cu(TCNQ) nanorod arrays for efficient solar desalination and salt production, ***Desalination***, 522 (2022) 115399.
153. P. Dash, S. Thirumurugan, C.L. Tseng, Y.C. Lin, S.L. Chen, U. Dhawan, R.J. Chung, Synthesis of Methotrexate-Loaded Dumbbell-Shaped Titanium Dioxide/Gold Nanorods Coated with Mesoporous Silica and Decorated with Upconversion Nanoparticles for Near-

Infrared-Driven Trimodal Cancer Treatment, ***ACS applied materials & interfaces***, 15 (2023) 33335-33347.
154. J. Liu, N. Liu, H. Wang, W. Shi, J. Zhuang, X. Wang, Hybrid MoO(3)-Polyoxometallate Sub-1 nm Nanobelt Superstructures, ***Journal of the American Chemical Society***, 142 (2020) 17557-17563.
155. S. Zhao, Y. Yuan, Q. Yu, B. Niu, J. Liao, Z. Guo, N. Wang, A Dual-Surface Amidoximated Halloysite Nanotube for High-Efficiency Economical Uranium Extraction from Seawater, ***Angewandte Chemie***, 58 (2019) 14979-14985.
156. Y. Du, P. Liu, H. Zhang, L. Zou, K. Deng, X. Li, W. Tian, J. Ji, Nature-Inspired Structure-Engineered TiN/TiO2 Nanotubes Array Toward Solar Desalination Synergy with Photothermal-Enhanced Degradation and Thermoelectric Generation, ***Advanced Functional Materials***, 34 (2023).
157. H. Ge, Z. Fang, S. Wu, J. Qiu, X. Liu, Z. Zhang, Efficient solar-driven interfacial evaporation and desalination using simple, salt-resistant, carbon nanotube-based Janus evaporators, ***Energy***, 306 (2024) 132354.
158. Q. Kang, T. Wang, P. Li, L. Liu, K. Chang, M. Li, J. Ye, Photocatalytic reduction of carbon dioxide by hydrous hydrazine over Au-Cu alloy nanoparticles supported on SrTiO3/TiO2 coaxial nanotube arrays, ***Angewandte Chemie***, 54 (2015) 841-845.
159. W. Wang, X. Yan, J. Geng, N. Zhao, L. Liu, T. Vogel, Q. Guo, L. Ge, B. Luo, Y. Zhao, Engineering a Copper@Polypyrrole Nanowire Network in the Near Field for Plasmon-Enhanced Solar Evaporation, ***ACS nano***, 15 (2021) 16376-16394.
160. Z. Zhang, H. Liu, Z. Kong, M. Fang, M. Wang, Y. Zhu, Mushroom-like Graphene Nanosheets/Copper Sulfide Nanowires Foam with Janus-Type Wettability for Solar Steam Generation, ***ACS Applied Nano Materials***, 5 (2022) 4931-4937.
161. L. Guo, J. Gong, C. Song, Y. Zhao, B. Tan, Q. Zhao, S. Jin, Donor–Acceptor Charge Migration System of Superhydrophilic Covalent Triazine Framework and Carbon Nanotube toward High Performance Solar Thermal Conversion, ***ACS Energy Letters***, 5 (2020) 1300-1306.
162. C. Ma, Q. Liu, Q. Peng, G. Yang, M. Jiang, L. Zong, J. Zhang, Biomimetic Hybridization of Janus-like Graphene Oxide into Hierarchical Porous Hydrogels for Improved Mechanical Properties and Efficient Solar Desalination Devices, ***ACS nano***, 15 (2021) 19877-19887.
163. Y. Xu, C. Li, X. Wu, M.X. Li, Y. Ma, H. Yang, Q. Zeng, J.L. Sessler, Z.X. Wang, Sheet-like 2D Manganese(IV) Complex with High Photothermal Conversion Efficiency, ***Journal of the American Chemical Society***, 144 (2022) 18834-18843.
164. Y. Wang, Z. Cao, A. Barati Farimani, Efficient water desalination with graphene nanopores obtained using artificial intelligence, ***npj 2D Materials and Applications***, 5 (2021).
165. S. Li, B. Xu, M. Lu, M. Sun, H. Yang, S. Liu, Z. Huang, H. Liu, Tensile-Strained Palladium Nanosheets for Synthetic Catalytic Therapy and Phototherapy, ***Advanced materials***, 34 (2022) e2202609.
166. D. Yang, Z. Deng, S. Wang, X. Yin, J. Xi, M. Andersson, J. Wang, D. Zhang, Polydopamine-coated two-dimensional nanomaterials as high-affinity photothermal signal tag towards dual-signal detection of Salmonella typhimurium by lateral flow immunoassay, ***Chemical Engineering Journal***, 472 (2023) 145110.
167. X. Yu, C. Zhao, L. Yang, J. Zhang, C. Chen, Photothermal catalytic oxidation of toluene over the Pt–Mn2O3/CN nanocomposite catalyst, ***EES Catalysis***, 2 (2024) 811-822.
168. G. Darabdhara, M.R. Das, V. Turcheniuk, K. Turcheniuk, V. Zaitsev, R. Boukherroub, S. Szunerits, Reduced graphene oxide nanosheets decorated with AuPd bimetallic nanoparticles: a multifunctional material for photothermal therapy of cancer cells, ***Journal of materials chemistry. B***, 3 (2015) 8366-8374.
169. J. Song, J. Li, X. Bai, L. Kang, L. Ma, N. Zhao, S. Wu, Y. Xue, J. Li, X. Ji, J. Sha, Cu nanoparticle-decorated two-dimensional carbon nanosheets with superior photothermal conversion efficiency of 65 % for highly efficient disinfection under near-infrared light, ***Journal of Materials Science & Technology***, 87 (2021) 83-94.

170. Y. Xu, J. Xu, J. Zhang, X. Li, B. Fu, C. Song, W. Shang, P. Tao, T. Deng, All-in-one polymer sponge composite 3D evaporators for simultaneous high-flux solar-thermal desalination and electricity generation, ***Nano Energy***, 93 (2022) 106882.
171. J. Xie, R. Liang, Q. Li, K. Wang, M. Hussain, L. Dong, C. Shen, H. Li, G. Shen, J. Zhu, J. Tao, Photosensitizer-loaded gold nanocages for immunogenic phototherapy of aggressive melanoma, ***Acta biomaterialia***, 142 (2022) 264-273.
172. X. Du, C. Zhao, M. Zhou, T. Ma, H. Huang, M. Jaroniec, X. Zhang, S.Z. Qiao, Hollow Carbon Nanospheres with Tunable Hierarchical Pores for Drug, Gene, and Photothermal Synergistic Treatment, ***Small***, 13 (2017).
173. G. Song, Y. Yuan, J. Liu, Q. Liu, W. Zhang, J. Fang, J. Gu, D. Ma, D. Zhang, Biomimetic Superstructures Assembled from Au Nanostars and Nanospheres for Efficient Solar Evaporation, ***Advanced Sustainable Systems***, 3 (2019).
174. S.I. In, D.D. Vaughn, 2nd, R.E. Schaak, Hybrid CuO-TiO(2-x)N(x) hollow nanocubes for photocatalytic conversion of CO2 into methane under solar irradiation, ***Angewandte Chemie***, 51 (2012) 3915-3918.
175. M. Gao, C.K. Peh, H.T. Phan, L. Zhu, G.W. Ho, Solar Absorber Gel: Localized Macro-Nano Heat Channeling for Efficient Plasmonic Au Nanoflowers Photothermic Vaporization and Triboelectric Generation, ***Advanced Energy Materials***, 8 (2018).
176. Y. Wang, X. Yuan, K. Ni, Y. Song, X. Li, X. Zeng, B. Shao, B. Sun, 3D dendritic hierarchically gradient nanoflowers in situ grown on conductive substrates for efficient hydrovoltaic power generation, ***Energy & Environmental Science***, 17 (2024) 4780-4793.
177. Q. Xue, P. Xiao, J. Gu, W. Wang, L. Yan, T. Chen, Superhydrophobic sand evaporator with core-shell structure for long-term salt-resistant solar desalination, ***Water research***, 253 (2024) 121290.
178. N. Arshad, M. Sohail Asghar, M. Sultan Irshad, N. Alwadie, J. Li, M.A.K. Yousaf Shah, Z. Ouyang, H. Wang, H. Xuan Nang, V.-D. Dao, X. Wang, Spatial patterned solar evaporator for effective thermal management in enhanced solar-driven evaporation, and wastewater remediation, ***Chemical Engineering Journal***, 500 (2024) 157068.
179. W. Zhang, L. Wang, K. Wang, M.U. Khan, M. Wang, H. Li, J. Zeng, Integration of Photothermal Effect and Heat Insulation to Efficiently Reduce Reaction Temperature of CO(2) Hydrogenation, ***Small***, 13 (2017).
180. H. Feng, T. Wang, W. Wang, L. Cao, X. Qin, J. Piao, P. Wang, W. Li, S. Chen, Photothermal insulation polydimethylsiloxane coating based on phase change materials for anticorrosion applications, ***Journal of Industrial and Engineering Chemistry***, 132 (2024) 161-171.
181. B. Wang, S. Zhao, S. Wang, Y. Fu, M. Liu, Coupling photothermal evaporation into photocatalysis for enhanced hydrogen production from water, ***The Innovation Energy***, 1 (2024) 100018.
182. Y. Lu, H. Zhang, D. Fan, Z. Chen, X. Yang, Coupling solar-driven photothermal effect into photocatalysis for sustainable water treatment, ***Journal of hazardous materials***, 423 (2022) 127128.
183. Y. Zhou, T. Ding, M. Gao, K.H. Chan, Y. Cheng, J. He, G.W. Ho, Controlled heterogeneous water distribution and evaporation towards enhanced photothermal water-electricity-hydrogen production, ***Nano Energy***, 77 (2020) 105102.
184. B. Xu, H. Li, H. Yang, W. Xiang, G. Zhou, Y. Wu, X. Wang, Colloidal 2D-0D Lateral Nanoheterostructures: A Case Study of Site-Selective Growth of CdS Nanodots onto Bi(2) Se(3) Nanosheets, ***Nano letters***, 15 (2015) 4200-4205.
185. Y. Chen, S. Zhao, X. Wang, Q. Peng, R. Lin, Y. Wang, R. Shen, X. Cao, L. Zhang, G. Zhou, J. Li, A. Xia, Y. Li, Synergetic Integration of Cu1.94S-ZnxCd1-xS Heteronanorods for Enhanced Visible-Light-Driven Photocatalytic Hydrogen Production, ***Journal of the American Chemical Society***, 138 (2016) 4286-4289.
186. X. Luo, R. Li, K.P. Homewood, X. Chen, Y. Gao, Hybrid 0D/2D Ni2P quantum dot loaded TiO2(B) nanosheet photothermal catalysts for enhanced hydrogen evolution, ***Applied Surface Science***, 505 (2020) 144099.

187. Y. Xu, J.X. Ma, Y. Han, J.J. Zhang, F.Y. Cui, Y. Zhao, X. Li, W. Wang, Multifunctional CuO Nanowire Mesh for Highly Efficient Solar Evaporation and Water Purification, ***Acs Sustainable Chemistry & Engineering***, 7 (2019) 5476-5485.
188. J. Huang, J. Tao, G. Liu, L. Lu, H. Tang, G. Qiao, In situ construction of 1D CdS/2D Nb2CTx MXene Schottky heterojunction for enhanced photocatalytic hydrogen production activity, ***Applied Surface Science***, 573 (2022) 151491.
189. A. Pramanik, J.A. Dhar, R. Banerjee, M. Davis, K. Gates, J. Nie, D. Davis, F.X. Han, P.C. Ray, WO(3) Nanowire-Attached Reduced Graphene Oxide-Based 1D-2D Heterostructures for Near-Infrared Light-Driven Synergistic Photocatalytic and Photothermal Inactivation of Multidrug-Resistant Superbugs, ***ACS applied bio materials***, 6 (2023) 919-931.
190. J. Pan, Y. Guan, Y. Zhang, Z. Xu, S. Han, H. Tang, X. Yan, H. Liu, Q. Lu, Near-Infrared-Induced Photothermal Enhanced Photocatalytic H(2) Production for 3D/2D Heterojunctions of Snowflake-like CuS/g-C(3)N(4) Nanosheets, ***Inorganic chemistry***, 62 (2023) 624-635.
191. T.Y. Huang, Z. Yang, S.Y. Yang, Z.H. Dai, Y.J. Liu, J.H. Liao, G.Y. Zhong, Z.J. Xie, Y.P. Fang, S.S. Zhang, Construction of 2D/2D Ti3C2T MXene/CdS heterojunction with photothermal effect for efficient photocatalytic hydrogen production, ***Journal of Materials Science & Technology***, 171 (2024) 1-9.
192. Y. Tian, H. Yang, S. Wu, J. Yan, K. Cen, T. Luo, G. Xiong, Y. Hou, Z. Bo, K. Ostrikov, Beyond lotus: Plasma nanostructuring enables efficient energy and water conversion and use, ***Nano Energy***, 66 (2019) 104125.
193. H. Yu, M. Dai, J. Zhang, W. Chen, Q. Jin, S. Wang, Z. He, Interface Engineering in 2D/2D Heterogeneous Photocatalysts, ***Small***, 19 (2023) e2205767.
194. H. Liang, B.B. Laird, M. Asta, Y. Yang, In-plane characterization of structural and thermodynamic properties for steps at faceted chemically heterogeneous solid/liquid interfaces, ***Acta Materialia***, 143 (2018) 329-337.
195. Y. Cheng, J. Zhang, X. Xiong, C. Chen, J. Zeng, Z. Kong, H. Wang, J. Xi, Y.-J. Yuan, Z. Ji, High photocatalytic and photoelectrochemical performances of the CuSe/MoSe2 2D/2D face-to-face heterojunction photocatalyst, ***Journal of Alloys and Compounds***, 870 (2021) 159540.
196. D. Xu, Y. Ma, J. Wang, W. Chen, Y. Tang, X. Li, L. Li, Interfacial engineering of 2D/2D MXene heterostructures: face-to-face contact for augmented photodegradation of amoxicillin, ***Chemical Engineering Journal***, 426 (2021) 131246.
197. J. Wu, J. Qu, G. Yin, T. Zhang, H.Y. Zhao, F.Z. Jiao, J. Liu, X. Li, Z.Z. Yu, Omnidirectionally irradiated three-dimensional molybdenum disulfide decorated hydrothermal pinecone evaporator for solar-thermal evaporation and photocatalytic degradation of wastewaters, ***Journal of colloid and interface science***, 637 (2023) 477-488.
198. J.J. Richardson, M. Bjornmalm, F. Caruso, Multilayer assembly. Technology-driven layer-by-layer assembly of nanofilms, ***Science***, 348 (2015) aaa2491.
199. G.W.S. C. Jeffrey Brinker, Sol-Gel Science The Physics and Chemistry of Sol-Gel Processing ***Academic Press***, (1990) 463.
200. N.J. Wilkinson, M.A.A. Smith, R.W. Kay, R.A. Harris, A review of aerosol jet printing—a non-traditional hybrid process for micro-manufacturing, ***The International Journal of Advanced Manufacturing Technology***, 105 (2019) 4599-4619.
201. K. Kang, K.H. Lee, Y. Han, H. Gao, S. Xie, D.A. Muller, J. Park, Layer-by-layer assembly of two-dimensional materials into wafer-scale heterostructures, ***Nature***, 550 (2017) 229-233.
202. A.E. Danks, S.R. Hall, Z. Schnepp, The evolution of 'sol–gel' chemistry as a technique for materials synthesis, ***Materials horizons***, 3 (2016) 91-112.
203. H.W. Tan, Y.Y.C. Choong, C.N. Kuo, H.Y. Low, C.K. Chua, 3D printed electronics: Processes, materials and future trends, ***Progress in Materials Science***, 127 (2022) 100945.
204. R. Zouari, M. Michel, J. Di Martino, V. Ball, Production of free standing composite membranes or of patterned films after sol–gel reactions in an exponential layer-by-layer architecture, ***Materials Science and Engineering: C***, 30 (2010) 1291-1297.
205. H. Abdolmaleki, P. Kidmose, S. Agarwala, Droplet-Based Techniques for Printing of Functional Inks for Flexible Physical Sensors, ***Advanced materials***, 33 (2021) e2006792.

Chapter 4
Structure Design

4.1 Energy Management

During solar evaporation, it has been found that besides harvesting light energy, efficient thermal energy management via the design of the photothermal evaporator's structure and configuration enhances solar evaporation performance [1, 2]. By employing strategic energy management, it is possible to decrease heat losses (via conduction, convection and radiation) in the evaporation system, while simultaneously capturing environment energy from the ambient, thereby increasing the evaporation rate [3–5]. Over recent years, numerous effective strategies for energy management have been proposed.

4.1.1 Surface Topography

The purpose of surface topography is to enhance light trapping and expand the actual evaporation surface area within the area illuminated by projected light, thereby improving photothermal evaporation efficiency [6, 7]. Improving solar absorption and minimizing reflection losses by surface engineering is pivotal for optimizing the performance. Surface modifications, such as texturing and the application of anti-reflective (AR) coatings, are widely employed to improve light trapping and absorption efficiency [8, 9]. Texturing with the increasing surface area introduces multi-incident angles, thereby promoting light trapping and reducing specular reflection. For instance, in silicon solar cells, surface texturing has been demonstrated to significantly diminish reflective losses, resulting in enhanced solar absorption and improved photovoltaic efficiency [10]. AR coatings offer an effective strategy by harnessing the destructive interference. These coatings are designed with specific thicknesses and refractive indices to eliminate reflected waves at both

G. Liu, T. Chen, *Solar Evaporation for Water-Energy-Resource Nexus*,
https://doi.org/10.1007/978-981-95-3815-7_4

the air-coating and coating-substrate interfaces, thereby maximizing light transmission into the material [11]. Additionally, the integration of photonic crystals into or onto material surfaces offers advanced control over light-matter interactions. Photonic crystals can create photonic bandgaps, selectively reflecting specific wavelengths while enhancing absorption in desired spectral regions [12, 13]. Together, surface texturing, anti-reflective coatings, and photonic crystals offer a versatile toolkit for customizing material properties to suit specific operational wavelengths and environmental conditions. This enables improved performance in a wide range of applications, including photovoltaics and photothermal evaporation.

A 3D origami-based solar evaporator has been fabricated by adjusting the origami's folding angles. The surface areal density (the ratio of the light interactive area to the projected area) could be manipulated to optimize light absorption [14]. Additionally, the increased surface area contributed to a lower steady-state evaporation temperature, effectively minimizing radiative and convective heat losses from the evaporation surfaces [15, 16]. Notably, when the surface areal density exceeded 4.65, a high light-to-vapor energy efficiency (greater than 100%) was achieved. In a similar vein, a honeycomb structure was introduced to further improve light absorption. Upon exposure to light, the tightly arranged 3D honeycomb structure captured light through multiple reflections and absorptions [17, 18]. This, coupled with the increased air/water interfacial area, led to superior solar steam generation performance [19]. Another notable study was conducted by Yu's team, who created hydrogels with engineered surface topographies using a template-assisted gelation process (Fig. 4.1) [6]. These hydrogels, compared to those with a flat surface, exhibited

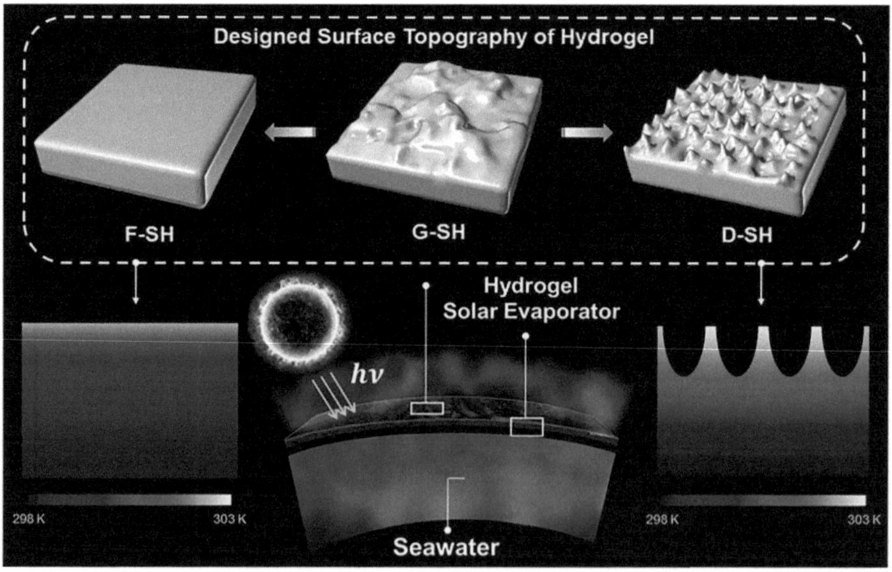

Fig. 4.1 Surface-modified hydrogel for solar vapor generation. (Reproduced with permission from [6] Copyright 2019 American Chemical Society)

increased heat flux on their dimpled surfaces. Due to the surface roughness, the evaporation surface area of the dimpled hydrogels was approximately five times greater than that of the flat surface, achieving an evaporation rate of 2.6 kg m^{-2} h^{-1} upon one sun irradiation. This concept of surface topography has been critical in development of advanced 3D evaporators.

4.1.2 Confined Water Path

In solar evaporation, two-dimensional self-floating architectures are predominantly employed, characterized by direct interfacial contact between the evaporator and the water reservoir [20–22]. This configuration facilitates a dual functionality: the water transport pathway not only supplies water to the evaporation interface during the solar-driven process but also serves as a thermal dissipation channel, transferring heat from the evaporation surface to the underlying water body [23, 24]. In this system, the incident solar energy (E_{in}) is absorbed by photothermal materials and subsequently transformed into thermal energy to drive the vaporization process at the evaporation interface. However, a fraction of the incident energy is inevitably lost to the surrounding environment through scattering and reflection phenomena. The thermal localization results in a temperature gradient between the photothermal evaporation surface and both the underlying water body and the ambient air. This thermal disparity induces energy dissipation through multiple pathways: conductive heat transfer to the bulk water, convective heat loss to the surrounding air, and radiative heat loss [25, 26]. Consequently, the effective energy harnessed for solar evaporation in a conventional 2D system can be expressed by the following energy balance equation:

$$E_{\mathrm{evaporation}} = \alpha E_{\mathrm{in}} - E_{\mathrm{conduction}} - E_{\mathrm{convection}} - E_{\mathrm{radiation}} \tag{4.1}$$

where α is the light absorption coefficient. The conduction, convection, and radiation losses can be calculated via Eqs. (2.7–2.9).

To reduce conductive heat loss, photothermal structures with low thermal conductivity are preferred. As documented by Zhou et al., naturally low-conductive wood (0.33 W m^{-1} K^{-1}) underwent flame treatment to function as a floating solar evaporator [27–29]. During solar evaporation, water ascends to the heated top surface via microchannels within the wood. The temperature differential (ΔT_w) between the evaporation surface and the submerged water (1 cm below the surface) was assessed to gauge heat dissipation. A ΔT_w of merely 0.5 °C was observed in the pure water setup, while the floating structure exhibited about 3.3 °C, highlighting improved thermal utilization due to the wood's low thermal conductivity. Nonetheless, a temperature rise in the bulk water indicated that heat loss persisted despite the floating evaporation structure, owing to the direct contact between the bulk water and the evaporator [30, 31]. In addressing this, floating evaporation systems with confined water transport paths were developed with 2D confined water path to reduce heat loss (Fig. 4.2a) [32]. A 1.6 cm thick polystyrene foam with a thermal conductivity of 0.04 W m^{-1} K^{-1} was used as an insulating layer between the evaporation surface and bulk water, wrapped in cellulose cloth for water wicking. This design greatly reduced conduction heat loss, resulting in only a 0.9 °C increase in surface water temperature, compared to 13.7 °C in a direct contact system, thereby boosting evaporation efficiency by 30%.

Moreover, 3D evaporators with a 1D water path were developed to reduce heat conduction loss [17, 33–35]. Mushrooms, with their naturally hydrophilic fibrous stipe, were optimized as low-cost solar thermal evaporators (Fig. 4.2b). Following carbonization treatment, the mushroom's light absorption improved to 96% across UV-vis-NIR spectrums. Leveraging its confined 1D water supply path, the system effectively suppressed conductive heat loss to water. Additionally, the mushroom's curved surface expanded the evaporation area, enhancing the evaporation rate (1.475 kg m^{-2} h^{-1}) and reducing the surface temperature compared to other flat light

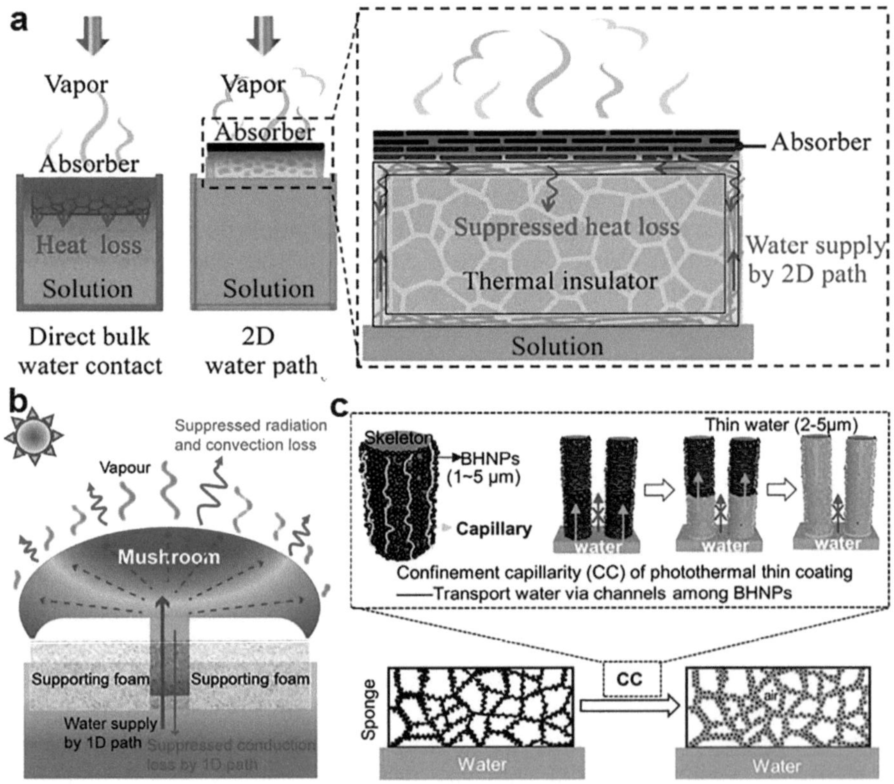

Fig. 4.2 (**a**) Solar desalination with 2D water supply. (Reproduced with permission from [32] Copyright 2016, National Academy of Sciences). (**b**) a mushroom-based device and the heat flow. (Reproduced with permission from [36] Copyright 2017, Wiley-VCH). (**c**) The confinement capillarity of photothermal thin coating for water skin formation to boost solar water evaporation. (Reproduced with permission from [41] Copyright 2021, Wiley-VCH)

absorbers, which in turn lowered convection and radiation losses [36]. An oil-lamp-like evaporator has also been developed using a cotton strand as a 1D wicking water path, which physically isolates the evaporation surface from the bulk water [37]. By adjusting the distance between the evaporation surface and the water level, they optimized water supply and minimized conductive heat loss, enhancing energy utilization for evaporation. This design achieved an evaporation rate exceeding 1.5 kg m^{-2} h^{-1} under one sun irradiation. Later developments saw the creation of self-contained solar evaporators [38, 39]. Common systems typically require external bulk water, the studies employed carbon sponge and cotton rods with high water retention capabilities as internal water reservoirs, thereby greatly eliminating conduction loss and boosting evaporation performance.

To further optimize water supply, controllable water transportation techniques were implemented. Qu and colleagues managed water supply at an injection rate of 8 μL/min, maintaining a capillary water state in the foam evaporator [40]. This approach effectively prevented energy wastage from excessive water heating while ensuring ample vapor escape channels, achieving equilibrium between water supply and evaporation, and resulting in an exceptional steam generation rate of 2.40 kg m^{-2} h^{-1} under one sun. Furthermore, Wang et al. aimed to establish an ultrathin liquid layer (2–5 μm) atop the evaporation framework using a confinement capillarity strategy [41, 42]. In such setup, hydrophobic sponges were coated with a rough layer of agglomerated black/hydrophilic nanoparticles (BHNPs), offering numerous nanochannels (100-500 nm) for capillary water transport (Fig. 4.2c). During evaporation, water remained within the BHNPs channels rather than saturating the sponge's pores, thus maximizes the surface area for evaporation and vapor release, leading to a solar vapor generation rate of 3.2 kg m^{-2} h^{-1}.

4.1.3 Environment Energy Harvesting

During solar evaporation, the temperature of the evaporation surface exceeds that of the ambient environment, significant energy losses through radiation and convection can occur, which undermines the efficiency of energy use [43, 44]. Notably, in 2D evaporators, the photothermal conversion increases the surface temperature of the evaporator above that of its surroundings, making energy loss inevitable [45, 46]. To mitigate these losses, the adoption of 3D evaporators with increased surface areas for evaporation has been proposed. Zhu's research team introduced an artificial transpiration device featuring a hollow cone absorber and a 1D confined water path for a comparative analysis against its 2D counterpart [25, 47]. Such 3D devices, within the same area exposed to light, offer a greater surface area which enhances the utilization of light energy for water evaporation. The resultant increase in evaporation leads to more effective cooling at the evaporation surface, which in turn lowers the surface temperature and reduces heat losses due to radiation and convection. These works pave the way for further improvements in evaporation performance through strategic structural design.

Hereby, 3D evaporators can incorporate additional light-absorbing side surfaces. These side surfaces, which do not directly absorb light, undergo evaporative cooling, resulting in temperatures lower than those of the surrounding environment. As a result, these side surfaces function as cold evaporation surfaces, capturing additional energy from the surrounding air via radiation and convection. By adjusting the proportional areas of the top (responsible for energy loss) and side evaporation surfaces (contributing to energy harvest), a net energy gain can be realized, potentially exceeding the theoretical evaporation efficiency limit (which assumes a 100% conversion of light energy to vapor). This innovative configuration differentiates the energy dynamics of a 3D system from a conventional 2D evaporation system, necessitating a revision of the energy balance equation as follows:

$$E_{\text{evaporation}} = \alpha E_{\text{in}} - E_{\text{conduction}} - E_{\text{convection,A1}} - E_{\text{radiation,A1}} + E_{\text{convection,A2}} + E_{\text{radiation,A2}} \tag{4.2}$$

$E_{\text{convection, A1}}$, $E_{\text{radiation, A1}}$ denote the convection and radiation losses from surface A1, while $E_{\text{convection, A2}}$ and $E_{\text{radiation, A2}}$ correspond to the energy gains via convection and radiation at the cooler surface A2. Subsequent research has focused on minimizing thermal losses by introduction of a 3D triangular structure [48]. By adjusting the apex angles, they could control the surface area density and temperature distribution across the evaporator. Utilizing a narrow apex angle of 8.4°, they achieved lower temperatures on most of the evaporative surfaces compared to the ambient, facilitating cold evaporation that draws energy from the surroundings.

A significant advancement was made by a heat sink-like evaporator with a top surface designed for light absorption and solar evaporation, and a series of vertical fins that supplied water to this top surface. Light exposure generated heat on the top surface, which was efficiently transferred to the underlying fins due to the temperature difference [49]. Adjusting the number of fins allowed the top surface's evaporation temperature to be reduced below the ambient temperature. The fins facilitated heat dissipation from the top surface through continuous water evaporation and its cooling effect, keeping it consistently cooler than both the environment and the water during the evaporation (Fig. 4.3). This was the first instance of achieving all-cold evaporation across all surfaces (top and fins) under standard solar conditions, effectively eliminating energy loss and significantly boosting energy gain from both the ambient environment and the bulk water. Calculations indicated that the net energy gain from the environment was about 1.7 times the received solar energy, resulting in an impressive evaporation rate of 4.1 kg m^{-2} h^{-1}.

Exploring energy gain from water via conduction is seldom and alternative approaches. Xu's group designed two specific evaporative structures. Initially, a solid cylinder structure was developed, featuring a top solar evaporation surface (SES) and side surfaces as cold evaporation surfaces (CES) [50]. During solar evaporation, thermal energy was transferred from the SES to the CES and the bulk water, driven by their temperature differences. By enlarging the CES area (e.g., >50.3 cm^2) relative to the SES (12.56 cm^2), the heat flow from the SES could be fully captured by the CES for cold evaporation, preventing loss to the bulk water, while simultaneously drawing thermal energy from the bulk water, thus achieving net energy

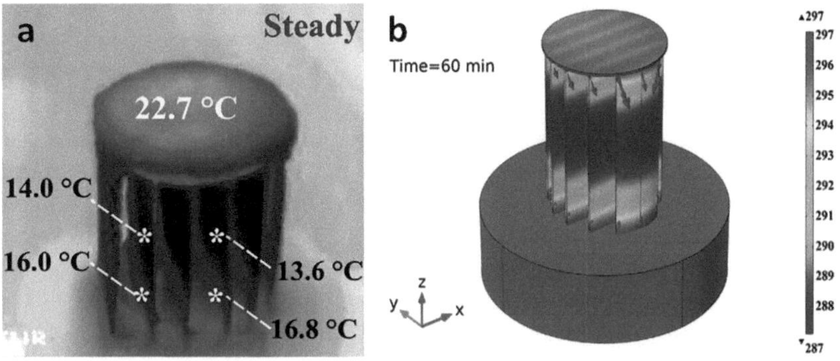

Fig. 4.3 IR images and heat transfer simulation of the evaporator. (Reproduced with permission from [49] Copyright 2021, Wiley-VCH)

harvest from the bulk water to the evaporator. This groundbreaking research revisited the conventional design of 2D evaporators, which typically use thermal insulating substrates to prevent heat conduction loss from evaporation surfaces to bulk water. By harnessing energy from the bulk water, a 3D evaporator with a highly thermally conductive aluminum support was further reported [39]. Compared to a 3D evaporator with a thermal insulator like PE foam, the aluminum support increased the solar evaporation rate by enhancing energy transfer from the bulk water to the evaporator. The energy harvested from the bulk water (0.66 W) surpassed the solar energy input (0.55 W), showing that bulk water can be a primary energy source for solar evaporation. Furthermore, all evaporation surfaces remained cooler than the ambient temperature, eliminating radiation and convection losses to the air and drawing additional energy from it. This approach could potentially address a critical issue in solar evaporation, achieving all-weather evaporation with substantial rates. The energy balances in this system are represented as follow [51]:

$$\begin{aligned} E_{\text{evaporation}} &= \alpha E_{\text{in}} + \left|E_{\text{conduction}}\right| + \left|E_{\text{convection,A1}}\right| + \left|E_{\text{radiation,A1}}\right| \\ &\quad + \left|E_{\text{convection,A2}}\right| + \left|E_{\text{radiation,A2}}\right| \end{aligned} \tag{4.3}$$

Based on this, it can be expected that with a strategically engineered 3D evaporation system, where all evaporating surfaces maintain temperatures below those of the surrounding ambient and bulk water, it is possible to reverse the direction of energy flow (i.e., from the water and surrounding air towards the evaporator) [26, 52]. This reversal significantly enhances the evaporation efficiency and performance.

4.1.4 Convective Flow

The local humidity levels above the evaporation surfaces significantly influence evaporation efficiency. Higher humidity levels increase water vapor content in the air, which in turn impedes the ongoing vapor release from the evaporation surfaces into the ambient environment [53, 54]. For instance, in a 2D evaporation system, convection typically initiates at the edges of the evaporation surface, facilitating rapid vapor dispersal. However, if the evaporation surface is larger than the reach of the convection currents, these currents do not extend to the central area, leading to the formation of a non-evaporative, vapor-saturated zone that constrains overall evaporation efficiency [55]. Consequently, smaller evaporation surfaces generally exhibit higher evaporation rates. To address this issue, Xu et al. firstly modified the evaporation surface by selectively removing its central part, which significantly boosted convection and thus enhanced the evaporation rate [55]. This approach not only increases vapor output but also reduces the required evaporation surface area, making it advantageous for large-scale applications. The same research group developed a transformable evaporator made from reduced graphene oxide and a cellulose sponge [56]. This evaporator can shift between a flat 2D form and a 3D spiral structure. Despite having the same mass and volume, transitioning from 2D to 3D significantly enhances the evaporation rate due to increased surface area and improved water transport. Numerical simulations indicate that the 3D spiral configuration effectively harnesses convective flows, maximizing evaporation across its surfaces. As a result, it achieves a substantially higher evaporation rate of up to 4.35 kg m^{-2} h^{-1}, which is 185.9% greater than that of the 2D flat configuration.

In practice, the simplest way to increase the evaporation rate is by enhancing thermal convection above the surface where evaporation occurs. This is the principle behind the effectiveness of hair dryers in rapidly drying wet hair or clothes. To capitalize on this concept, a donut-shaped hydrogel evaporator has been developed using 3D-printed molds [57]. This design aims to boost thermal convection and minimize areas where evaporation is less effective. Simulations indicate that these optimized structural designs significantly enhance heat distribution and air convection, leading to marked improvements in evaporation efficiency. Experimental data further supports these findings, showcasing a evaporation rate of 3.47 kg m^{-2} h^{-1}, which is a 66% increase over traditional evaporator designs. In contrast to natural convection, applying a strong, externally driven convective flow, which requires additional energy, can substantially enhance the evaporation rate. A groundbreaking study by Zhu's group introduced a 3D evaporator featuring an interconnected porous structure, extensive evaporation surface, and outstanding hydrophilicity for this purpose [58]. The use of external convective flow speeds up the phase transition and vapor diffusion, promoting sustained evaporation (Fig. 4.4). By optimizing the pore size and velocity of the convective flow, an impressive evaporation rate above 10 kg m^{-2} h^{-1} was achieved under one sun irradiation.

4.1.5 Energy Recycle

In addition to harvesting energy from environment, the energy loss from evaporation system can also be captured and reused to enhance evaporation efficiency [51, 59]. Wang et al. developed a photothermal device featuring a 2D flat bottom surface

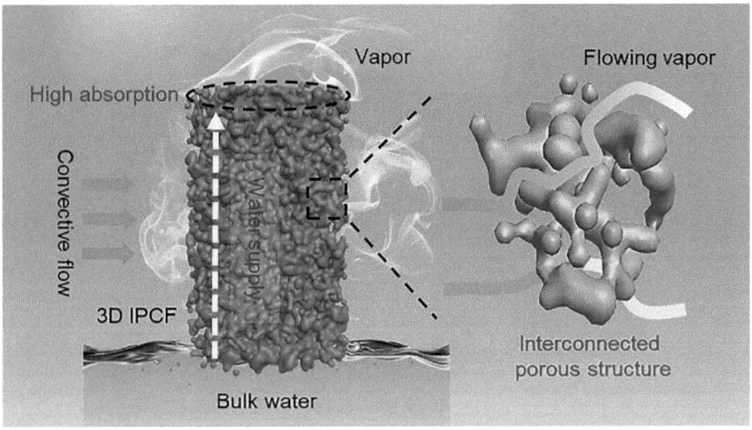

Fig. 4.4 The interconnected porous structure facilitates vapor diffusion alongside convective flow. (Reproduced with permission from [58] Copyright 2020, Cell Press)

and a curved wall. This design helps recover scattered light and radiative heat loss from the bottom during solar evaporation, thanks to the 3D wall structure, achieving nearly 100% net energy efficiency [60]. Subsequently, Deng et al. implemented energy recycling in a 2D evaporation system by selectively coating air-laid paper with carbon black to create a patterned surface with alternating black and white areas [61]. This configuration generates a temperature differential during solar evaporation, promoting lateral heat and mass transfer and thus boosting steam production. A similar concept has also been implemented in a waffle-shaped solar evaporator, characterized by additional sidewalls that enhance convection through synergistic solute and thermal Marangoni effects [62]. As a result, this design enables spontaneous salt rejection and maintains stable, efficient evaporation for over 60 days in brine without any fouling.

The solar-to-vapor conversion efficiency is calculated as the ratio of the enthalpy used for vapor generation to the solar energy input. An efficiency of 100% indicates that all solar energy is converted into water vapor. Theoretically, the heat released during steam condensation and cooling could be utilized to generate additional vapor. If all energy from the vapor were recycled, the theoretical maximum efficiency could reach as high as 417% [63]. However, only a few reported systems have achieved efficiencies over 100%. To recycle the latent heat from condensation, multi-stage solar evaporators were introduced, where multiple evaporation units were vertically stacked to form a passive multi-stage solar distiller [64–66]. The first unit included an outer thermal insulator, a light-absorbing layer, a wicking layer, an air gap, and an aluminum condenser. In subsequent units, the wicking/evaporation layer was attached to the backside of the aluminum condenser from the previous unit, followed by an air gap [2, 63, 67]. During desalination, the latent heat from vapor condensation is transferred to the evaporative layer of the next unit, driving water evaporation and maximizing energy utilization (Fig. 4.5). Unlike 2D

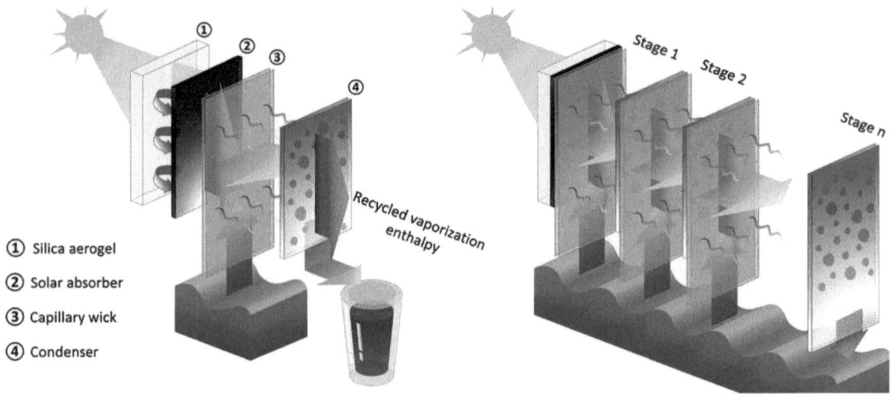

Fig. 4.5 A multistage solar still system. (Reproduced with permission from [63] Copyright 2020, Royal Society of Chemistry)

evaporators, where vapor begins to condense and release latent heat upon encountering cold air above the evaporation surfaces, placing vertically aligned cold evaporation surfaces above the solar evaporation surface (i.e., in the vapor condensation zone) can also effectively recycle latent heat to improve solar evaporation efficiency.

4.1.6 Reduce Equivalent Enthalpy

Reducing the energy required for water evaporation can speed up the process. This energy, called evaporation enthalpy, aids in converting water from liquid to gas. At 100 °C, the evaporation enthalpy is approximately 2260 kJ/kg, largely due to strong molecular interactions like hydrogen bonding [4, 51, 68]. Modifying the phase transition behavior and state of water by disrupting the original hydrogen-bond network among water molecules can adjust the evaporation enthalpy, thus enhancing evaporation efficiency. The energy necessary to sever these hydrogen bonds determines the latent heat of water evaporation. Although the total energy for solar evaporation comprises both sensible heat and latent heat, research efforts are aimed at reducing the latent heat component to decrease the overall energy requirement, given that sensible heat is significantly less than latent heat. Enthalpy (H) is the total of internal energy (E) and the product of volume (V) and pressure (p). Theoretically, the latent heat of evaporation (ΔH_{vap}) is the difference between the enthalpy of the gas phase and the liquid phase [51].

$$\Delta H_{\mathrm{vap}} = H_{\mathrm{g}} - H_{\mathrm{l}} = E_{\mathrm{g}} + p_{\mathrm{g}} V_{\mathrm{g}} - E_{\mathrm{l}} - p_{\mathrm{l}} V_{\mathrm{l}} \tag{4.4}$$

For an ideal gas, the equation $p_{\mathrm{g}} V_{\mathrm{g}} = N k_{\mathrm{B}} T$ holds true, where $E = K + U$ presents the total energy, comprising kinetic energy (K) and potential energy. Potential energy is further divided into intermolecular potential (U^{ie}) and intramolecular potential (U^{ia}).

In both liquid and gas phases, the kinetic energy of water molecules and the vibrational energy of bonds (U^{ia}) within each molecule remain constant [4, 69]. Consequently, Eq. (4.4) can be written as follows:

$$\Delta H_{\text{vap}} = -U_1^{\text{ie}} + Nk_{\text{B}}T \tag{4.5}$$

Additionally, the surface-driven pressure impacts the additional pressure on the evaporation enthalpy [70], which can be expressed as follows:

$$\Delta H_{\text{vap}} = -U_1^{\text{ie}} + \left[Nk_{\text{B}}T - \left(p + 4\sigma \frac{\cos\theta}{r} V_1 \right) \right] \tag{4.6}$$

where σ denotes the surface tension at the water/gas interface, θ represents the contact angle at this interface, and r is the radius of curvature at the water/gas interface. The evaporation enthalpy is clearly associated with the intermolecular potential U_1^{ie} and the surface tension σ of the water/gas interface. Therefore, disrupting the interactions between water molecules is an effective way to reduce the energy needed for evaporation.

To quantify changes in evaporation enthalpy, two primary methods have been utilized: practical dark evaporation measurement and differential scanning calorimetry (DSC) characterization [4, 71]. The equivalent evaporation enthalpy (H_{equ}) is determined through these methods. In the dark evaporation approach, evaporation experiments are conducted in a controlled environment within an enclosed chamber, maintaining specific humidity and temperature conditions [20, 72]. These experiments measure the evaporation rates from both pure water and a specifically designed evaporator, each having the same projection surface area. The equivalent evaporation enthalpy for the evaporative system is then derived using the equation:

$$E_{\text{in}} = H_0 m_0 = H_{\text{equ}} m_{\text{g}} \tag{4.7}$$

Here, E_{in} represents the incident light energy, while H_0 and m_0 denote the evaporation enthalpy and mass change of water. Additionally, m_{g} refers to the mass change observed in the photothermal evaporator. As in Eq. (4.7), it becomes apparent that for a fixed amount of incident light energy used for evaporation, reducing the evaporation enthalpy is crucial for achieving an enhanced evaporation rate.

However, it is important to recognize that the H_{equ} value derived from the dark evaporation method provides only a basic understanding of the enthalpy changes occurring during an ideal dehydration process [73]. By contrast, DSC provides a comprehensive heat curve that illustrates water vaporization as temperature-dependent, outlining the entire dehydration process. Water molecules vibrate in picoseconds, with hydrogen bonds relaxing in 0.5–1.0 ps and their rotational relaxation occurring in 2–5 ps [74]. Raman spectroscopy, with its characteristic time parameters nearing 10–15 s, emerges as the primary method for analyzing liquid water structure [75]. Four Raman peaks are identified, categorized into two groups: peaks at 3200 and 3350 cm^{-1} represent free water with tetrahedral hydrogen bonding, while peaks at 3450 and 3610 cm^{-1} indicate intermediate water with weaker or partial bonds (Fig. 4.6a).

Water molecule vibrations include translational, bending, and stretching movements of –OH, influenced by both intra and intermolecular forces. The translation mode is depicted by the 162 cm^{-1} Raman peak, illustrating the whole molecule's motion [4]. The 1595 cm^{-1} peak signifies the –OH bending vibration, crucial for water evaporation and sensitive to hydrogen bond strength, indicated by shifts in peak position [76]. The –OH stretching vibration, most intense and directly related to hydrogen bonding, varies with the local molecular environment, affecting water's evaporation dynamics. Hydrogen bond structures are categorized into DDAA, DAA, DDA, and DA types, with corresponding spectral sub-peaks indicating the strength and interaction of these bonds (Fig. 4.6b) [77]. Among the five Gaussian deconvoluted sub-peaks from the –OH stretching mode, the low and high frequency sub-peaks, mainly defined by the DDAA and DA structures, are crucial. A slight red-shift in these sub-peaks indicates reduced energy and weaker hydrogen bonds, which facilitate molecular dissociation at the water-air interface and enhance evaporation.

4.1.6.1 Intermediate Water Adjustment

Numerous strategies have been explored to weaken the noncovalent interactions among water molecules. These strategies include increasing the effective surface area, adjusting the intermediate water content, and engineering water clusters or capillary structures [4, 78]. Materials with strong interactions with water molecules can modify the state and phase-transition behavior of water. Yu's team conducted the pioneering study on the relationship between water states and the energy

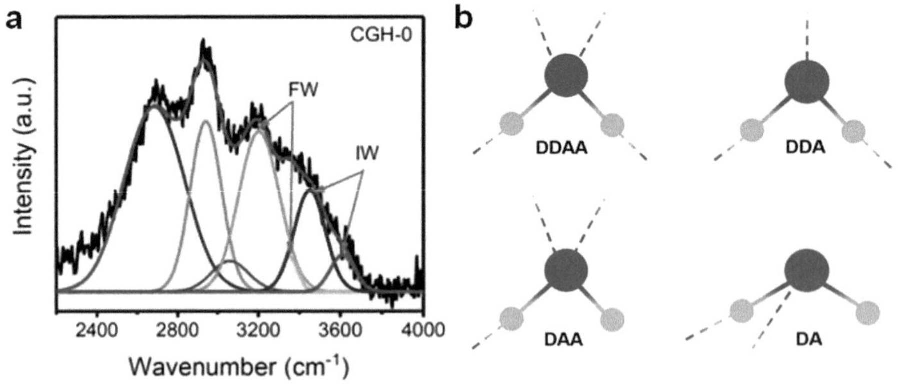

Fig. 4.6 (**a**) Raman spectra by fitting curves for CGH-0. (Reproduced with permission from [75], Copyright 2022 American Chemical Society). (**b**) Four bond structures associated with the -OH stretching mode. (Reproduced with permission from [77], Copyright 2009 Elsevier)

required for water evaporation in hydrogels [79, 80]. They found that polar groups in hydrophilic hydrogels can form weak hydrogen bonds with adjacent water molecules. For example, abundant sulfonate ($-SO_3H$) and hydroxyl (−OH) groups in hydrophilic polymer gels can disrupt the hydrogen bonding network among water molecules (Fig. 4.7a). Additionally, functional groups like amine ($-NH_2$), carboxylic acid (–COOH), epoxy (–CH(O)CH–), aldehyde (–CHO), and carbonyl (C=O) can also enhance hydrophilicity. For example, chitosan with $-NH_2$ groups was used to construct hydratable polymer networks that regulate water-polymer interactions in a light-absorbing hydrogel (Fig. 4.7b) [81]. In this design, the water state was modulated by increasing the chitosan/PVA ratio, which significantly reduced the energy demand for water vaporization and increased the evaporation rate to 3.6 kg m^{-2} h^{-1}.

Water can be classified into three types based on its interactions with hydrophilic materials: bound water (BW), intermediate water (IW), and free water (FW). BW refers to water molecules tightly held by the functional groups in polymer chains due to hydration. FW resembles bulk water with minimal interactions with materials. The fewer and weaker hydrogen bonds in IW result in a lower evaporation enthalpy. During solar evaporation, IW often resides in confined areas near the surface of a nanoporous light absorber, characterized by a continuous thermal energy supply, higher localized temperatures, and an increased evaporation surface area. As a result, evaporating IW requires less energy, providing a basis for enhancing solar evaporation [82–85]. Yu et al. incorporated polystyrene sulfonate into a polyvinyl alcohol network to influence the water state and phase change behaviors. The polystyrene sulfonate promotes the formation of more intermediate water (IW), reducing the evaporation enthalpy through electrostatic interactions and hydrogen bonding among water molecules [86]. The crosslinked polyvinyl alcohol limits

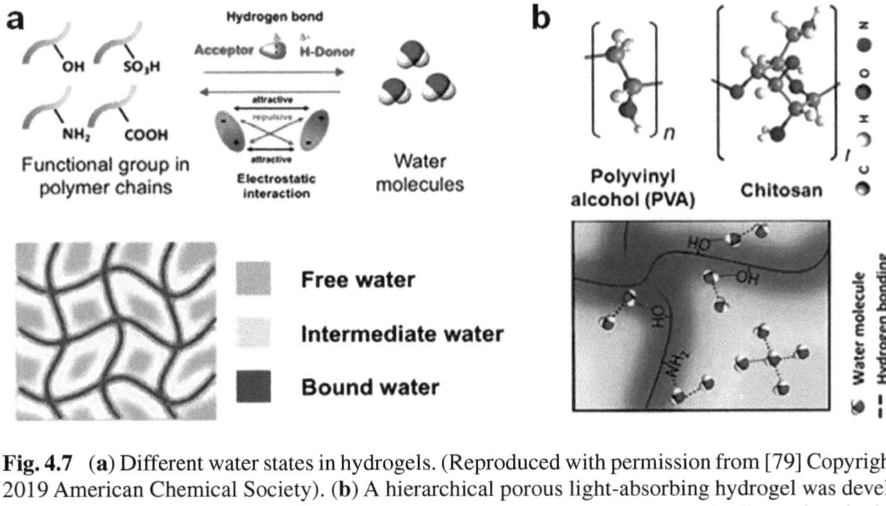

Fig. 4.7 (**a**) Different water states in hydrogels. (Reproduced with permission from [79] Copyright 2019 American Chemical Society). (**b**) A hierarchical porous light-absorbing hydrogel was developed by integrating polypyrrole into the polyvinyl alcohol polymeric network. (Reproduced with permission from [81] Copyright 2019 American Association for the Advancement of Science)

water saturation in the hydrogel, reducing energy loss and increasing the solar vapor generation rate to 3.9 kg m^{-2} h^{-1} with an energy utilization efficiency of 92% under one-sun irradiation.

In addition, stacked atomic-thick 2D $MoN_{1.2}$ and rGO nanosheets with heterointerfaces can create heterogeneous interactions with surrounding water, leading to an unstable water state, which facilitates the breaking of hydrogen bonds, considerably decreases evaporation enthalpy, and boosts the evaporation rate to 2.6 kg m^{-2} h^{-1} [87]. Other Janus structured materials also influence evaporation enthalpy [88–90]. For instance, Zhang designed a hydrophilic/hydrophobic membrane that accelerates water flux for efficient evaporation by adjusting the state of confined water and ensuring timely water replenishment [91]. These examples demonstrate how material and structural designs can create weak interactions with nearby water molecules to generate more intermediate water, thereby reducing energy consumption for evaporation and increasing the solar evaporation rate.

4.1.6.2 Formation of More Water Clusters

In nature, water, whether in ice or liquid form, forms clusters of multiple molecules connected by hydrogen bonds. These clusters can evaporate either as single molecules or as groups of a few to several tens of molecules. Within hydrogel meshes, water molecules typically escape as smaller clusters rather than larger ones [72, 80, 92]. The evaporation enthalpy of water ($\Delta H_{p-water}$) and for water confined in hydrogels ($\Delta H_{cluster}$) can be calculated using the equations [51, 81]:

$$\Delta H_{p-water} = -U_{1-water}^{inter} + NkT \tag{4.8}$$

$$\Delta H_{cluster} = -\left[xU_{1-water}^{inter} + (1-x)U_{cluster-water} \right] + NkT \tag{4.9}$$

$U_{1-water}^{inter}$ and $U_{cluster-water}$ represent the bond energy in liquid water and the energy required for cluster evaporation, respectively. Here, N denotes the number of water molecules, T represents temperature, k is the Boltzmann constant, and x is the fraction of a gas.

$$\Delta H_{cluster} - \Delta H_{p-water} = (1-x)\left[-U_{cluster-water} - \left(-U_{1-water}^{inter} \right) \right] \tag{4.10}$$

Fewer hydrogen bonds need to be broken for the evaporation of smaller water clusters, such as intermediate water (IW), compared to larger clusters like free water (FW). Consequently, $-U_{cluster-water}$ is smaller than $-U_{1-water}^{inter}$, indicating that $\Delta H_{cluster}$ is less than $\Delta H_{p-water}$. This indicates that smaller water clusters can lower the enthalpy of evaporation. Techniques like Raman spectroscopy and lithium-ion tracking can qualitatively and quantitatively confirm the presence of different water states and small clusters in polymers. Yu's team showed that the incorporation of PPy into the PVA polymeric gel network alters water cluster sizes, resulting in a solar evaporation rate of 3.2 kg m^{-2} h^{-1}, surpassing the physic limit of normal evaporators [93]. Moreover, MoS_{2-x} nanosheets facilitate water transport and lower vaporization enthalpy by promoting cluster evaporation, achieving a evaporation efficiency of 89.6% upon standard sunlight conditions [94].

Modifying materials and nanostructures to interact with water molecules is essential for reducing the energy required for water evaporation. Porous hydrophilic materials can increase the content of smaller water clusters within networks, which require less energy due to weaker hydrogen bonds. Additionally, optimizing microstructural designs can help vaporize water as smaller clusters, thus lowering the enthalpy of evaporation. Furthermore, capillary water in porous structures exposes more evaporation surfaces and facilitates unimpeded water transport (Fig. 4.8) [4]. For example, Gong et al. developed a porous carbon structure that promotes the formation of numerous small water clusters, thus reducing the enthalpy of evaporation [95]. Guo et al. introduced hydrophilic hydrogel evaporators with hydrophobic island-shaped patches, achieving a record evaporation rate of 4.0 kg m^{-2} h^{-1} with 93% efficiency under one-sun irradiation [96]. This high rate stems from synergistic effects in the wetting regions: the thicker water layer in the hydrophilic area allows for quick water molecule escape, and extended contact lines in the hydrophobic area, collectively boosting the process. Lu and Wang et al. developed a hybrid hydrogel embedded with a 2D nanostructure, which, due to the Marangoni effect and the presence of small water clusters, achieved a high evaporation rate of 3.62 kg m^{-2} h^{-1} [84, 97, 98]. Chen et al. introduced a photomolecular effect in which a strong electric field at the water-air interface drives the separation of water clusters, enabling smaller clusters to escape rapidly into the air. This results in an evaporation rate of 2.4 kg m^{-2} h^{-1}, exceeding the thermal evaporative limit [99]. Developing

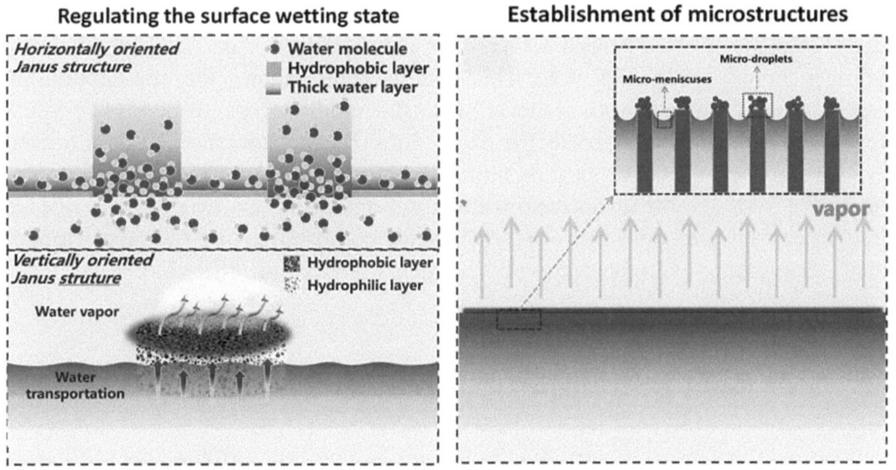

Fig. 4.8 The regulation of surface wetting state and the establishment of microstructure in enthalpy reduction. (Reproduced with permission from [4], Copyright 2024 Elsevier)

more efficient and competitive strategies to decrease latent enthalpy remains essential for further improving the evaporation performance.

4.1.6.3 Water Activation

Previous studies showed that using an alternating electric field can aid water evaporation by disrupting the hydrogen bonding network through polarization and Lorentz forces, thereby reducing the energy needed for evaporation [4, 100]. However, the deployment of large-scale equipment for this purpose is limited by high costs and energy consumption, making smaller, portable power devices like piezoelectric fabrics a more feasible option. The kinetic and potential energy from the ocean are high-density blue energy sources. These can be harnessed using piezoelectric materials to generate clean energy that activates water molecules [101, 102]. A novel approach employs a wave-driven piezoelectric interface solar evaporator, featuring a photothermal layer made of carbon felt and a piezoelectric layer consisting of a cellulose-based ultra-thin nanofilm coated with barium titanate ($BaTiO_3$) nanoparticles [101]. The $BaTiO_3$ nanoparticles are optimally dispersed on the cellulose matrix to convert wave energy into electrical energy, enabling continuous water activation and enhanced evaporation. The hydrophilic, porous cellulose structure ensures effective water transport and nanoparticle-water contact, optimizing energy transfer. Under 1 V AC from the piezoelectric material, FTIR analysis reveals a pronounced blue shift and intensity decrease in the hydroxyl absorption peak, indicating O-H bond contraction and weakened hydrogen bonding. This reduces water's evaporation enthalpy from 2100 to 1455 kJ/kg, significantly boosting evaporation performance.

To optimize electro-activated water efficiency, it is essential to address the omnidirectional energy of ocean waves. One solution is a composite hydrogel incorporating multiple anisotropic poly(fluoroethylene-trifluoroethylene) piezoelectric layers arranged uniformly before gelling. This composite is divided into three sections, with the end sections oriented perpendicularly to the middle section and bonded via strong hydrogen bonds [102] (Fig. 4.9). This configuration enables the end sections to respond to wave impacts from sea level, while the middle section responds to perpendicular wave forces. The piezoelectric fibers within the co-generation evaporator convert wave impacts from various directions into electrical energy. This energy activates water and reduces the evaporation enthalpy, achieving an evaporation rate of 3.545 kg m^{-2} h^{-1} under one Sun—a 28% increase compared to non-piezoelectric activated evaporators—with the evaporation enthalpy reduced to 1180 kJ/kg.

4.2 Anti-fouling Strategies

In practical deployment, it is important to address salt accumulation and biofouling [103–105]. The operational dynamics of these systems entail the continual introduction of seawater or brine to the evaporation surface. Here, a phase transition

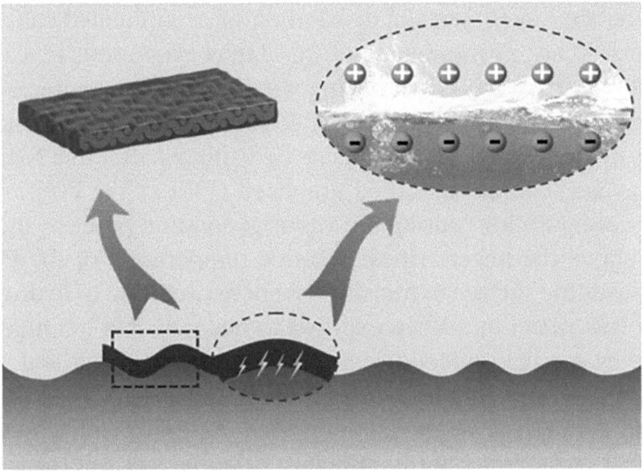

Fig. 4.9 The water activation of electricity in enthalpy reduction. (Reproduced with permission from [4], Copyright 2024 Elsevier)

facilitates the vapor generation, concurrently leaving behind dissolved salts and organic contaminants. In the absence of effective antifouling measures, these residual substances progressively build up, precipitating two primary deleterious effects: 1) the optical obstruction of incident solar radiation, and 2) the physical obstruction of water transport channels and vapor release pathways [106]. Such fouling phenomena invariably lead to a decline in both the evaporation efficiency and the overall performance [46, 107, 108]. Therefore, the antifouling strategies are indispensable for maintaining the long-term efficiency and stability of solar evaporation system.

4.2.1 Anti-salt Fouling

Previous research addressed the salt-accumulation issue by periodic water washing following each evaporation cycle [104, 109, 110]. To mitigate this challenge, subsequent studies pioneered three more efficacious strategies: (1) facilitating the salt diffusion back into the bulk water, (2) implementing contactless evaporation techniques and (3) removing salt from evaporation surfaces via dynamic evaporation.

4.2.1.1 Salt Diffusion

Inspired by natural self-cleaning in lotus and water lily plants, a Janus double-layered evaporation structure was engineered [35]. This structure incorporates a hydrophobic photothermal top layer paired with a hydrophilic substrate. This Janus configuration effectively restricts water transport to the hydrophobic photothermal surface, thereby localizing salt accumulation to the hydrophilic bottom layer. Here, continuous water flux facilitates the dissolution of accumulated salts back into the bulk liquid. Performance assessments of this Janus evaporator revealed consistent evaporation rates over 16 days, contrasting with a control evaporator that exhibited a 15% reduction in evaporation rate within 45 minutes due to salt deposition. Geng and colleagues made parallel advancements by using a melamine foam coated with poly(*N*-isopropylacrylamide)-modified graphene (PG) as the evaporative medium [111]. In their configuration, during the vapor generation process, the surface temperature rises above the lower critical solution temperature of the PG membrane. This change causes the surface to transition from hydrophilic to hydrophobic, facilitating effective salt rejection. When exposed to one sun, this intelligently designed structure achieves a water collection rate of 4.2 kg/m^2 per hour and an ionic rejection rate of over 99%, utilizing both transpiration and guttation mechanisms to process brine feed through a single unit. Further, Asinari et al. capitalized on the Marangoni effect at the air-water interface to expedite salt rejection [112]. During evaporation, the salinity gradient at the interface induces a surface tension differential, thereby creating a Marangoni flow that significantly augments salt rejection capabilities. This phenomenon resulted in a 75% reduction in surface salt concentration within two hours under dark conditions.

Salt accumulation can also be managed through rapid water replenishment of the evaporation surface [113–115]. Han et al. developed a 3D solar evaporator with vertically aligned mass transfer bridges that improve water supply and facilitate salt diffusion back into the bulk water [116]. By increasing the number of these bridges, they optimized the flow of water through hydrophilic microchannels to the evaporation surface, thereby also enhancing the diffusion of salt away from it. Hu et al. reported a self-regenerating solar evaporator with outstanding antifouling properties, engineered with a strategically designed artificial channel-array embedded in a natural wood substrate [113]. Concentration gradients enable spontaneous salt exchange between channels through 1–2 μm pits, effectively diluting salt concentrations in the micro-wood channels. These drilled channels, with high hydraulic conductivity, serve as efficient pathways for salt rejection and facilitate rapid salt exchange with the bulk solution, allowing for continuous self-regeneration of the evaporator (Fig. 4.10). Zhu's group designed a hierarchical evaporator that includes a central water supply layer, sandwiched between a top light-harvesting layer and a bottom insulation layer [117]. Leveraging the Donnan effect, the mesoporous structure of the water supply channels is endowed with negative charges, effectively trapping Na^+ ions within the microchannels and establishing a chemical potential. This innovative design significantly restricts cation movement into the water supply layer, preventing salt crystallization and deposition on the evaporation surfaces. These advancements represent substantial progress in addressing salt buildup in solar evaporation systems.

4.2.1.2 Contactless Evaporation

Complete isolation of salt ions from the solar evaporator can be achieved by introducing an air gap between the evaporator and seawater, a notable advancement seen in the development of contactless solar evaporation systems [103, 118–121]. These

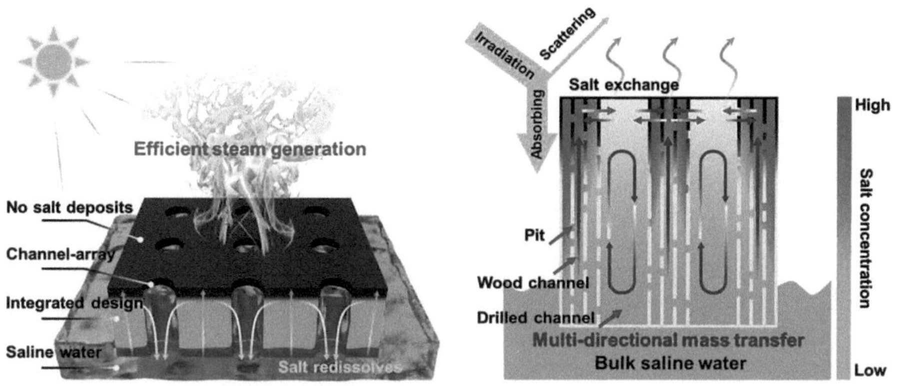

Fig. 4.10 The diffusion evaporator and multidirectional mass transfer. (Reproduced with permission from [113], Copyright 2019 Wiley-VCH)

systems are distinguished by separating the light-absorbing materials from direct contact with water, thereby enabling the efficient transfer of accumulated thermal energy to the evaporation surface via mechanisms such as thermal radiation or conduction [89, 122]. As detailed by Chen et al., the system features a dual-coated aluminum shell, where a high-efficiency light absorber is applied to the front surface and a material with significant thermal emittance is applied to the back [119]. When exposed to solar radiation, the generated heat is transferred to the emittance material, which then re-emits it as thermal radiation. This radiative energy is primarily absorbed by the water's uppermost layer (less than 100 μm thick), enhancing evaporation. While this configuration effectively prevents salt deposition, its solar-vapor conversion efficiency is somewhat constrained at 24.6% under standard solar irradiation, due to thermal losses to the bulk liquid and the inherent inefficiencies associated with multi-step energy transformations.

To address these challenges, Prasher's group proposed an innovative noncontact method that converts sunlight into mid-infrared radiation, which is directly absorbed by water within a penetration depth of less than 100 μm (Fig. 4.11) [120]. This approach not only prevents fouling but also achieves an evaporation rate of 0.49 kg m^{-2} h^{-1} and a solar-to-vapor efficiency of about 43% when processing saturated brine under standard solar conditions. Although this efficiency is generally lower than that of traditional floatable solar absorbers, it offers a promising new method for water-solute separation and recovery from complex aqueous sources [123]. To enhance the energy efficiency, Wang integrated the contactless approach with conventional heat localization strategies by employing a floating thermal insulation structure with vertically aligned macrochannels [103]. This setup confines the solar-thermal conversion to a defined water layer, improving energy management and achieving an evaporation efficiency exceeding 80% for 20 wt% saline solutions.

Further innovations aimed at optimizing water productivity were introduced by Qu et al., who developed a Janus evaporator employing a high thermal conductivity metal framework (copper foil/foam). This design uniquely separates the light absorption surface from the evaporative surface. The foil side of the metal

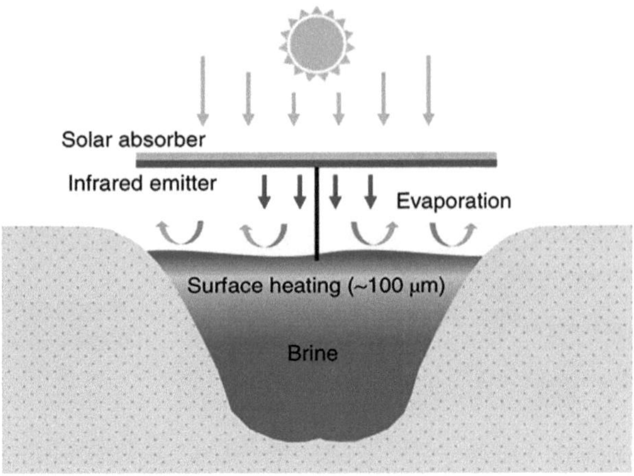

Fig. 4.11 Surface heating using a photothermal umbrella. (Reproduced with permission from [120], Copyright 2020 Nature Publishing Group)

framework is laser-treated to enhance anti-reflective properties, while the foam side is coated with a superhydrophilic polymer membrane to facilitate water transport [89]. The Janus evaporator is engineered to capture a broad range of natural sunlight angles (from −75 to +75 degrees). The metal structure's high thermal conductivity ensures rapid heat transfer from the light-absorbing surface to the evaporative side, leading to high evaporation rates and reducing salt accumulation issues. Existing research mainly addresses the evaporation of NaCl brine, whereas the thermodynamic precipitation of seawater's real salt content, particularly divalent ions like Mg^{2+} and Ca^{2+}, typically results in dense salt layers that obstruct water transport and vapor release [124]. Addressing this challenge necessitates seawater pre-treatment to remove these ions or the addition of salt crystallization inhibitors such as nitrilotriacetic acid.

4.2.1.3 Dynamic Evaporation

Solar absorbers with self-cleaning capabilities enhance the extraction of metal salts from high-salinity brines by autonomously achieving water-solute separation without external intervention. Innovatively designed evaporators featuring self-rotating mechanisms have been developed to enable autonomous regeneration of the evaporation surface [125–127]. Xu et al. developed a self-rotating spherical photothermal evaporator with dual evaporation zones (high-temperature and low-temperature) [128]. This design is sensitive to weight imbalances of less than 15 mg, allowing it to quickly rotate and refresh the evaporation surface in response to salt accumulation (Fig. 4.12a). The dual zones optimize energy dynamics during solar evaporation, achieving excellent salt resistance and a high evaporation rate of 2.6 kg m^{-2} h^{-1}. Bai et al. improved the design by using spherical polystyrene foam as structural support and applying black Ti_4O_7 as a photothermal coating to significantly enhance sunlight absorption. Additionally, the integration of sodium alginate imparts

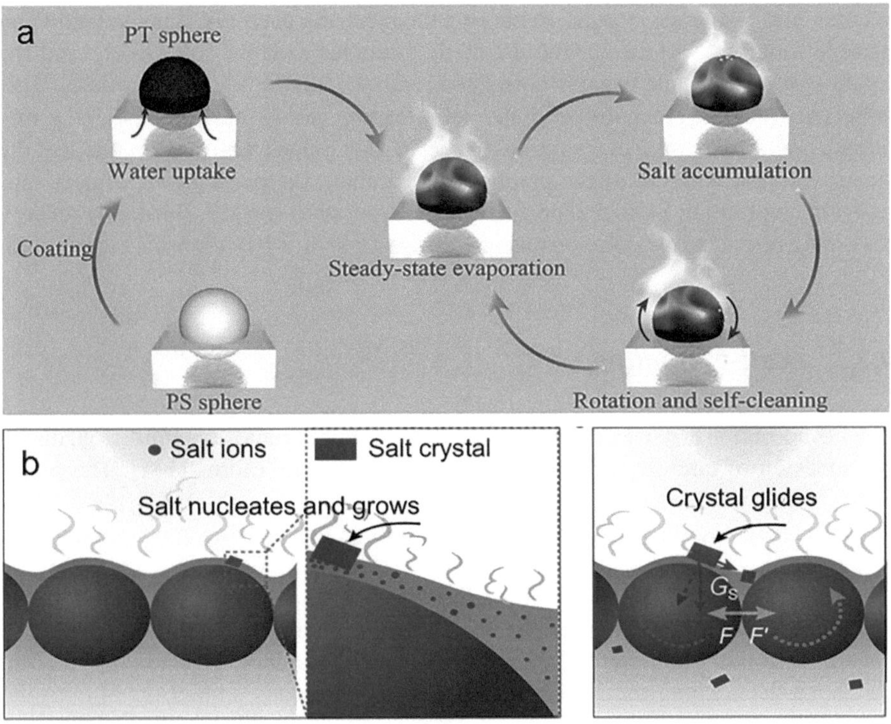

Fig. 4.12 (**a**) Dynamic evaporation of the photothermal sphere. (Reproduced with permission from [128], Copyright 2021 Wiley-VCH). (**b**) Fish-school-inspired system by particles for solar-powered autonomous water-solute separation. (Reproduced with permission from [129], Copyright 2021 Oxford University Press)

hydrophilicity [125]. This strategic combination not only boosts the evaporation rate to 2.46 kg m^{-2} h^{-1} but also excels in water purification. These spherical designs use self-rotation to periodically clean salt deposits from the evaporator surface, offering a promising solution to salt deposition challenges.

A self-cleaning cylindrical absorber, made from an expanded polyethylene cylinder coated with CNT/PVA cloth, was designed for solar evaporation and salt extraction. As evaporation proceeds, salt crystals nucleate and grow at the top of the cylinder, eventually dropping off as the cylinder rotates under the weight of the accumulated salt, thus accomplishing self-cleaning [126, 130]. This system achieves an evaporation rate of 1.41 kg m^{-2} h^{-1} and a solar-to-vapor efficiency of 88.5% when processing hypersaline brine (300 g/L). To achieve large-scale application, Zhu et al. engineered an autonomous system comprising particles with an expanded polystyrene core and a graphene oxide shell (Fig. 4.12b) [123]. These particles placed on water, form a thin, albeit uneven, water film on their surface due to surface tension. Under light irradiation, enhanced evaporation occurs at the top where the water film is thinner, leading to salt nucleation and deposition. The salt accumulation generates torque, disrupting the mechanical equilibrium and inducing particle rotation. Surface tension drives the particles to bond collectively, and the rotation of one particle triggers correlated rotations among adjacent particles. This collective dynamic behavior facilitates self-cleaning across the system, thereby sustaining continuous solar evaporation. However, due to the small size ~3 mm and the superhydrophilic nature of the graphene oxide shell, the particles are almost submerged, resulting in limited exposure of the evaporation surface, substantial energy loss, and a restricted evaporation rate of 1.26 kg m^{-2} h^{-1} for 10 wt% brine evaporation.

4.2.2 *Anti-biofouling*

Solar evaporation are susceptible to biofouling and/or organic contamination, due to the adhesion of microorganisms and bacteria that form biofilms [131]. These biofilms impede water transport and light absorption, thereby undermining the stability and efficiency of the system. To address biofouling, a promising approach involves the integration of antibacterial materials into the evaporator design [132]. Despite the critical nature of this issue, biofouling has not yet garnered the same level of attention as salt fouling. Due to the abundance of microorganisms and bacteria in natural seawater, photothermal evaporators often experience adhesion and biofilm formation during long-term operation [107, 133]. This causes blockage of essential structures and reduces evaporation performance. Therefore, developing solar evaporators with anti-biofouling capabilities is becoming a crucial advancement in the field.

4.2.2.1 Photocatalytic Oxidation

Photocatalytic oxidation is a pivotal strategy in antibacterial applications, particularly by the integration of photocatalysts such as ZnO [134], CuO [135], MXene [136], WO_3 [137] and MoS_2, [138, 139], alongside photothermal evaporation. This method utilizes the intrinsic properties of photocatalysts to generate reactive oxygen species (ROS), which are essential for targeting and neutralizing bacterial pathogens [140, 141]. The process begins when photocatalysts, upon exposure to light, absorb photons and generate electron-hole pairs. These pairs interact with surrounding water and oxygen molecules, forming reactive oxygen species. These highly reactive molecules, including hydroxyl radicals, superoxide anions, and hydrogen peroxide, possess potent oxidative capabilities that can disrupt bacterial cellular structures, such as cell walls, membranes, and DNA, leading to cell damage or destruction [142]. Inspiration from the design of desalination membranes, metallic biocidal nanomaterials such as silver can also be incorporated onto photothermal surfaces [143]. Silver nanoparticles, in particular, are renowned for their broad-spectrum antimicrobial properties, attributed primarily to the sustained release of silver ions which can interact with thiol groups in bacterial proteins, leading to protein inactivation and cell death [144, 145]. The integration of photothermal materials in this context serves a dual purpose. Firstly, these materials can convert light energy into heat, elevating the local temperature upon irradiation. This temperature increase can enhance the kinetic energy of the photocatalysts, potentially increasing the rate of ROS generation. Secondly, the localized heating can directly contribute to bacterial inactivation, as many pathogens are sensitive to temperature changes [146].

The synergistic of photocatalytic and photothermal activities can be optimized by careful selection and structuring of the materials [142, 144]. For instance, the bandgap of photocatalysts can be tuned to match the solar spectrum more effectively, enhancing light absorption and ROS production [147, 148]. The morphology and surface properties of the catalysts can be engineered to maximize their interaction with light and microbial cells [139, 144, 149, 150]. Recent studies demonstrated that coatings or films containing these photocatalysts can be applied to various surfaces prone to bacterial contamination, such as medical devices, water treatment equipment, and even textiles [137, 140, 141, 151] (Fig. 4.13). This approach not only offers a passive and continuous antibacterial defense but also reduces the reliance on chemical disinfectants, which can lead to resistant bacterial strains. Furthermore, photocatalytic oxidations are generally clean, relying only on light (often solar) and naturally abundant (water and oxygen). This eco-friendly aspect, combined with the potential for solar energy utilization, aligns well with sustainability goals. Future research could focus on optimizing these systems for specific applications, enhancing their stability, and scaling up their production to meet industrial demands.

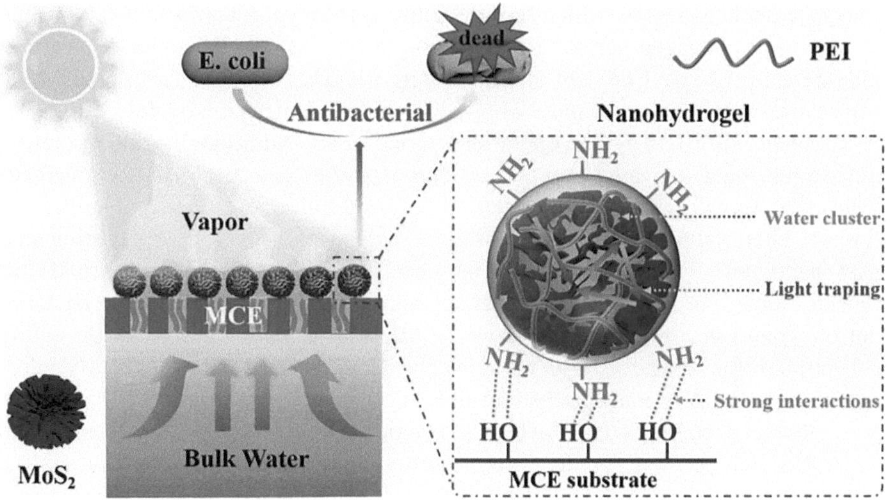

Fig. 4.13 The antibacterial process of a evaporator incorporated with photocatalysts. (Reproduced with permission from [139] Copyright 2019 Royal Society of Chemistry)

4.2.2.2 Molecular Engineering

Molecular engineering plays a pivotal role in imparting anti-biofouling properties of photothermal evaporators [152, 153]. One compelling approach involves the incorporation of zwitterionic hydrogel coatings [154]. These coatings utilize the ionic solvation effect to create a robust hydration layer on the evaporator's surface. This layer serves as a physical barrier, preventing bacterial cells from directly contacting the surface and thereby reducing the risk of biofouling. For instance, the use of zwitterionic materials in the hydrogel formation creates a hydrated interface that is inhospitable to bacterial adhesion due to its highly hydrophilic nature [155]. Further advancements in molecular engineering are exemplified in the work of Gong and colleagues, who introduced sodium lignosulfonate into photothermal hydrogels [78]. The strategic incorporation of sodium lignosulfonate capitalizes on its abundant sulfonate groups to deter bacterial adhesion, while its phenolic hydroxyl groups actively inhibit bacterial growth. This dual-functionality results in a hydrogel with exceptional antibacterial properties, showing efficacy in reducing both bacterial adhesion and proliferation. Additionally, the use of cationic polymers such as chitosan and its derivatives has been prevalent in the development of anti-biofouling evaporators [156]. These polymers are known for their inherent antimicrobial properties, which are primarily attributed to their ability to disrupt bacterial cell membranes. A notable contribution by Zhang et al. involves the synthesis of a cationic photothermal hydrogel using [2-(methacryloyloxy) ethyl] trimethylammonium chloride [157]. This polymer is enriched with quaternary ammonium groups that enhance its bactericidal activity, thereby enabling the hydrogel to achieve nearly 100% efficiency in bacterial inactivation.

Recent innovations also include the development of catechol-functionalized chitosan networks by Yu and co-workers [153]. This design utilizes the oxidative properties of catechol groups, which, upon exposure to oxygen and water, lead to the production of hydrogen peroxide. Hydrogen peroxide is a well-known disinfectant that offers broad-spectrum antimicrobial activity. The catechol functionality not only ensures the sustained release of H_2O_2 but also enhances the overall stability of the antibacterial effect under operational conditions. Additionally, incorporating quinone-modified activated carbon particles into the hydrogel network enhances bactericidal functionality. The interaction between quinone groups and sulfhydryl groups in bacterial proteins disrupts essential metabolic processes, accelerating bacterial inactivation (Fig. 4.14). This synergistic effect from both catechol and quinone modifications offers a robust mechanism for continuous and effective bacterial control. Therefore, molecular engineering offers a versatile toolkit for designing and optimizing evaporators with superior anti-fouling capabilities [152]. By introducing specific functional groups and utilizing synergistic material combinations, these advanced coatings and materials not only prevent bacterial adhesion and growth but also enhance the operation stability and efficiency of solar evaporation systems.

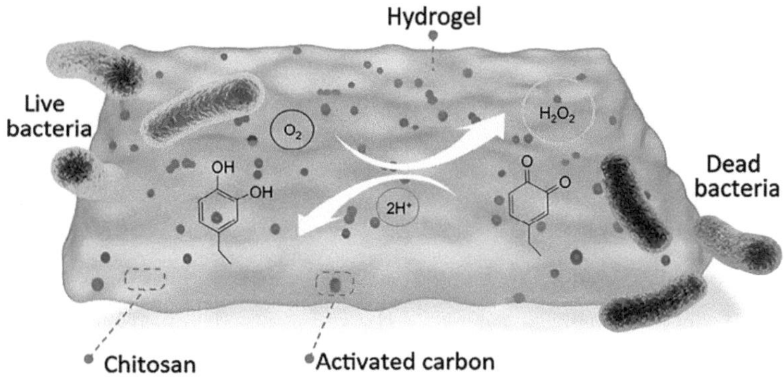

Fig. 4.14 Antibacterial mechanism of hydrogel for water disinfection. (Reproduced with permission from [153], Copyright 2021 Wiley-VCH)

4.2.3 Other Contamination Resistance

The performance of evaporators in wastewater treatment is hindered by the presence of organic pollutants, including non-volatile organic compounds like oils and volatile organic compounds (VOCs) [158, 159]. High-viscosity oil contaminants can lead to the clogging of pores and hinder the flow of water through evaporators. An effective strategy to mitigate oil fouling involves the engineering of superoleophobic surfaces on evaporators (Fig. 4.15a) [160]. Nanostructures on these surfaces induce Cassie contact, enhancing oleophobicity and achieving a contact angle of 143° with soybean oil underwater. This significant angle allows oil droplets to retain their spherical shape, preventing them from adhering to and blocking the evaporator surfaces.

VOCs, on the other hand, can co-evaporate with water, posing significant health risks [161]. To address this, both physical and chemical methods are being developed to remove VOCs during solar-driven evaporation processes [162]. The predominant chemical approach involves the integration of photocatalysis with interfacial solar evaporation (Fig. 4.15b) [163]. This technique utilizes solar energy to activate photocatalysts, which subsequently produce highly reactive hydroxyl radicals that effectively degrade VOCs into carbon dioxide and other benign compounds. For instance, a composite of reduced graphene oxide and polypyrrole in an aerogel form has demonstrated an evaporation rate of 2.08 kg m^{-2} h^{-1} and a VOC removal efficiency of 94.8% at a concentration of 20 ppm [164]. However, the efficacy of photocatalysis diminishes at higher VOC concentrations (>20 ppm), thus limiting its application [165].

Physical removal methods, such as molecular sieving (Fig. 4.15c) [166] and semipermeable membrane filtration (Fig. 4.15d) [167–169], have proven effective for managing high VOC concentrations in the condensation water during evaporation. These methods selectively permit water molecules to pass while blocking VOCs through distinct mechanisms. Molecular sieving uses the size-exclusion

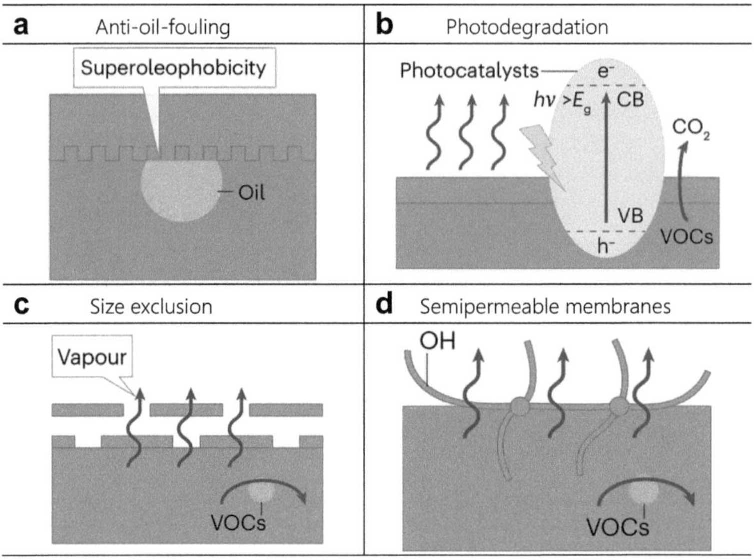

Fig. 4.15 Resistance to other contamination: (**a**) Integration of superoleophobic properties in absorbers for anti-oil fouling; removal of volatile organic compounds through (**b**) Photocatalytic degradation, (**c**) Size exclusion, and (**d**) selective permeable membrane. (Adapted with permission from [159], Copyright 2025 Nature)

principle, whereas semipermeable membranes rely on differing affinities for water and VOCs. These physical approaches can handle VOC concentrations up to several hundred parts per million, achieving retention rates of 99% at 400 ppm [166]. However, a major limitation is their inability to degrade VOCs, which can lead to secondary pollution. Therefore, developing integrated physicochemical strategies that combine the advantages of both physical and chemical removal techniques is essential for efficiently treating high-concentration VOCs in wastewater.

4.3 Steam Release and Condensation

4.3.1 Steam Release

Solar evaporation technology uses solar energy to convert water into vapor, which is then condensed into fresh water. A crucial factor in optimizing these systems is improving vapor transport from the evaporation site to the condensation site [170]. Here, we explore strategies to enhance vapor transport, emphasizing structural design, airflow dynamics, and pressure modulation within the system. Optimizing Structural Design for Vapor Transport: Traditional designs often involve simple flat surfaces, recent advancements suggest that three-dimensional structures with specific microstructural features can significantly enhance performance (Fig. 4.16a)

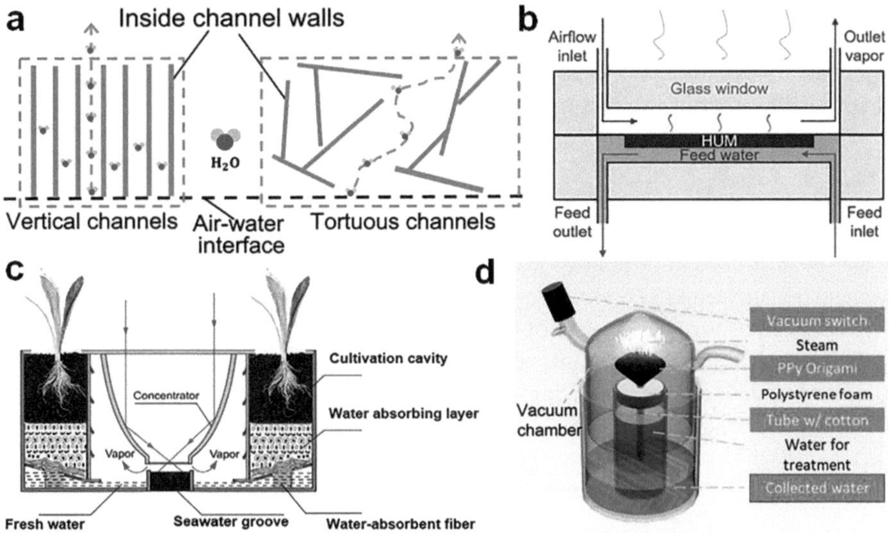

Fig. 4.16 Strategies for promoting vapor diffusion. (**a**) Vapor escape in different channels. (Reproduced with permission from [170], Copyright 2023 Elsevier). (**b**) Airflow directional drives. (Reproduced with permission from [175], Copyright 2021 Wiley-VCH). (**c**) Thermal gradient driving the steam. (Reproduced with permission from [177] Copyright 2018 Elsevier). (**d**) Low-pressure water collection system. (Reproduced with permission from [178] Copyright 2019 Wiley-VCH)

[171]. (1) Straight versus Tortuous Channels: Vapor diffusion is more efficient in straight channels as they provide a direct path for vapor transport, minimizing resistance and enhancing flow rate. For instance, graphene-based evaporators with vertically aligned graphene pillar arrays have demonstrated evaporation rates substantially higher than those of traditional designs, reaching up to 2.1 kg m^{-2} h^{-1} upon solar irradiation [172]. (2) Interconnected Pore Structures: These structures facilitate horizontal vapor transport, which can be particularly effective under windy conditions. A notable implementation is in carbon foam evaporators, where interconnected porosity has achieved evaporation rates exceeding 10 kg m^{-2} h^{-1} when external convective airflow is introduced [58]. (3) Multi-layer Evaporators: By stacking multiple evaporative layers, the surface area for evaporation is significantly increased without expanding the solar footprint [173]. This approach has more than doubled evaporation rates compared to single-layer systems, though it also introduces challenges such as increased heat loss and lower vapor temperatures, which can impede condensation efficiency.

Airflow Dynamics and Vapor Transport: The manipulation of airflow within the evaporator can dramatically influence vapor transport dynamics and overall system efficiency [174]. (1) Unidirectional Airflow Systems: These systems direct airflow across the evaporative surface to enhance vapor transport and prevent the accumulation of non-condensable gases, which can insulate the surface and reduce evaporation rates (Fig. 4.16b) [175]. Implementations using simple mechanical devices or natural airflow patterns to create unidirectional flow have shown improved water production rates. (2) Mechanical Airflow Regulation: In more complex systems, mechanical devices such as fans or wave-driven cantilever beams have been used to direct airflow within multi-dimensional structures [176]. This not only enhances vapor transport but also helps in maintaining optimal temperature and pressure conditions across the evaporative surface. (3) Researchers have employed steam for directional transport by creating substantial thermal gradients. Wang et al. intensified these gradients on condensing and evaporating surfaces using a light-gathering device (Fig. 4.16c) to direct vapor from the hotter evaporating surface to the cooler condensing surface [177]. However, the wide spacing between these surfaces led to a reduced water production rate of 0.196 kg m^{-2} h^{-1}.

Pressure Modulation Techniques: Adjusting the ambient pressure around the evaporative surface can significantly impact the vapor pressure gradient, a critical driver of evaporation [170, 174]. (1) Vacuum Systems: By reducing the pressure in the condensation chamber (e.g., through vacuum pumps), the vapor pressure gradient increases, accelerating the evaporation process [179]. Studies have shown that reducing the chamber pressure can increase evaporation rates by up to 50%. (2) Low-Pressure Solar Harvesting Systems: Innovations in material and structural design allow for the maintenance of lower pressures over extended periods, enhancing sustained evaporation rates. For example, systems utilizing chemically polymerized surfaces have demonstrated increased water collection efficiency under reduced pressure conditions [178]. Improvements in evaporator structural design, alongside strategic airflow management and pressure modulation, offer promising ways to enhance the efficiency of solar evaporation systems. These strategies not only increase water production rates but also support the overall sustainability of water resource management. Future research should continue to explore these areas, focusing on integrating these strategies into scalable and economically viable systems. Further development of materials and technologies that support efficient vapor transport and condensation will be crucial for the advancement of this field.

4.3.2 Steam Condensation

The steam condensation is pivotal in regulating water production [43, 174]. In the front-side solar evaporation system, the upward diffusion of steam initiates upon photothermal evaporation, rapidly saturating the entire transparent cover, which leads to supersaturation and exothermic condensation at the periphery of the cover. Prior studies showed that the evaporator configuration, along with the thermal characteristics of steam and air, governs the steam transport dynamics, independent of materials [180, 181]. Transparent covers in front-side systems are generally constructed from polymers or glass, chosen for their high optical transmittance. Nevertheless, these materials typically exhibit low thermal conductivity (less than 5 W m^{-1} K^{-1}), which impedes the steam condensation process [182]. Moreover, the accumulation of condensed water intensifies thermal resistance and optical attenuation. The properties of the inner surface of the transparent cover are essential for efficient steam condensation. Research indicates that steam condenses more rapidly on hydrophobic surfaces, and some studies have reported enhanced heat transfer characteristics on hydrophobic or superhydrophobic surfaces compared to nature metal surfaces during the condensation process [183]. In addition, hydrophobic surfaces aid in the rapid removal of condensate, thereby minimizing light loss due to droplet retention [184, 185].

4.3.2.1 Single-Slope Surface

The first device for water harvesting through interfacial evaporation was introduced in 2015. Made from transparent, lightweight plastic, it features separate zones for evaporation and condensation collection. The upper section includes a solar-powered fan to induce air convection, and the device is designed to float independently on a water body [186]. Since then, numerous single-slope evaporators using materials like acrylic, PET, and glass for condensing surfaces have been developed [174]. Liu et al. designed a condensing device with an acrylic single-slope top and an interfacial evaporation layer coated with carbon black on hydrophilic fibers. While the evaporation rate in an open setting reached 1.28 kg m^{-2} h^{-1}, the actual water yield within a sealed system was only 0.83 kg m^{-2} h^{-1}, significantly lower than the rate in an open environment during the day [187]. This disparity prompted further research into methods to enhance evaporation rates, though often at the expense of actual water production metrics.

The limitations of two-dimensional evaporation areas and theoretical evaporation caps paved the way for the exploration of three-dimensional evaporators, which offer extensive evaporation surfaces and lower surface temperatures. Wu et al. developed a compact 3D bionic structure featuring a CNT-based evaporator within a single-slope light-transmitting design, aimed at optimizing water transport paths and evaporation sites for enhanced rate and energy efficiency [188]. This configuration achieved an evaporation rate of 2.63 kg m^{-2} h^{-1} in an open environment, with an internal water yield of 1.72 kg m^{-2} h^{-1}. The high cost of photothermal absorbers, often due to complex and scarce production methods, has prompted innovative use of biowaste [189–191]. For instance, Zhang et al. carbonized cattails and integrated them into a monoclinic condensing device. Despite a high evaporation rate of 4.12 kg m^{-2} h^{-1} in an open environment, the actual water yield was only 8.32 kg m^{-2} day^{-1}, due to the high aspect ratio of the carbonized cattails leading to inefficient vapor condensation [192].

Recently, Yang et al. developed an internally recessed 3D evaporator array on a single-slope condensing surface, achieving an impressive evaporation rate of 6.09 kg m^{-2} h^{-1} in an open environment (Fig. 4.17) [193]. However, the wide spacing between condensing and evaporating surfaces posed challenges in maintaining the necessary temperature gradient, affecting the condensation rate. Furthermore, Mu et al. explored the dual application of water evaporation and thermoelectricity by harnessing photothermal conversion for thermoelectric collection and utilizing

Fig. 4.17 A 3D evaporator with the single-slope condensing device. (Reproduced with permission from [193], Copyright 2023 Nature Publishing Group)

thermoelectric waste heat to enhance water evaporation. This innovative approach not only improved the thermoelectric yield but also promoted efficient water evaporation, achieving a collection rate of 0.92 kg m^{-2} h^{-1} [194]. Despite these advancements, it is regrettable that the focus remains on the evaporation rate, with less attention given to the condensation efficiency. In real applications, the variability of solar incidence due to changing angles affects the efficiency of single-slope condensing surfaces, underscoring the need for further research to optimize these systems under variable solar conditions.

4.3.2.2 Symmetrical Surface

The nature limitations in single-slope surfaces prompte researchers to develop symmetrical surfaces that offering a reduced incidence angle sensitivity to sunlight. Ni et al. designed a dual-sloped condensation surface using a polymer film with a hydrophilic black fiber fabric as the photothermal absorption layer, optimizing the insulation-to-fiber ratio to enhance mass and thermal transfer while preventing salt crystallization (Fig. 4.18a). This setup achieved a water production rate of 2.5 L m^{-2} day^{-1}. Raihananda et al. improved this by adjusting the slopes to a symmetrical 45° angle, resulting in 4.9 g/h over an area of 16.5 cm × 25.5 cm [195]. Other researchers have explored alternative designs, such as pyramidal and house-like condensation surfaces. Liang et al. created a pyramidal surface that maximizes energy use with a simple mechanical water supply device [40]. Chen et al. developed a house-like collector mimicking plant root and stem water supply, achieving 0.24 kg m^{-2} h^{-1} in actual production and 1.5 kg/m^2 on a sunny day outdoors [196]. Similarly, Yang et al. used PMMA to create a house-like condensing surface, yielding 175 mL in one day, equivalent to 5.0 m^{-2} day^{-1} [116].

Despite the advantages of symmetrical surfaces in reducing shadow areas and refraction, the incidence of sunlight can still be affected by multiple angles. Xu et al. developed a prism-free condensing device featuring a laterally placed semi-cylindrical polypropylene surface, which enabled salt crystallization on low-cost

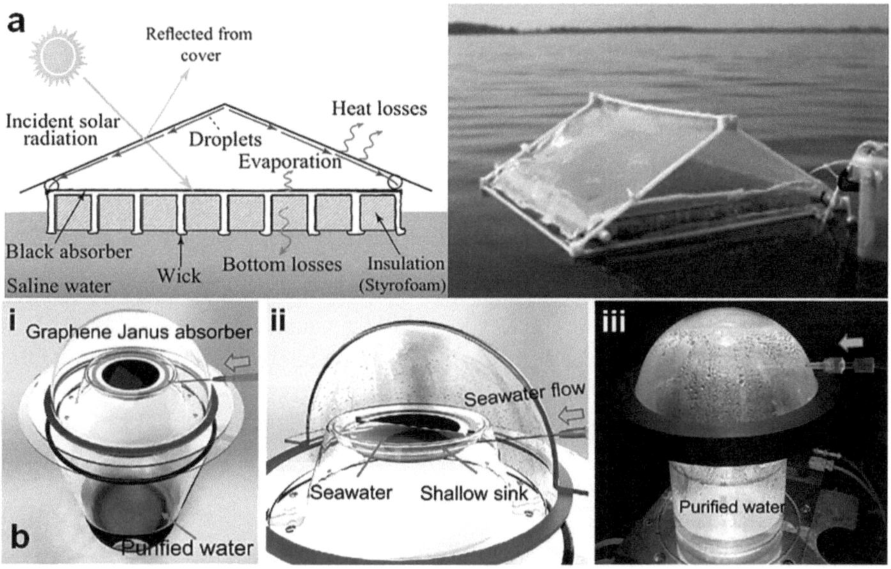

Fig. 4.18 (**a**) A salt-rejecting evaporator is designed according to the Fick diffusion law. (Adapted from with permission from [114], Copyright 2018 Royal Society of Chemistry). (**b**) Omnidirectional semispherical condensing surface. (Reproduced with permission from [199], Copyright 2022 Elsevier)

2D printing paper [197]. This design achieved a water production rate of 2 L m^{-2} h^{-1} and 400 g m^{-2} day^{-1} of salt. However, the 3D evaporator was not well-suited for the semicylindrical condensation collector, indicating the need for vertical cylindrical condensation collectors for improved efficiency. Wu et al. developed a bullet-shaped condensation collector made of glass, which, by reducing the distance between evaporating and condensing surfaces, facilitated a consistent temperature gradient essential for efficient condensation [198]. Additionally, spherical and hemispherical condensing devices have been engineered to fully eliminate refraction losses (Fig. 4.18b) [199]. Zhang et al. designed a hemispherical glass condensing surface with a vertically aligned MXene aerogel as the evaporator, achieving a water yield of 6 L m^{-2} day^{-1} [200]. Yang et al. developed a one-way water circulation system using a hemispherical condensing surface with carbonized coconut, achieving a high evaporation rate of 7.28 kg m^{-2} h^{-1} in an open environment, but the actual water yield was only 8 L m^{-2} day^{-1} [201]. Despite the promise of innovative condensing surface designs in enhancing water collection efficiency, challenges remain regarding light transmittance, angle sensitivity, and non-condensable gas accumulation at the gas-liquid interface. Further research is needed to address these issues and optimize the condensation process to improve freshwater yield.

4.3.2.3 Improved Surface Treatment

Water harvesting by improvements to the condensation surface received significant interest. The adhesion of water droplets on condensing surfaces significantly impedes water harvesting. To address this, researchers have advocated for reducing droplet adhesion through the application of hydrophobic treatments to these surfaces [186, 202, 203]. Nonetheless, such hydrophobic surfaces typically exhibit high roughness, which can lead to substantial scattering of solar light. Minimizing roughness can reduce the intensity of light scattering and achieve a balance between hydrophobicity and light transmission, although the high cost associated with this approach limits its large-scale application in solar water extraction systems. In a promising development, researchers have devised an alternative by decoupling the condensation and light transmittance surfaces through reverse evaporation, a technique that facilitates hydrophobic treatment without compromising light transmission efficiency [204, 205]. This advancement simplifies the implementation of hydrophobic treatments to enhance vapor condensation and increase water yield.

Yao et al. utilized copper's high thermal conductivity to achieve photothermal separation and reverse evaporation [89]. They modified an aluminum sheet's surface with hydrophobic fluorine and silane groups, creating a reverse evaporation structure that generated a significant temperature and humidity difference between the evaporating and condensing layers. This effectively directed vapor toward the condensing surface, resulting in a water yield of 1.95 kg m^{-2} h^{-1} over a large evaporating area. Xu et al. applied a hydrophobic fluorine coating to the condensation surface, achieving a rolling angle of 5 degrees, which allowed condensate removal by gravity [63]. Their multistage evaporation device produced 10–20 L m^{-2} day^{-1}, sufficient to meet the water needs of several households. Furthermore, nanostructured treatments have been shown to impart hydrophobic properties, enhancing condensation performance by reducing the thermal barrier and increasing vapor turbulence in the diffusion layer (Fig. 4.19a) [182]. Peng et al. utilized a capillary

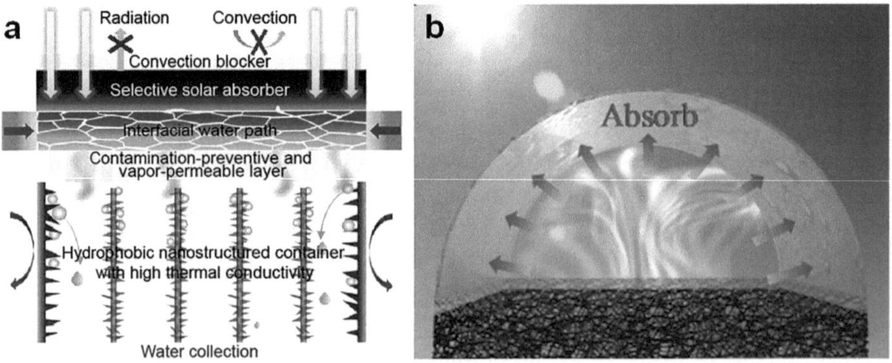

Fig. 4.19 (**a**) Nanohydrophobic copper oxide condensation surface. (Reproduced with permission from [182], Copyright 2021 Cell Press). (**b**) Moisture-absorbing materials retrofitted to collect cold steam. (Reproduced with permission from [208], Copyright 2023 American Chemical Society)

grid in a photothermal device to improve the hydrophilicity of the condensing surface, which resulted in a thin film of water forming on super-hydrophilic glass rather than droplets, thus preserving light transmission and reducing energy loss [206]. This approach yielded 7 kg m^{-2} day^{-1} in a two-stage evaporator. Alternative methods, including hydrophilic treatments and atmospheric water harvesting strategies, have been explored. Dong et al. utilized a high-aspect-ratio evaporator with an RGO film rolled onto a cold evaporation surface, capturing ambient energy for evaporation and achieving a water production efficiency of 2.1 kg m^{-2} h^{-1} [207]. Zhan et al. developed highly translucent hydrogels that directly adsorbed vapors from photothermal evaporation, achieving an impressive evaporation rate and showing the potential for efficient long-term water collection and recycling from storage hydrogels (Fig. 4.19b) [208].

References

1. L. Cheng, L. Zuo, C. Li, J. Shen, L. Wang, Y. Li, D. Dukashin, P. Senin, S. Yan, T. Bian, Sandwiched solar interfacial evaporation system: Integrating phase change material/hydrogel/NiCo2O4 nanoflower for superior desalination, ***Separation and Purification Technology***, 348 (2024) 127816.
2. E. Chiavazzo, M. Morciano, F. Viglino, M. Fasano, P. Asinari, Passive solar high-yield seawater desalination by modular and low-cost distillation, ***Nature Sustainability***, 1 (2018) 763-772.
3. C. Ge, D. Xu, X. Feng, H. Du, Z. Chen, H.L. Ong, C. Gao, G. Yan, J. Fang, 3D spacer fabric structured airflow channel for enhanced solar desalination with efficient multi-energy harvesting, ***Separation and Purification Technology***, 354 (2025) 128849.
4. X. Geng, P. Yang, Y. Wan, How to reduce enthalpy in the interfacial solar water generation system for enhancing efficiency?, ***Nano Energy***, 123 (2024) 109434.
5. Z. Guo, W. Zhou, N. Arshad, Z. Zhang, D. Yan, M.S. Irshad, L. Yu, X. Wang, Excellent energy capture of hierarchical MoS2 nanosheets coupled with MXene for efficient solar evaporators and thermal packs, ***Carbon***, 186 (2022) 19-27.
6. Y. Guo, F. Zhao, X. Zhou, Z. Chen, G. Yu, Tailoring Nanoscale Surface Topography of Hydrogel for Efficient Solar Vapor Generation, ***Nano letters***, 19 (2019) 2530-2536.
7. J. Huang, Q. Wang, Z. Wu, Z. Ma, C. Yan, Y. Shi, B. Su, 3D-Printed Underwater Super-Oleophobic Shark Skin toward the Electricity Generation through Low-Adhesion Sliding of Magnetic Nanofluid Droplets, ***Advanced Functional Materials***, 31 (2021).
8. B. Wang, C. Wang, Y. Li, J. Jin, X. Lin, C. Shi, Bionic design: insights from nature for solar interfacial evaporators, ***Energy & Environmental Science***, 18 (2025) 3432-3461.
9. K. Bae, G. Kang, S.K. Cho, W. Park, K. Kim, W.J. Padilla, Flexible thin-film black gold membranes with ultrabroadband plasmonic nanofocusing for efficient solar vapour generation, ***Nature communications***, 6 (2015) 10103.
10. Y. Wang, F. Xi, K. Wei, Z. Tong, S. Li, Z. Ding, W. Ma, Bridging efficiency and scalability: A systematic evaluation of diamond wire sawn silicon wafer texturing technologies for high-performance photovoltaics, ***Applied Energy***, 386 (2025) 125591.
11. H.K. Raut, V.A. Ganesh, A.S. Nair, S. Ramakrishna, Anti-reflective coatings: A critical, in-depth review, ***Energy & Environmental Science***, 4 (2011) 3779.
12. J.D. Joannopoulos, P.R. Villeneuve, S. Fan, Photonic crystals, ***Solid State Communications***, 102 (1997) 165-173.

13. R.K. Cersonsky, J. Antonaglia, B.D. Dice, S.C. Glotzer, The diversity of three-dimensional photonic crystals, ***Nature communications***, 12 (2021) 2543.
14. S. Hong, Y. Shi, R. Li, C. Zhang, Y. Jin, P. Wang, Nature-Inspired, 3D Origami Solar Steam Generator toward Near Full Utilization of Solar Energy, ***ACS applied materials & interfaces***, 10 (2018) 28517-28524.
15. X. Liu, Y. Tian, F. Chen, Y. Mu, A. Caratenuto, M.L. Minus, Y. Zheng, A waterbomb origami tower for convertible photothermal evaporation, ***Journal of Materials Chemistry A***, 10 (2022) 18657-18670.
16. Y. Xu, J. Ma, D. Liu, H. Xu, F. Cui, W. Wang, Origami system for efficient solar driven distillation in emergency water supply, ***Chemical Engineering Journal***, 356 (2019) 869-876.
17. Z. Lei, X. Sun, S. Zhu, K. Dong, X. Liu, L. Wang, X. Zhang, L. Qu, X. Zhang, Nature Inspired MXene-Decorated 3D Honeycomb-Fabric Architectures Toward Efficient Water Desalination and Salt Harvesting, ***Nano-micro letters***, 14 (2021) 10.
18. X. Liu, L. Hou, R. Hu, H. Zhang, X. Zhang, X. Ge, Y. Zhang, G. Yue, Z. Cui, J. Bai, J. Liu, N. Wang, Y. Li, Y. Zhao, Honeycomb-like Microthermal Traps on a Photothermal Surface for Highly Efficient Solar Evaporation, ***ACS applied materials & interfaces***, 16 (2024) 62710-62720.
19. S. Qi, T. Jia, Y. Zhang, Y. Zhao, Y. Xing, M. Zhang, F. Wang, C. Chen, W.Y. Wong, Efficient "Core–Shell" Carbon-Based Solar Evaporator with Honeycomb Structure for Large-Scale Solar-Driven Water Purification and Energy Harvesting, ***Advanced Functional Materials***, (2025).
20. N. Ding, J. Zhu, Y. Qiao, D. Yao, X. Gao, J. Chen, C. Lu, X. Pang, Janus-structured self-floating hydrogel via thermal localization and arched evaporation interface design toward all-weather efficient solar desalination, ***Chemical Engineering Journal***, 513 (2025) 162851.
21. B. Lv, C. Gao, Y. Xu, X. Fan, J. Xiao, Y. Liu, C. Song, A self-floating, salt-resistant 3D Janus radish-based evaporator for highly efficient solar desalination, ***Desalination***, 510 (2021) 115093.
22. M. Tan, J. Wang, W. Song, J. Fang, X. Zhang, Self-floating hybrid hydrogels assembled with conducting polymer hollow spheres and silica aerogel microparticles for solar steam generation, ***Journal of Materials Chemistry A***, 7 (2019) 1244-1251.
23. C. Chen, Y. Kuang, L. Hu, Challenges and Opportunities for Solar Evaporation, ***Joule***, 3 (2019) 683-718.
24. P. Tao, G. Ni, C. Song, W. Shang, J. Wu, J. Zhu, G. Chen, T. Deng, Solar-driven interfacial evaporation, ***Nature Energy***, 3 (2018) 1031-1041.
25. G. Ni, G. Li, Svetlana V. Boriskina, H. Li, W. Yang, T. Zhang, G. Chen, Steam generation under one sun enabled by a floating structure with thermal concentration, ***Nature Energy***, 1 (2016).
26. M. Wang, Y. Wei, R. Li, X. Wang, C. Wang, N. Ren, S.H. Ho, Sustainable Seawater Desalination and Energy Management: Mechanisms, Strategies, and the Way Forward, ***Research***, 6 (2023) 0290.
27. H. Liu, C. Chen, G. Chen, Y. Kuang, X. Zhao, J. Song, C. Jia, X. Xu, E. Hitz, H. Xie, S. Wang, F. Jiang, T. Li, Y. Li, A. Gong, R. Yang, S. Das, L. Hu, High-Performance Solar Steam Device with Layered Channels: Artificial Tree with a Reversed Design, ***Advanced Energy Materials***, 8 (2017).
28. Y. Dong, Y. Tan, K. Wang, Y. Cai, J. Li, C. Sonne, C. Li, Reviewing wood-based solar-driven interfacial evaporators for desalination, ***Water research***, 223 (2022) 119011.
29. C. Jia, Y.J. Li, Z. Yang, G. Chen, Y.G. Yao, F. Jiang, Y.D. Kuang, G. Pastel, H. Xie, B. Yang, S. Das, L.B. Hu, Rich Mesostructures Derived from Natural Woods for Solar Steam Generation, ***Joule***, 1 (2017) 588-599.
30. X. Zhu, M. Li, L. Song, X.-F. Zhang, J. Yao, Metal organic framework enabled wood evaporator for solar-driven water purification, ***Separation and Purification Technology***, 281 (2022) 119912.

31. C. Yan, J. Huang, Z. Liang, Z. Fang, D. Yang, Z. Li, Fabrication of a Highly Efficient Wood-Based Solar Interfacial Evaporator with Self-Desalting and Sterilization Performance, ***Langmuir : the ACS journal of surfaces and colloids***, 38 (2022) 12813-12821.
32. X. Li, W. Xu, M. Tang, L. Zhou, B. Zhu, S. Zhu, J. Zhu, Graphene oxide-based efficient and scalable solar desalination under one sun with a confined 2D water path, ***Proceedings of the National Academy of Sciences of the United States of America***, 113 (2016) 13953-13958.
33. Y. Xia, Q. Hou, H. Jubaer, Y. Li, Y. Kang, S. Yuan, H. Liu, M.W. Woo, L. Zhang, L. Gao, H. Wang, X. Zhang, Spatially isolating salt crystallisation from water evaporation for continuous solar steam generation and salt harvesting, ***Energy & Environmental Science***, 12 (2019) 1840-1847.
34. Q. Hu, X. Lin, G. Ren, J. Lü, W. Wang, D. Zhang, S. Zhou, Hydrovoltaic electricity generation induced by living leaf transpiration, ***Nature Water***, 2 (2024) 988-998.
35. N. Xu, J. Li, Y. Wang, C. Fang, X. Li, Y. Wang, L. Zhou, B. Zhu, Z. Wu, S. Zhu, J. Zhu, A water lily-inspired hierarchical design for stable and efficient solar evaporation of high-salinity brine, ***Science advances***, 5 (2019) eaaw7013.
36. N. Xu, X. Hu, W. Xu, X. Li, L. Zhou, S. Zhu, J. Zhu, Mushrooms as Efficient Solar Steam-Generation Devices, ***Advanced materials***, 29 (2017).
37. X. Wu, L.M. Wu, J. Tan, G.Y. Chen, G. Owens, H.L. Xu, Evaporation above a bulk water surface using an oil lamp inspired highly efficient solar-steam generation strategy, ***Journal of Materials Chemistry A***, 6 (2018) 12267-12274.
38. L. Zhu, M. Gao, C.K.N. Peh, X. Wang, G.W. Ho, Self-Contained Monolithic Carbon Sponges for Solar-Driven Interfacial Water Evaporation Distillation and Electricity Generation, ***Advanced Energy Materials***, 8 (2018).
39. Y. Wang, X. Wu, P. Wu, J. Zhao, X. Yang, G. Owens, H. Xu, Enhancing solar steam generation using a highly thermally conductive evaporator support, ***Science bulletin***, 66 (2021) 2479-2488.
40. H. Liang, Q. Liao, N. Chen, Y. Liang, G. Lv, P. Zhang, B. Lu, L. Qu, Thermal Efficiency of Solar Steam Generation Approaching 100 % through Capillary Water Transport, ***Angewandte Chemie***, 58 (2019) 19041-19046.
41. Z. Wang, X. Wu, F. He, S. Peng, Y. Li, Confinement Capillarity of Thin Coating for Boosting Solar-Driven Water Evaporation, ***Advanced Functional Materials***, 31 (2021).
42. Z. Wang, M. Hu, L. Zhu, J. Zhou, F. He, Y. Liu, Y. Li, Y. Li, Z. Lin, F. Caruso, Cracked Metal-Phenolic Networks with Durable Confinement Capillarity for Enhanced Solar Desalination, ***Advanced materials***, (2025) e2503896.
43. C. Dang, Y. Cao, H. Nie, W. Lang, J. Zhang, G. Xu, M. Zhu, Structure integration and architecture of solar-driven interfacial desalination from miniaturization designs to industrial applications, ***Nature Water***, 2 (2024) 115-126.
44. J. Wang, Y. Kong, Z. Liu, H. Wang, Solar-driven interfacial evaporation: Design and application progress of structural evaporators and functional distillers, ***Nano Energy***, 108 (2023) 108115.
45. S. Tian, Z. Huang, J. Tan, X. Cui, Y. Xiao, Y. Wan, X. Li, Q. Zhao, S. Li, C.-S. Lee, Manipulating Interfacial Charge-Transfer Absorption of Cocrystal Absorber for Efficient Solar Seawater Desalination and Water Purification, ***ACS Energy Letters***, 5 (2020) 2698-2705.
46. F. Wang, S. Zhao, X. Zhang, Z. Su, Interfacial solar evaporation based on Janus films: An effective strategy to improve salt tolerance and antifouling performance, ***Desalination***, 543 (2022) 116085.
47. X.Q. Li, R.X. Lin, G. Ni, N. Xu, X.Z. Hu, B. Zhu, G.X. Lv, J.L. Li, S.N. Zhu, J. Zhu, Three-dimensional artificial transpiration for efficient solar waste-water treatment, ***National science review***, 5 (2018) 70-77.
48. H. Song, Y. Liu, Z. Liu, M.H. Singer, C. Li, A.R. Cheney, D. Ji, L. Zhou, N. Zhang, X. Zeng, Z. Bei, Z. Yu, S. Jiang, Q. Gan, Cold Vapor Generation beyond the Input Solar Energy Limit, ***Advanced science***, 5 (2018) 1800222.

49. X. Wu, Z. Wu, Y. Wang, T. Gao, Q. Li, H. Xu, All-Cold Evaporation under One Sun with Zero Energy Loss by Using a Heatsink Inspired Solar Evaporator, ***Advanced science***, 8 (2021) 2002501.
50. Y.D. Wang, X. Wu, X.F. Yang, G. Owens, H.L. Xu, Reversing heat conduction loss: Extracting energy from bulk water to enhance solar steam generation, ***Nano Energy***, 78 (2020).
51. X. Wu, Y. Lu, X. Ren, P. Wu, D. Chu, X. Yang, H. Xu, Interfacial Solar Evaporation: From Fundamental Research to Applications, ***Advanced materials***, 36 (2024) e2313090.
52. H.T. Kim, L. Philip, A. McDonagh, M. Johir, J. Ren, H.K. Shon, L.D. Tijing, Recent Advances in High-Rate Solar-Driven Interfacial Evaporation, ***Advanced science***, 11 (2024) e2401322.
53. X. Liu, Z. Liu, D. Devadutta Mishra, Z. Chen, J. Zhao, C. Hu, Evaporation rate far beyond the input solar energy limit enabled by introducing convective flow, ***Chemical Engineering Journal***, 429 (2022) 132335.
54. C. Onggowarsito, S. Mao, X.S. Zhang, A. Feng, H. Xu, Q. Fu, Updated perspective on solar steam generation application, ***Energy & Environmental Science***, 17 (2024) 2088-2099.
55. T. Gao, Y. Wang, X. Wu, P. Wu, X. Yang, Q. Li, Z. Zhang, D. Zhang, G. Owens, H. Xu, More from less: improving solar steam generation by selectively removing a portion of evaporation surface, ***Science bulletin***, 67 (2022) 1572-1580.
56. Y. Wang, X. Wu, T. Gao, Y. Lu, X. Yang, G.Y. Chen, G. Owens, H. Xu, Same materials, bigger output: A reversibly transformable 2D–3D photothermal evaporator for highly efficient solar steam generation, ***Nano Energy***, 79 (2021) 105477.
57. X. Li, J. Li, L. Zou, M. Chen, H. Wang, L. Yang, R. Yang, H. Zhang, Y. Zhou, Less is More: Reducing Evaporation Dead Zones to Tune Thermal Convection in Donut-Structure Evaporators for High-Efficiency Interfacial Solar Steam Generation, ***Advanced Functional Materials***, (2025).
58. J. Li, X. Wang, Z. Lin, N. Xu, X. Li, J. Liang, W. Zhao, R. Lin, B. Zhu, G. Liu, L. Zhou, S. Zhu, J. Zhu, Over 10 kg m−2 h−1 Evaporation Rate Enabled by a 3D Interconnected Porous Carbon Foam, ***Joule***, 4 (2020) 928-937.
59. Z. Huang, Y. Liu, S. Li, C.S. Lee, X.H. Zhang, From Materials to Devices: Rationally Designing Solar Steam System for Advanced Applications, ***Small Methods***, 6 (2022) e2200835.
60. Y. Shi, R. Li, Y. Jin, S. Zhuo, L. Shi, J. Chang, S. Hong, K.-C. Ng, P. Wang, A 3D Photothermal Structure toward Improved Energy Efficiency in Solar Steam Generation, ***Joule***, 2 (2018) 1171-1186.
61. Y. Luo, B. Fu, Q. Shen, W. Hao, J. Xu, M. Min, Y. Liu, S. An, C. Song, P. Tao, J. Wu, W. Shang, T. Deng, Patterned Surfaces for Solar-Driven Interfacial Evaporation, ***ACS applied materials & interfaces***, 11 (2019) 7584-7590.
62. Y. Wang, T. Wei, Y. Wang, J. Zeng, T. Wang, Q. Wang, S. Zhang, M. Zeng, F. Wang, P. Dai, X. Jiang, M. Hu, J. Zhao, Z. Hu, J. Zhu, X. Wang, Quasi-waffle solar distiller for durable desalination of seawater, ***Science advances***, 10 (2024) eadk1113.
63. Z. Xu, L. Zhang, L. Zhao, B. Li, B. Bhatia, C. Wang, K.L. Wilke, Y. Song, O. Labban, J.H. Lienhard, R. Wang, E.N. Wang, Ultrahigh-efficiency desalination via a thermally-localized multistage solar still, ***Energy & Environmental Science***, 13 (2020) 830-839.
64. L. Yang, T. Sun, J. Tang, Y. Shao, N. Li, A. Shen, J. Chen, Y. Zhang, H. Liu, G. Xue, Photovoltaic-multistage desalination of hypersaline waters for simultaneous electricity, water and salt harvesting via automatic rinsing, ***Nano Energy***, 87 (2021) 106163.
65. W. Wang, Y. Shi, C. Zhang, S. Hong, L. Shi, J. Chang, R. Li, Y. Jin, C. Ong, S. Zhuo, P. Wang, Simultaneous production of fresh water and electricity via multistage solar photovoltaic membrane distillation, ***Nature communications***, 10 (2019) 3012.
66. J. Gao, L. Zhang, J. You, Z. Ye, Y. Zhong, R. Wang, E.N. Wang, Z. Xu, Extreme salt-resisting multistage solar distillation with thermohaline convection, ***Joule***, 7 (2023) 2274-2290.
67. Y. Li, S. Cheng, Z. Yu, R. Gu, X. He, Stable, zero liquid discharge, and highly efficient solar-driven multistage distillation device based on tree-inspired radial water transfer, ***Journal of Cleaner Production***, 375 (2022) 134025.

68. F. Zhao, Y. Guo, X. Zhou, W. Shi, G. Yu, Materials for solar-powered water evaporation, ***Nature Reviews Materials***, 5 (2020) 388-401.
69. S. Lv, Y. Guo, W. Lv, Z. Wang, Y. Pan, J. Ren, Modeling and performance evaluation of radiative cooling-assisted interfacial evaporation for all-day freshwater harvesting, ***Desalination***, 593 (2025) 118241.
70. A. Luzar, D. Chandler, Hydrogen-bond kinetics in liquid water, ***Nature***, 379 (1996) 55-57.
71. J. Liu, S. Zhang, J. Wang, Q. Lan, Recent progress in solar-driven interfacial evaporation: Evaporators, condensers, applications and prospects, ***Desalination***, 597 (2025) 118356.
72. Y. Guo, X. Zhou, F. Zhao, J. Bae, B. Rosenberger, G. Yu, Synergistic Energy Nanoconfinement and Water Activation in Hydrogels for Efficient Solar Water Desalination, ***ACS nano***, 13 (2019) 7913-7919.
73. A. Caratenuto, Y. Zheng, Critical assessment of water enthalpy characterization through dark environment evaporation, ***Science advances***, 10 (2024).
74. C.P. Lawrence, J.L. Skinner, Vibrational spectroscopy of HOD in liquid D2O. III. Spectral diffusion, and hydrogen-bonding and rotational dynamics, ***The Journal of Chemical Physics***, 118 (2003) 264-272.
75. C. Li, S. Cao, J. Lutzki, J. Yang, T. Konegger, F. Kleitz, A. Thomas, A Covalent Organic Framework/Graphene Dual-Region Hydrogel for Enhanced Solar-Driven Water Generation, ***Journal of the American Chemical Society***, 144 (2022) 3083-3090.
76. D.N. Whiteman, G.E. Walrafen, W.H. Yang, S.H. Melfi, Measurement of an isosbestic point in the Raman spectrum of liquid water by use of a backscattering geometry, ***Applied optics***, 38 (1999) 2614-2615.
77. Q. Sun, The Raman OH stretching bands of liquid water, ***Vibrational Spectroscopy***, 51 (2009) 213-217.
78. L. Hao, N. Liu, H. Bai, P. He, R. Niu, J. Gong, High-performance solar-driven interfacial evaporation through molecular design of antibacterial, biomass-derived hydrogels, ***Journal of colloid and interface science***, 608 (2022) 840-852.
79. X. Zhou, Y. Guo, F. Zhao, G. Yu, Hydrogels as an Emerging Material Platform for Solar Water Purification, ***Accounts of chemical research***, 52 (2019) 3244-3253.
80. Y. Guo, J. Bae, Z. Fang, P. Li, F. Zhao, G. Yu, Hydrogels and Hydrogel-Derived Materials for Energy and Water Sustainability, ***Chemical reviews***, 120 (2020) 7642-7707.
81. X. Zhou, F. Zhao, Y. Guo, B. Rosenberger, G. Yu, Architecting highly hydratable polymer networks to tune the water state for solar water purification, ***Science advances***, 5 (2019) eaaw5484.
82. L. Zang, L. Sun, S. Zhang, C. Finnerty, A. Kim, J. Ma, B. Mi, Nanofibrous hydrogel-reduced graphene oxide membranes for effective solar-driven interfacial evaporation and desalination, ***Chemical Engineering Journal***, 422 (2021) 129998.
83. Q. Xia, Y. Pan, B. Liu, X. Zhang, E. Li, T. Shen, S. Li, N. Xu, J. Ding, C. Wang, C.D. Vecitis, G. Gao, Solar-driven abnormal evaporation of nanoconfined water, ***Science advances***, 10 (2024) eadj3760.
84. Y. Lu, D. Fan, Y. Wang, H. Xu, C. Lu, X. Yang, Surface Patterning of Two-Dimensional Nanostructure-Embedded Photothermal Hydrogels for High-Yield Solar Steam Generation, ***ACS nano***, 15 (2021) 10366-10376.
85. Z. Yu, Y. Su, R. Gu, W. Wu, Y. Li, S. Cheng, Micro-Nano Water Film Enabled High-Performance Interfacial Solar Evaporation, ***Nano-micro letters***, 15 (2023) 214.
86. X. Zhou, Y. Guo, F. Zhao, W. Shi, G. Yu, Topology-Controlled Hydration of Polymer Network in Hydrogels for Solar-Driven Wastewater Treatment, ***Advanced materials***, 32 (2020) e2007012.
87. H. Yu, D. Wang, H. Jin, P. Wu, X. Wu, D. Chu, Y. Lu, X. Yang, H. Xu, 2D MoN1.2-rGO Stacked Heterostructures Enabled Water State Modification for Highly Efficient Interfacial Solar Evaporation, ***Advanced Functional Materials***, 33 (2023).

88. P. Zhou, K. Yang, L. Liu, Q. Liu, N. Zhang, J. Xu, Janus-structured MXene-PA/MS with an ultrathin intermediate layer for high-salinity water desalination and wastewater purification, ***Journal of colloid and interface science***, 682 (2024) 795-803.
89. H. Yao, P. Zhang, C. Yang, Q. Liao, X. Hao, Y. Huang, M. Zhang, X. Wang, T. Lin, H. Cheng, J. Yuan, L. Qu, Janus-interface engineering boosting solar steam towards high-efficiency water collection, ***Energy & Environmental Science***, 14 (2021) 5330-5338.
90. B. Nie, Y. Meng, S. Niu, L. Gong, Y. Chen, L. Guo, X. Li, Y.-C. Wu, H.-J. Li, W. Zhang, Janus solar evaporators: a review of innovative technologies and diverse applications, ***Energy & Environmental Science***, 18 (2025) 3502-3525.
91. Y. Zhang, F. Shen, W. Cao, Y. Wan, Hydrophilic/hydrophobic Janus membranes with a dual-function surface coating for rapid and robust membrane distillation desalination, ***Desalination***, 491 (2020) 114561.
92. W.H. Lee, C.W. Lee, G.D. Cha, B.H. Lee, J.H. Jeong, H. Park, J. Heo, M.S. Bootharaju, S.H. Sunwoo, J.H. Kim, K.H. Ahn, D.H. Kim, T. Hyeon, Floatable photocatalytic hydrogel nanocomposites for large-scale solar hydrogen production, ***Nature nanotechnology***, 18 (2023) 754-762.
93. F. Zhao, X. Zhou, Y. Shi, X. Qian, M. Alexander, X. Zhao, S. Mendez, R. Yang, L. Qu, G. Yu, Highly efficient solar vapour generation via hierarchically nanostructured gels, ***Nature nanotechnology***, 13 (2018) 489-495.
94. Q. Lu, W. Shi, H. Yang, X. Wang, Nanoconfined Water-Molecule Channels for High-Yield Solar Vapor Generation under Weaker Sunlight, ***Advanced materials***, 32 (2020) e2001544.
95. N. Liu, L. Hao, B. Zhang, R. Niu, J. Gong, T. Tang, Rational Design of High-Performance Bilayer Solar Evaporator by Using Waste Polyester-Derived Porous Carbon-Coated Wood, ***Energy & Environmental Materials***, 5 (2021) 617-626.
96. Y. Guo, X. Zhao, F. Zhao, Z. Jiao, X. Zhou, G. Yu, Tailoring surface wetting states for ultrafast solar-driven water evaporation, ***Energy & Environmental Science***, 13 (2020) 2087-2095.
97. X. Liu, Y. Tian, Y. Wu, F. Chen, Y. Mu, M.L. Minus, Y. Zheng, Fully Biomass-Based Hybrid Hydrogel for Efficient Solar Desalination with Salt Self-Cleaning Property, ***ACS applied materials & interfaces***, 13 (2021) 42832-42842.
98. R. Li, Y. Shi, M. Alsaedi, M. Wu, L. Shi, P. Wang, Hybrid Hydrogel with High Water Vapor Harvesting Capacity for Deployable Solar-Driven Atmospheric Water Generator, ***Environmental science & technology***, 52 (2018) 11367-11377.
99. Y. Tu, J. Zhou, S. Lin, M. Alshrah, X. Zhao, G. Chen, Plausible photomolecular effect leading to water evaporation exceeding the thermal limit, ***Proceedings of the National Academy of Sciences of the United States of America***, 120 (2023) e2312751120.
100. Y. Cao, J. Wang, W. Guan, M. An, P. Yan, Z. Li, C. Zhao, G. Yu, Spatially regulated water-heat transport by fluidic diode membrane for efficient solar-powered desalination and electricity generation, ***Nature communications***, 16 (2025) 5050.
101. S. Meng, C.Y. Tang, J. Jia, J. Yang, M.B. Yang, W. Yang, A Wave-Driven Piezoelectric Solar Evaporator for Water Purification, ***Advanced Energy Materials***, 12 (2022).
102. S. Meng, Z. Xu, L. Bai, J. Yang, M.-B. Yang, W. Yang, A customized kinetic energy harvesting system with multilayer piezoelectric membrane for solar interfacial vapor generation, ***Nano Energy***, 104 (2022) 107996.
103. L. Zhang, X. Li, Y. Zhong, A. Leroy, Z. Xu, L. Zhao, E.N. Wang, Highly efficient and salt rejecting solar evaporation via a wick-free confined water layer, ***Nature communications***, 13 (2022) 849.
104. G. Liu, T. Chen, J. Xu, G. Yao, J. Xie, Y. Cheng, Z. Miao, K. Wang, Salt-Rejecting Solar Interfacial Evaporation, ***Cell Reports Physical Science***, 2 (2021) 100310.
105. Y. Zhang, T. Xiong, D.K. Nandakumar, S.C. Tan, Structure Architecting for Salt-Rejecting Solar Interfacial Desalination to Achieve High-Performance Evaporation With In Situ Energy Generation, ***Advanced science***, 7 (2020) 1903478.

106. H. Li, Z. Yan, Y. Li, W. Hong, Latest development in salt removal from solar-driven interfacial saline water evaporators: Advanced strategies and challenges, ***Water research***, 177 (2020) 115770.
107. P. Huang, C. Huang, Y. Sun, Z. Zhao, L. Yang, H. Yang, L. Gong, W. Yang, H. Zeng, Bioinspired magnetically induced self-assembling Janus solar evaporator with antifouling and antiscaling properties, ***Chemical Engineering Journal***, 501 (2024) 157824.
108. A.M. Saleque, S. Ma, A.K. Thakur, R. Saidur, T.K. Han, M.I. Hossain, W. Qarony, Y. Ma, R. Sathyamurthy, Y.H. Tsang, MXene/MnO2 nanocomposite coated superior salt-rejecting biodegradable luffa sponge for efficient solar steam generation, ***Desalination***, 554 (2023) 116488.
109. Z. Yang, D. Li, Y. Zhu, X. Zhu, W. Yu, K. Yang, B. Chen, Developing Salt-Rejecting Evaporators for Solar Desalination: A Critical Review, ***Environmental science & technology***, 58 (2024) 8610-8630.
110. M. Sheng, Y. Yang, X. Bin, S. Zhao, C. Pan, F. Nawaz, W. Que, Recent advanced self-propelling salt-blocking technologies for passive solar-driven interfacial evaporation desalination systems, ***Nano Energy***, 89 (2021) 106468.
111. H. Geng, Q. Xu, M. Wu, H. Ma, P. Zhang, T. Gao, L. Qu, T. Ma, C. Li, Plant leaves inspired sunlight-driven purifier for high-efficiency clean water production, ***Nature communications***, 10 (2019) 1512.
112. M. Morciano, M. Fasano, S.V. Boriskina, E. Chiavazzo, P. Asinari, Solar passive distiller with high productivity and Marangoni effect-driven salt rejection, ***Energy & Environmental Science***, 13 (2020) 3646-3655.
113. Y. Kuang, C. Chen, S. He, E.M. Hitz, Y. Wang, W. Gan, R. Mi, L. Hu, A High-Performance Self-Regenerating Solar Evaporator for Continuous Water Desalination, ***Advanced materials***, 31 (2019) e1900498.
114. G. Ni, S.H. Zandavi, S.M. Javid, S.V. Boriskina, T.A. Cooper, G. Chen, A salt-rejecting floating solar still for low-cost desalination, ***Energy & Environmental Science***, 11 (2018) 1510-1519.
115. S. He, C. Chen, Y. Kuang, R. Mi, Y. Liu, Y. Pei, W. Kong, W. Gan, H. Xie, E. Hitz, C. Jia, X. Chen, A. Gong, J. Liao, J. Li, Z.J. Ren, B. Yang, S. Das, L. Hu, Nature-inspired salt resistant bimodal porous solar evaporator for efficient and stable water desalination, ***Energy & Environmental Science***, 12 (2019) 1558-1567.
116. K. Yang, T. Pan, S. Dang, Q. Gan, Y. Han, Three-dimensional open architecture enabling salt-rejection solar evaporators with boosted water production efficiency, ***Nature communications***, 13 (2022) 6653.
117. W. Zhao, H. Gong, Y. Song, B. Li, N. Xu, X. Min, G. Liu, B. Zhu, L. Zhou, X.X. Zhang, J. Zhu, Hierarchically Designed Salt-Resistant Solar Evaporator Based on Donnan Effect for Stable and High-Performance Brine Treatment, ***Advanced Functional Materials***, 31 (2021).
118. S. Zheng, J. Yu, H. Shan, J. Gao, R. Wang, Z. Xu, High-performance and scalable contactless solar evaporation with 3D structure, ***Science bulletin***, 70 (2025) 563-572.
119. T.A. Cooper, S.H. Zandavi, G.W. Ni, Y. Tsurimaki, Y. Huang, S.V. Boriskina, G. Chen, Contactless steam generation and superheating under one sun illumination, ***Nature communications***, 9 (2018) 5086.
120. A.K. Menon, I. Haechler, S. Kaur, S. Lubner, R.S. Prasher, Enhanced solar evaporation using a photo-thermal umbrella for wastewater management, ***Nature Sustainability***, 3 (2020) 144-151.
121. L. Wang, Y. Feng, Q. Liu, G. Liu, Solar-thermo-radiative evaporator for continuous steam generation and salt harvesting, ***Solar Energy***, 250 (2023) 347-354.
122. L. Wang, Z. Liu, J. Xu, K. Wang, Q. Wang, G. Liu, Conical solar-thermo-radiative evaporator for sustainable desalination and salt recovery, ***Desalination***, 567 (2023) 116993.
123. P. Shi, J. Li, Y. Song, N. Xu, J. Zhu, Cogeneration of Clean Water and Valuable Energy/Resources via Interfacial Solar Evaporation, ***Nano Letters***, 24 (2024) 5673-5682.

124. D.M. Warsinger, J. Swaminathan, E. Guillen-Burrieza, H.A. Arafat, J.H. Lienhard V, Scaling and fouling in membrane distillation for desalination applications: A review, ***Desalination***, 356 (2015) 294-313.
125. Y. Ge, Z. Su, L. Wang, C. Wang, P. Cai, S. Lin, Y. Liu, S. Xu, G. Bai, Self-Rotating Spherical Evaporator Based on Hydrogel and Black Titanium Oxide for Continuous Desalination of Seawater, ***ACS Materials Letters***, 5 (2023) 2576-2583.
126. Y. Xia, Y. Li, S. Yuan, Y. Kang, M. Jian, Q. Hou, L. Gao, H. Wang, X. Zhang, A self-rotating solar evaporator for continuous and efficient desalination of hypersaline brine, ***Journal of Materials Chemistry A***, 8 (2020) 16212-16217.
127. J. Choi, S. Jeon, Simultaneous evaporation and vapor condensation enhancement in 3D solar steam generators via rotational evaporation, ***Chemical Engineering Journal***, 505 (2025) 159372.
128. X. Wu, Y. Wang, P. Wu, J. Zhao, Y. Lu, X. Yang, H. Xu, Dual-Zone Photothermal Evaporator for Antisalt Accumulation and Highly Efficient Solar Steam Generation, ***Advanced Functional Materials***, 31 (2021).
129. N. Xu, H. Zhang, Z. Lin, J. Li, G. Liu, X. Li, W. Zhao, X. Min, P. Yao, L. Zhou, Y. Song, B. Zhu, S. Zhu, J. Zhu, A scalable fish-school inspired self-assembled particle system for solar-powered water-solute separation, ***National science review***, 8 (2021) nwab065.
130. R. Zhang, W. Zhang, H. Ren, K. Liu, S. Yang, B. Chen, W. Li, C. Ge, D. Xu, W. Xu, A Current-Adaptive Evaporator with Optimized Energy Utilization for Self-Cleaning High-Salinity Desalination, ***Advanced Functional Materials***, (2024).
131. R. Kolter, E.P. Greenberg, Microbial sciences: the superficial life of microbes, ***Nature***, 441 (2006) 300-302.
132. I. Banerjee, R.C. Pangule, R.S. Kane, Antifouling coatings: recent developments in the design of surfaces that prevent fouling by proteins, bacteria, and marine organisms, ***Advanced materials***, 23 (2011) 690-718.
133. J. Zhou, Z. Cui, X. Liu, Q. Wang, B. Chen, S. Zeng, J. Si, Multifunctional nanofiber membrane with antifouling properties for efficient seawater desalination and wastewater purification, ***Desalination***, 586 (2024) 117812.
134. X.Y. Wang, J.Z. Xue, C.F. Ma, T. He, H.S. Qian, B. Wang, J.W. Liu, Y. Lu, Anti-biofouling double-layered unidirectional scaffold for long-term solar-driven water evaporation, ***Journal of Materials Chemistry A***, 7 (2019) 16696-16703.
135. Y. Xu, J.X. Ma, Y. Han, J.J. Zhang, F.Y. Cui, Y. Zhao, X. Li, W. Wang, Multifunctional CuO Nanowire Mesh for Highly Efficient Solar Evaporation and Water Purification, ***Acs Sustainable Chemistry & Engineering***, 7 (2019) 5476-5485.
136. W. Xiao, J. Yan, S. Gao, X. Huang, J. Luo, L. Wang, S. Zhang, Z. Wu, X. Lai, J. Gao, Superhydrophobic MXene based fabric composite for high efficiency solar desalination, ***Desalination***, 524 (2022) 115475.
137. A. Pramanik, J.A. Dhar, R. Banerjee, M. Davis, K. Gates, J. Nie, D. Davis, F.X. Han, P.C. Ray, WO(3) Nanowire-Attached Reduced Graphene Oxide-Based 1D-2D Heterostructures for Near-Infrared Light-Driven Synergistic Photocatalytic and Photothermal Inactivation of Multidrug-Resistant Superbugs, ***ACS applied bio materials***, 6 (2023) 919-931.
138. T. Wu, B. Liu, C. Liu, J. Wan, A. Yang, K. Liu, F. Shi, J. Zhao, Z. Lu, G. Chen, A. Pei, H.Y. Hwang, Y. Cui, Solar-driven efficient heterogeneous subminute water disinfection nanosystem assembled with fingerprint MoS2, ***Nature Water***, 1 (2023) 462-470.
139. Y. Li, M. Zhao, Y. Xu, L. Chen, T. Jiang, W. Jiang, S. Yang, Y. Wang, Manipulating light trapping and water vaporization enthalpy via porous hybrid nanohydrogels for enhanced solar-driven interfacial water evaporation with antibacterial ability, ***Journal of Materials Chemistry A***, 7 (2019) 26769-26775.
140. X. Meng, X. Wang, K. Yin, Y. Jing, L. Gu, Z. Tao, X. Ren, M. Tang, X. Shao, L. Sun, Y. Sun, Y. Dai, Y. Xiong, Integration of photothermal water evaporation with photocatalytic microplastics upcycling via nanofluidic thermal management, ***Proceedings of the National Academy of Sciences of the United States of America***, 121 (2024) e2317192121.

141. X. Li, R. Yang, L. Zou, S. Zheng, M. Chen, J. Wen, H. Zhang, C. Wu, Y. Zhang, Y. Zhou, Reassessing the Role of Thermal Convection in Simultaneous Water Production and Pollutant Degradation in Interfacial Photothermal-Photocatalytic Systems, ***Advanced materials***, (2024) e2416283.
142. R. Djellabi, L. Noureen, V.-D. Dao, D. Meroni, E. Falletta, D.D. Dionysiou, C.L. Bianchi, Recent advances and challenges of emerging solar-driven steam and the contribution of photocatalytic effect, ***Chemical Engineering Journal***, 431 (2022) 134024.
143. Y. Xu, J. Ma, Y. Han, H. Xu, Y. Wang, D. Qi, W. Wang, A simple and universal strategy to deposit Ag/polypyrrole on various substrates for enhanced interfacial solar evaporation and antibacterial activity, ***Chemical Engineering Journal***, 384 (2020) 123379.
144. J. Wu, J. Qu, G. Yin, T. Zhang, H.Y. Zhao, F.Z. Jiao, J. Liu, X. Li, Z.Z. Yu, Omnidirectionally irradiated three-dimensional molybdenum disulfide decorated hydrothermal pinecone evaporator for solar-thermal evaporation and photocatalytic degradation of wastewaters, ***Journal of colloid and interface science***, 637 (2023) 477-488.
145. H. Li, F. Zhai, D. Gui, X. Wang, C. Wu, D. Zhang, X. Dai, H. Deng, X. Su, J. Diwu, Z. Lin, Z. Chai, S. Wang, Powerful uranium extraction strategy with combined ligand complexation and photocatalytic reduction by postsynthetically modified photoactive metal-organic frameworks, ***Applied Catalysis B: Environmental***, 254 (2019) 47-54.
146. Q. Kang, T. Wang, P. Li, L. Liu, K. Chang, M. Li, J. Ye, Photocatalytic reduction of carbon dioxide by hydrous hydrazine over Au-Cu alloy nanoparticles supported on SrTiO3/TiO2 coaxial nanotube arrays, ***Angewandte Chemie***, 54 (2015) 841-845.
147. J. Huang, J. Tao, G. Liu, L. Lu, H. Tang, G. Qiao, In situ construction of 1D CdS/2D Nb2CTx MXene Schottky heterojunction for enhanced photocatalytic hydrogen production activity, ***Applied Surface Science***, 573 (2022) 151491.
148. J. Pan, Y. Guan, Y. Zhang, Z. Xu, S. Han, H. Tang, X. Yan, H. Liu, Q. Lu, Near-Infrared-Induced Photothermal Enhanced Photocatalytic H(2) Production for 3D/2D Heterojunctions of Snowflake-like CuS/g-C(3)N(4) Nanosheets, ***Inorganic chemistry***, 62 (2023) 624-635.
149. Y. Cheng, J. Zhang, X. Xiong, C. Chen, J. Zeng, Z. Kong, H. Wang, J. Xi, Y.-J. Yuan, Z. Ji, High photocatalytic and photoelectrochemical performances of the CuSe/MoSe2 2D/2D face-to-face heterojunction photocatalyst, ***Journal of Alloys and Compounds***, 870 (2021) 159540.
150. B. Yuan, Y. Sui, J. Dong, X. Lv, M. Guo, Y. Xiao, L. Yang, H. Yang, L. Bai, W. Wang, D. Wei, Y. Liang, H. Chen, A high-efficiency solar water evaporation-photocatalysis system achieved by manipulating surface wettability and constructing heterojunction, ***Applied Surface Science***, 611 (2023) 155678.
151. Y. Lu, Z. Li, J. Liang, X. Xu, G. Zhang, H. Min, Vermiculite-derived MgFe nanosheet-enhanced hydrogel with dual photothermal and photocatalytic Fenton functions for solar water purification, ***Desalination***, 592 (2024) 118196.
152. B. Guo, Z. Sheng, D. Hu, A. Li, S. Xu, P.N. Manghnani, C. Liu, L. Guo, H. Zheng, B. Liu, Molecular Engineering of Conjugated Polymers for Biocompatible Organic Nanoparticles with Highly Efficient Photoacoustic and Photothermal Performance in Cancer Theranostics, ***ACS nano***, 11 (2017) 10124-10134.
153. Y. Guo, C.M. Dundas, X. Zhou, K.P. Johnston, G. Yu, Molecular Engineering of Hydrogels for Rapid Water Disinfection and Sustainable Solar Vapor Generation, ***Advanced materials***, 33 (2021) e2102994.
154. X. Su, D. Hao, M. Sun, T. Wei, D. Xu, X. Ai, X. Guo, T. Zhao, L. Jiang, Nature Sunflower Stalk Pith with Zwitterionic Hydrogel Coating for Highly Efficient and Sustainable Solar Evaporation, ***Advanced Functional Materials***, 32 (2021).
155. C. Wen, H. Guo, J. Yang, Q. Li, X. Zhang, X. Sui, M. Cao, L. Zhang, Zwitterionic hydrogel coated superhydrophilic hierarchical antifouling floater enables unimpeded interfacial steam generation and multi-contamination resistance in complex conditions, ***Chemical Engineering Journal***, 421 (2021) 130344.

156. N. Li, L. Luo, C. Guo, J. He, S. Wang, L. Yu, M. Wang, P. Murto, X. Xu, Shape-controlled fabrication of cost-effective, scalable and anti-biofouling hydrogel foams for solar-powered clean water production, ***Chemical Engineering Journal***, 431 (2022) 134144.
157. B. Peng, Y. Gao, Q. Lyu, Z. Xie, M. Li, L. Zhang, J. Zhu, Cationic Photothermal Hydrogels with Bacteria-Inhibiting Capability for Freshwater Production via Solar-Driven Steam Generation, ***ACS applied materials & interfaces***, 13 (2021) 37724-37733.
158. L. Cui, P. Wang, H. Che, J. Chen, B. Liu, Y. Ao, Solar-driven interfacial water evaporation for wastewater purification: Recent advances and challenges, ***Chemical Engineering Journal***, 477 (2023) 147158.
159. Y. Song, S. Fang, N. Xu, J. Zhu, Solar-driven interfacial evaporation technologies for food, energy and water, ***Nature Reviews Clean Technology***, 1 (2025) 55-74.
160. X. Hao, H. Yao, P. Zhang, Q. Liao, K. Zhu, J. Chang, H. Cheng, J. Yuan, L. Qu, Multifunctional solar water harvester with high transport selectivity and fouling rejection capacity, ***Nature Water***, 1 (2023) 982-991.
161. P. Zhang, F. Zhao, W. Shi, H. Lu, X. Zhou, Y. Guo, G. Yu, Super Water-Extracting Gels for Solar-Powered Volatile Organic Compounds Management in the Hydrological Cycle, ***Advanced materials***, 34 (2022) e2110548.
162. J. Ma, Y. Xu, F. Sun, X. Chen, W. Wang, Perspective for removing volatile organic compounds during solar-driven water evaporation toward water production, ***EcoMat***, 3 (2021).
163. C. Song, D. Qi, Y. Han, Y. Xu, H. Xu, S. You, W. Wang, C. Wang, Y. Wei, J. Ma, Volatile-Organic-Compound-Intercepting Solar Distillation Enabled by a Photothermal/Photocatalytic Nanofibrous Membrane with Dual-Scale Pores, ***Environmental science & technology***, 54 (2020) 9025-9033.
164. S. Yan, H. Song, Y. Li, J. Yang, X. Jia, S. Wang, X. Yang, Integrated reduced graphene oxide/polypyrrole hybrid aerogels for simultaneous photocatalytic decontamination and water evaporation, ***Applied Catalysis B: Environmental***, 301 (2022) 120820.
165. W. Zhao, M. Adeel, P. Zhang, P. Zhou, L. Huang, Y. Zhao, M.A. Ahmad, N. Shakoor, B. Lou, Y. Jiang, I. Lynch, Y. Rui, A critical review on surface-modified nano-catalyst application for the photocatalytic degradation of volatile organic compounds, ***Environmental Science: Nano***, 9 (2022) 61-80.
166. Y. Peng, X. Wei, Y. Wang, W. Li, S. Zhang, J. Jin, Metal-Organic Framework Composite Photothermal Membrane for Removal of High-Concentration Volatile Organic Compounds from Water via Molecular Sieving, ***ACS nano***, 16 (2022) 8329-8337.
167. D. Qi, Y. Liu, Y. Liu, Z. Liu, Y. Luo, H. Xu, X. Zhou, J. Zhang, H. Yang, W. Wang, X. Chen, Polymeric Membranes with Selective Solution-Diffusion for Intercepting Volatile Organic Compounds during Solar-Driven Water Remediation, ***Advanced materials***, 32 (2020) e2004401.
168. X. Zhao, T. Wang, R. Wang, Y. Lan, Y. Jiang, J. Pan, Superwetting photothermal membranes enabled by polyphenol-mediated nanostructured coating with raspberry-like architectures for solar-driven interfacial evaporation, ***Desalination***, 542 (2022) 116046.
169. C. Violet, A. Ball, M. Heiranian, L.F. Villalobos, J. Zhang, B. Uralcan, H. Kulik, A. Haji-Akbari, M. Elimelech, Designing membranes with specific binding sites for selective ion separations, ***Nature Water***, 2 (2024) 706-718.
170. C. Du, X. Zhao, X. Qian, C. Huang, R. Yang, Heat-localized solar evaporation: Transport processes and applications, ***Nano Energy***, 107 (2023) 108086.
171. P. Zhang, J. Li, L. Lv, Y. Zhao, L. Qu, Vertically Aligned Graphene Sheets Membrane for Highly Efficient Solar Thermal Generation of Clean Water, ***ACS nano***, 11 (2017) 5087-5093.
172. P. Zhang, Q. Liao, H. Yao, H. Cheng, Y. Huang, C. Yang, L. Jiang, L. Qu, Three-dimensional water evaporation on a macroporous vertically aligned graphene pillar array under one sun, ***Journal of Materials Chemistry A***, 6 (2018) 15303-15309.
173. P. Xiao, J. Gu, C. Zhang, F. Ni, Y. Liang, J. He, L. Zhang, J. Ouyang, S.-W. Kuo, T. Chen, A scalable, low-cost and robust photo-thermal fabric with tunable and programmable 2D/3D structures towards environmentally adaptable liquid/solid-medium water extraction, ***Nano Energy***, 65 (2019) 104002.

174. Z. Zhao, C. Wang, D. Wei, F. Wang, Condensation device design represents a critical step for solar-driven water evaporation toward practical applications, ***Cell Reports Physical Science***, 5 (2024) 101794.
175. H. Lu, W. Shi, F. Zhao, W. Zhang, P. Zhang, C. Zhao, G. Yu, High-Yield and Low-Cost Solar Water Purification via Hydrogel-Based Membrane Distillation, ***Advanced Functional Materials***, 31 (2021).
176. W. Deng, T. Fan, Y. Li, Water wave vibration-promoted solar evaporation with super high productivity, ***Nano Energy***, 92 (2022) 106745.
177. Q. Wang, Z. Zhu, G. Wu, X. Zhang, H. Zheng, Energy analysis and experimental verification of a solar freshwater self-produced ecological film floating on the sea, ***Applied Energy***, 224 (2018) 510-526.
178. W. Li, Z. Li, K. Bertelsmann, D.E. Fan, Portable Low-Pressure Solar Steaming-Collection Unisystem with Polypyrrole Origamis, ***Advanced materials***, 31 (2019) e1900720.
179. M. Mohsenzadeh, L. Aye, P. Christopher, Development and experimental analysis of an innovative self-cleaning low vacuum hemispherical floating solar still for low-cost desalination, ***Energy Conversion and Management***, 251 (2022) 114902.
180. M. Gao, L. Zhu, C.K. Peh, G.W. Ho, Solar absorber material and system designs for photothermal water vaporization towards clean water and energy production, ***Energy & Environmental Science***, 12 (2019) 841-864.
181. S.K. Patel, C.L. Ritt, A. Deshmukh, Z. Wang, M. Qin, R. Epsztein, M. Elimelech, The relative insignificance of advanced materials in enhancing the energy efficiency of desalination technologies, ***Energy & Environmental Science***, 13 (2020) 1694-1710.
182. F. Wang, N. Xu, W. Zhao, L. Zhou, P. Zhu, X. Wang, B. Zhu, J. Zhu, A high-performing single-stage invert-structured solar water purifier through enhanced absorption and condensation, ***Joule***, 5 (2021) 1602-1612.
183. S. Zhang, S. Xu, Y. Liu, R. Lei, T. Guo, Y. Yao, S. Gao, J. Ding, Z. Zhang, The effect of surface-free energy and microstructure on the condensation mechanism of water vapor, ***Progress in Natural Science: Materials International***, 33 (2023) 37-46.
184. J. Chen, J.L. Yin, B. Li, Z. Ye, D. Liu, D. Ding, F. Qian, N.V. Myung, Q. Zhang, Y. Yin, Janus Evaporators with Self-Recovering Hydrophobicity for Salt-Rejecting Interfacial Solar Desalination, ***ACS nano***, 14 (2020) 17419-17427.
185. X. Li, B. Li, Y. Li, J. Sun, Nonfluorinated, transparent, and spontaneous self-healing superhydrophobic coatings enabled by supramolecular polymers, ***Chemical Engineering Journal***, 404 (2021) 126504.
186. L. Zhang, B. Tang, J. Wu, R. Li, P. Wang, Hydrophobic Light-to-Heat Conversion Membranes with Self-Healing Ability for Interfacial Solar Heating, ***Advanced materials***, 27 (2015) 4889-4894.
187. Z. Liu, H. Song, D. Ji, C. Li, A. Cheney, Y. Liu, N. Zhang, X. Zeng, B. Chen, J. Gao, Y. Li, X. Liu, D. Aga, S. Jiang, Z. Yu, Q. Gan, Extremely Cost-Effective and Efficient Solar Vapor Generation under Nonconcentrated Illumination Using Thermally Isolated Black Paper, ***Global challenges***, 1 (2017) 1600003.
188. L. Wu, Z. Dong, Z. Cai, T. Ganapathy, N.X. Fang, C. Li, C. Yu, Y. Zhang, Y. Song, Highly efficient three-dimensional solar evaporator for high salinity desalination by localized crystallization, ***Nature communications***, 11 (2020) 521.
189. H. Cao, D. Li, J. Fu, Y. Yang, Y. Zheng, Y. Que, X. He, H. Wang, J. Liu, H. Zhou, Multifunctional biomass-based solar evaporator with enhanced sterilization performance for solar-driven clean water evaporation, ***Desalination***, 591 (2024) 118052.
190. Y. Geng, W. Sun, P. Ying, Y. Zheng, J. Ding, K. Sun, L. Li, M. Li, Bioinspired Fractal Design of Waste Biomass-Derived Solar–Thermal Materials for Highly Efficient Solar Evaporation, ***Advanced Functional Materials***, 31 (2020).
191. H.-S. Guan, T.-T. Fan, H.-Y. Bai, Y. Su, Z. Liu, X. Ning, M. Yu, S. Ramakrishna, Y.-Z. Long, A waste biomass-derived photothermic material with high salt-resistance for efficient solar evaporation, ***Carbon***, 188 (2022) 265-275.

192. C. Zhang, B. Yuan, Y. Liang, L. Yang, L. Bai, H. Yang, D. Wei, W. Wang, H. Chen, Solar vapor generator: A natural all-in-one 3D system derived from cattail, ***Solar Energy Materials and Solar Cells***, 227 (2021) 111127.
193. B. Yang, Z. Zhang, P. Liu, X. Fu, J. Wang, Y. Cao, R. Tang, X. Du, W. Chen, S. Li, H. Yan, Z. Li, X. Zhao, G. Qin, X.Q. Chen, L. Zuo, Flatband lambda-Ti(3)O(5) towards extraordinary solar steam generation, ***Nature***, 622 (2023) 499-506.
194. X. Mu, J. Zhou, P. Wang, H. Chen, T. Yang, S. Chen, L. Miao, T. Mori, A robust starch–polyacrylamide hydrogel with scavenging energy harvesting capacity for efficient solar thermoelectricity–freshwater cogeneration, ***Energy & Environmental Science***, 15 (2022) 3388-3399.
195. F.A. Raihananda, E. Philander, A.F. Lauvandy, T.A.F. Soelaiman, B.A. Budiman, F.B. Juangsa, P. Sambegoro, Low-cost floating solar still for developing countries: Prototyping and heat-mass transfer analysis, ***Results in Engineering***, 12 (2021) 100300.
196. S. Chen, P. Zhao, G. Xie, Y. Wei, Y. Lyu, Y. Zhang, T. Yan, T. Zhang, A floating solar still inspired by continuous root water intake, ***Desalination***, 512 (2021) 115133.
197. J. Xu, Z. Wang, C. Chang, B. Fu, P. Tao, C. Song, W. Shang, T. Deng, Solar-driven interfacial desalination for simultaneous freshwater and salt generation, ***Desalination***, 484 (2020) 114423.
198. X. Wu, T. Gao, C. Han, J. Xu, G. Owens, H. Xu, A photothermal reservoir for highly efficient solar steam generation without bulk water, ***Science bulletin***, 64 (2019) 1625-1633.
199. T. Yang, H. Lin, K.-T. Lin, D. Mesa Saldarriaga, G. Yang, C. Guo, H. Zhang, J. Zhang, S. Fraser, A.K.-T. Lau, T. Ma, B. Jia, Single-shot production of Janus graphene thin film for solar steam generation with 94.5% efficiency, ***Carbon***, 199 (2022) 469-478.
200. Q. Zhang, G. Yi, Z. Fu, H. Yu, S. Chen, X. Quan, Vertically Aligned Janus MXene-Based Aerogels for Solar Desalination with High Efficiency and Salt Resistance, ***ACS nano***, 13 (2019) 13196-13207.
201. Y. Yang, H. Feng, W. Que, Y. Qiu, Y. Li, L. Guo, Q. Li, A Diode-like Scalable Asymmetric Solar Evaporator with Ultra-high Salt Resistance, ***Advanced Functional Materials***, 33 (2022).
202. D. Weng, F. Xu, X. Li, Y. Li, J. Sun, Bioinspired photothermal conversion coatings with self-healing superhydrophobicity for efficient solar steam generation, ***Journal of Materials Chemistry A***, 6 (2018) 24441-24451.
203. Q. Xue, P. Xiao, J. Gu, W. Wang, L. Yan, T. Chen, Superhydrophobic sand evaporator with core-shell structure for long-term salt-resistant solar desalination, ***Water research***, 253 (2024) 121290.
204. M.S. Irshad, N. Arshad, G. Maqsood, I. Ahmed, B. Shakoor, M.S. Asghar, U. Ghazanfar, L. Lin, M.A.K.Y. Shah, I. Ahmed, X. Chen, J. Wang, C. Yi, J. Li, J. Qian, W. Li, Z. Said, H. Li, N.X. Ho, H. Wang, X. Fan, N. Mushtaq, V.-D. Dao, T. Mei, X. Wang, Advancing water collection efficiency in hybrid solar evaporators: key factors, strategic innovations, and synergistic applications, ***Materials Science and Engineering: R: Reports***, 165 (2025) 101018.
205. T. Wang, F. Bai, D. Lei, L. Wang, X. Zhang, G. Yuan, Z. Wang, Experimental investigation of an asymmetric CPC concentrated solar interfacial evaporation device (CSIED), ***Desalination***, 604 (2025) 118739.
206. G. Peng, S.W. Sharshir, Z. Hu, R. Ji, J. Ma, A.E. Kabeel, H. Liu, J. Zang, N. Yang, A compact flat solar still with high performance, ***International Journal of Heat and Mass Transfer***, 179 (2021) 121657.
207. S. Dong, Y. Xu, C. Wang, C. Liu, J. Zhang, Y. Di, L. Yu, L. Dong, Z. Gan, Atmospheric water harvester-assisted solar steam generation for highly efficient collection of distilled water, ***Journal of Materials Chemistry A***, 10 (2022) 1885-1890.
208. Z. Zhan, L. Chen, C. Wang, Y. Shuai, H. Duan, Z. Wang, Super Water-Storage Self-Adhesive Gel for Solar Vapor Generation and Collection, ***ACS applied materials & interfaces***, 15 (2023) 8181-8189.

Chapter 5
Simulation and AI Modeling

Solar evaporation is an intricate process coupled across multiple physical fields [1, 2]. Efficient design of solar-driven interfacial evaporators necessitates a thorough understanding of thermal management and water-salt transport dynamics during evaporation [3, 4]. Numerical simulations provide a dynamic and continuous representation of the entire evaporation process, aiding in the comprehension of both macroscopic changes and microscopic details [5]. This enables the optimization and design of evaporators through material modifications and structural improvements. In this section, we highlight recent advancements in using numerical simulations and modeling techniques for solar-driven interfacial evaporation.

Firstly, macroscopic scale simulations elucidate the complex interactions between heat and mass transfer during evaporation, along with the distribution of water, salt, and vapor [6]. These simulations offer theoretical insights that guide the enhancement of material properties and structural optimization. Secondly, microscale simulations enable the analysis of particle dynamics during interfacial evaporation and the calculation of material properties [7]. This approach offers a detailed atomic-level understanding of the evaporation process and predicts potential challenges and constraints, aiding in the development of optimized photothermal materials and evaporation structures from a microscopic perspective. Additionally, the integration of AI techniques in assessing solar desalination system performance is highlighted, promoting the development of intelligent, multifunctional materials and systems [8, 9], thereby enhancing the efficiency and applicability of solar evaporation technologies.

5.1 Macroscopic Simulation

Macroscopic simulations employing finite volume/element methods, are pivotal for cross-scale modeling of multi-physical fields in solar desalination, including temperature and flow fields. These simulations facilitate the analysis of temperature distribution, salt concentration, humidity levels, and vapor dynamics throughout the evaporation process [5]. Such insights assist researchers in understanding salt resistance, thermal management, and water transport mechanisms during evaporation. This understanding is crucial for optimizing distillation units to improve

G. Liu, T. Chen, *Solar Evaporation for Water-Energy-Resource Nexus*,
https://doi.org/10.1007/978-981-95-3815-7_5

evaporation efficiency. Commonly utilized software for macroscopic simulations includes COMSOL Multiphysics [10], ANSYS [11], among others.

5.1.1 COMSOL Multiphysics

COMSOL Multiphysics facilitates the integrated interaction among diverse multiphysical fields. Utilizing the finite element method, this software models singular physical phenomena by resolving individual partial differential equations and addresses complex multiphysical phenomena through coupled systems of partial differential equations [12]. Its applications span across various domains including fluid dynamics, thermal conductivity, structural mechanics, and electromagnetic analyses, among others. In COMSOL Multiphysics, the configuration of physical field couplings is extensively customizable, imposing no constraints on the physical fields, coupling modalities, or necessitating the creation of user-defined subroutines. A cohesive interface within the software supports comprehensive data analysis and the generation of dynamic, visually engaging displays of simulation outcomes [13]. Solar desalination represents a typical example of a coupled multiphysical process involving fluid dynamics and heat transfer, making COMSOL Multiphysics an invaluable simulation tool for examining water flow, salt ion transport, and thermal dynamics [5, 14, 15].

Photothermal materials capture solar radiation and convert it into thermal energy, but only a portion of this heat is used for steam generation, with the rest lost to the environment. Exploring heat transfer dynamics and thermal management strategies during evaporation and condensation is crucial. A novel approach to enhancing the evaporation rate involves examining condensation mechanisms beginning with the vapor generation process [16–18]. Simulations were conducted to evaluate the effects of vapor diffusion distance and nail array configurations across various parameters. Optimization experiments revealed that a brass nail array condenser with a 5 mm vapor diffusion distance and an optimal 5×5 nail array configuration achieved a water vapor collection rate of 90.5% [19]. Furthermore, the performance of solar evaporators is influenced by the surface patterning. The application of carbon black on aerated paper to create distinct surface patterns was studied using a 3D COMSOL Multiphysic model to analyze the surface temperature distribution across these patterns [20]. Results support that patterned solar absorbers enhance evaporation efficiency by modifying heat conduction pathways.

Water transport dynamics play a crucial role in efficient evaporation, with any disparity between water transport and evaporation rates significantly affecting the system's effectiveness. Overly rapid transport rates may increase heat transfer losses, whereas inadequate rates can lead to salt accumulation at the interface, obstructing water channels and impeding evaporation. Numerical simulations provide a visual representation of water and salt transport, aiding in the understanding of the water-salt separation mechanism and improving evaporation performance. Recent advancements in photothermal materials, particularly hydrogels, have shown promise in interfacial solar desalination due to their high tunability and potential for post-processing modifications. The 3D hydrogel evaporator models with different configurations were simulated to compare their performance [21]. The hydrogel evaporator with its vertically aligned structure, exhibited superior water transport rates compared to other models, effectively matching the water transport rate with the evaporation rate, thereby achieving optimal thermal management and salt resistance. The introduction of 2D nanostructure-embedded hybrid hydrogel evaporators has surpassed traditional evaporation rates, as demonstrated by the hydrogel evaporator with a rate of 3.62 kg m^{-2} h^{-1} [10]. The hydrogel's groove structure not only induces a temperature gradient but also a surface tension gradient, fostering a Marangoni effect that enhances water flow and increases the evaporation surface area. Additionally, porous hydrogels, prepared through self-assembled templating methods without the need for freeze-drying, exhibit superior elasticity and stability [22]. Simulations predicting the impact of pore size on water transport rates corroborate experimental findings, indicating that larger pore sizes reduce diffusion barriers and enhance water transport, thereby meeting diverse evaporation demands more effectively.

During solar evaporation, a concentration gradient of salt is established between the evaporation interface and the underlying water body. This gradient facilitates the spontaneous diffusion of salt ions back to the water body, partially mitigating salt accumulation at the evaporation interface [23]. Consequently, salt deposition is contingent upon the dynamic equilibrium between ion convection at the interface and their diffusion back into the water. When convection predominates over diffusion, ion concentration at the interface escalates to saturation, precipitating salt and consequently diminishing evaporation efficiency due to increased light reflection and reduced light absorption [24]. Severe salt deposition may obstruct the water supply channels, impeding vapor release. Xu et al. engineered an efficient, cost-effective Solar Evaporation-based Crystallizer with distinct evaporation and non-evaporation zones, facilitating simultaneous evaporation and salt collection [25]. This system maintains robust evaporation and consistent salt extraction even under one-sun radiation. The design incorporates Water Transport Layer beneath the photothermal layer to enhance water transport and ion diffusion, promoting salt crystallization away from the water inlet. Simulations of salt ion distribution reveal preferential salt precipitation at the lateral edges of the evaporation surface and the base of the vertical Water Transport Layer, ensuring high evaporation performance without salt deposition interference. A T-shaped synthetic tree with an asymmetric structure optimizes salt collection and solar steam generation. The tree features an evaporation disk and transport columns directing brine to the disk for evaporation. Simulated studies of salt concentration distribution and water transport elucidate the dynamics of salt convection and diffusion during evaporation. Results indicate that a critical disk diameter enables convection to surpass diffusion, leading to salt precipitation primarily at the disk's edge [26, 27]. This mechanism is experimentally validated, showing crystalline salt formation that detaches under gravitational forces, achieving concurrent salt collection and water evaporation. To further enhance salt resistance, an asymmetric double-layer evaporator with a trapezoidal evaporation surface and a wide-leg water supply path has been developed, demonstrating prolonged operational stability under high salinity and irradiation conditions [28]. This design reduces salt accumulation rates compared to conventional systems, as confirmed by both numerical simulations and experimental tests.

The evaporation rate, a critical performance metric for solar evaporation systems, is influenced by phase changes and diffusion processes. Enhanced vapor diffusion, which diminishes vapor pressure and ambient relative humidity, is pivotal for sustained evaporation [29, 30]. Novel approaches in evaporator design, such as Song et al. introduced a three-dimensional conical evaporator to achieve an evaporation efficiency of 2.63 kg m^{-2} h^{-1} under standard solar illumination, directly address challenges such as salt crystallization which impedes evaporation efficiency. The spatial dependency of evaporation indicated by simulations reveals higher vapor diffusion flux and lower temperatures at the apex compared to the base, suggesting optimal conditions for evaporation at the top and dominant energy absorption at the bottom [31]. Further simulations explored the impact of structural configurations, like the umbrella evaporator with its unique double-sided evaporation design, on enhancing evaporation rates. This design doubles the mass transfer area and maintains lower surface temperatures conducive to improved evaporation rates. Adjustments in the opening angle and surface height of the umbrella evaporator were experimentally and computationally analyzed to optimize vapor transfer and evaporation rates [6]. Moreover, introducing convection has been identified as a beneficial strategy for promoting vapor diffusion. Xu et al. demonstrated this by designing a spiral structure evaporator capable of alternating between two-dimensional and three-dimensional configurations [32]. This design not only facilitates rapid water supply and compact storage in its two-dimensional state but also augments the evaporative surface area. Simulations highlight the advantages of structural transformation and convection in elevating evaporation performance [33, 34], showing that external convection effectively mitigates vapor accumulation, thereby enhancing evaporation rates (Fig. 5.1).

In practical applications, such as on islands and ships, integrated multi-stage passive solar stills simplify water collection and minimize nocturnal salt accumulation. These systems leverage latent heat from steam condensation to enhance steam recycling and improve overall solar efficiency [35]. Additionally, the incorporation of Marangoni effect-driven liquid transport, which exploits surface tension gradients induced by salt concentration variations, presents a novel approach for enhancing salt rejection in solar desalination processes [36]. These advancements in solar evaporation highlight the importance of innovative design and simulation in overcoming the challenges posed by salt deposition, thereby enhancing the scalability of solar evaporation systems.

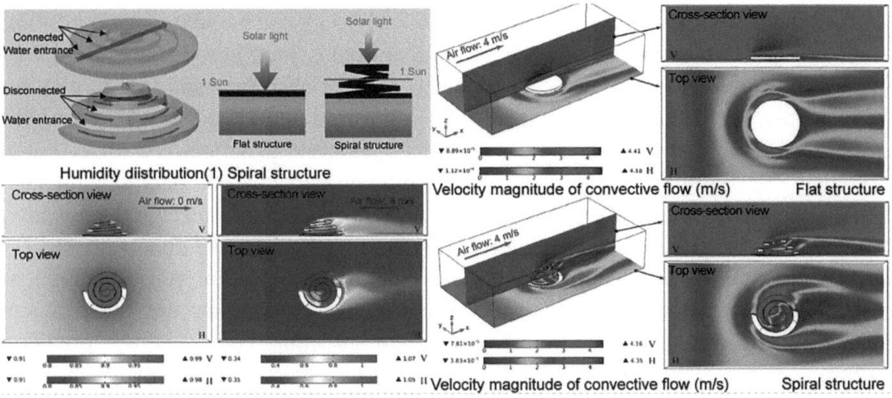

Fig. 5.1 COMSOL Simulation of Vapor Convective Flow of 2D flat and 3D spiral evaporation structures. (Reproduced with permission from [32] Copyright 2021 Elsevier)

5.1.2 ANSYS Fluent

ANSYS Fluent is a computational fluid dynamics (CFD) software used to simulate fluid flow, heat transfer, chemical reactions, and multiphase systems in various industries. It leverages finite volume methods to solve complex fluid dynamics problems, offering high-fidelity simulations with turbulence modeling (k-ε, k-ω, LES, DES), multiphysics coupling (fluid-structure interaction (FSI), magnetohydrodynamics), and adaptive meshing for accurate results [37]. Engineers apply Fluent in aerospace (aircraft aerodynamics), automotive (engine cooling, fuel efficiency), energy (wind turbines, combustion), and biomedical (blood flow, respiratory analysis). Its user-friendly interface supports custom scripting (UDFs, Python), enabling advanced automation and optimization, making it a leading tool for fluid dynamics research and industrial applications [38]. ANSYS is capable of simulating the temperature distribution, water flow rate, and salt concentration during the evaporation process at a solar interface. By numerical simulations, insights can be gained into material modifications and structural designs aimed at enhancing evaporation efficiency and salt rejection [5]. This understanding aids in overcoming limiting factors from a mechanistic perspective, offering innovative approaches for advancing solar evaporation technology.

Wood, as a renewable resource, is characterized by low thermal conductivity, low density, ease of processing, and natural water transport microchannels, making it suitable for solar interfacial evaporation applications [39–41]. A 3D solar evaporator constructed from wood demonstrates high evaporation performance through surface carbonization, adjustments in hydrophilicity, and optimal positioning above the water surface. ANSYS Fluent was utilized to simulate the temperature distribution and water transport within the wood structure. The simulation results showed a significant difference in temperature between larger and smaller channels within the wood, with surface temperatures closely aligning with experimental observations. The efficient heat confinement to the evaporator's surface and the rapid water transport through microchannels enabled a high evaporation rate of 1.93 kg m^{-2} h^{-1} under standard solar irradiation [42]. A novel fluidic photothermal structure, incorporating metal-organic framework-derived mesoporous carbon sheets as solar absorbers and thin fabrics for water supply, demonstrated the capability for salt rejection via fluid-driven salt migration [43]. This system not only produces clean water and generates electricity but is also adaptable to multi-stage interface evaporation processes. It can operate with various water sources including salt water, human sweat, groundwater, and rainwater. Numerical simulations of this structure provided insights into the conditions necessary for preventing salt crystallization at the outlet, thereby guiding practical applications. The design of a flat disc evaporator effectively addresses the issue of salt deposition impeding evaporation. Comprising an absorbing layer, a water spreading layer, and a heat preservation layer, the evaporator features a central brine injection that creates a radial concentration gradient. This design spatially separates the zones of highest salt concentration from the evaporation zones, enhancing evaporation efficiency. Computational Fluid Dynamics models validated the experimental findings, showing that the placement and number of brine inlets can control salt distribution, with simulation results consistent across various inlet configurations [44] (Fig. 5.2). This research underscores the pivotal role of numerical simulations in optimizing evaporator design and operational strategies.

5.1.3 SOLIDWORKS Flow Simulation

SOLIDWORKS Flow Simulation facilitates the simulation of fluid dynamics, heat transfer, and fluid forces with remarkable speed and ease. Utilizing the finite volume method, this software allows for the detailed analysis of how fluid flow,

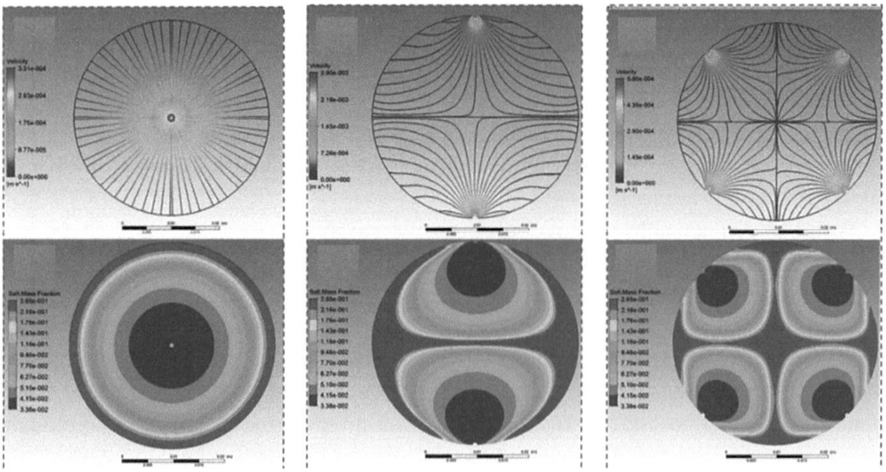

Fig. 5.2 ANSYS Fluent simulation of water transport and velocity distribution at different inlets. (Reproduced with permission from [44]. Copyright 2019 Royal Society of Chemistry)

heat transfer, and associated forces impact engineered devices [45]. One of the noteworthy aspects of SOLIDWORKS Flow Simulation is its user-friendly interface, which enables users to commence simulations directly after creating models in SOLIDWORKS, significantly streamlining the design and analysis process. This simulation tool finds its application in a variety of contexts, including the development of cost-effective and efficient evaporative technologies. The use of inexpensive raw materials and straightforward manufacturing processes for evaporators can be a game-changer for such areas. Wood, being a readily available and renewable resource, serves as an excellent base material for these devices. Advances in technology have enhanced the functionality of wood in evaporative processes; notable developments include surface treatments such as coating and carbonization, and precision modifications like laser engraving [46, 47].

An innovative application involves the creation of a high-efficiency wood-based evaporator with a unique double-layer structure. This structure consists of natural wood as the substrate and a carbonized wood surface that acts as a light-absorbing layer. The carbonization of the wood surface significantly enhances its light absorption capabilities, which, when combined with the natural transport and evaporation capabilities of wood, greatly increases the evaporator's efficiency. Experimental results have demonstrated that such an evaporator can achieve an evaporation rate of 11 $kg/m^2/h$. This rate enables the device to extract water efficiently from various substrates like soil and sand and to operate effectively in saline environments over extended periods. The integration of SOLIDWORKS Flow Simulation provides deeper insights into the interactions between light absorption and water flow within the wood structure (Fig. 5.3). Simulations have shown that the carbonized wood layer utilizes internal reflections within the wood's large-sized channels to enhance light absorption. This optimization of light absorption is critical for maximizing the efficiency of the

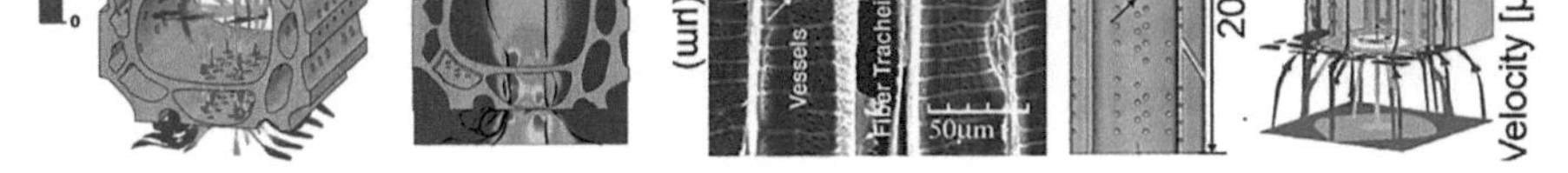

Fig. 5.3 Solidworks simulation of water transport (left) trajectory and (right) velocity contour in porous layered wood structure. (Reproduced with permission from [48] Copyright 2017 Wiley-VCH). Light absorption simulation in the wood's internal structure; SEM image of carbonized wood and its cross-sectional view; simulation results of water velocity cross-sections along the wood channel. (Reproduced with permission from [46] Copyright 2017 Wiley-VCH)

evaporative process [46]. Moreover, the software has helped elucidate the dynamics of water transport within the wood. The presence of a water film on the wood surface, sustained by the capillary action within the internal structure, facilitates a higher water distribution capacity than the physical height of the wood might suggest. However, the water flow is somewhat impeded in smaller tracheids, whereas larger pits connecting these tracheids to the central vessels do not significantly affect water transport. The simulation results indicate that the efficiency of water transport within the wood primarily depends on the size of the lumen in the open-ended vessels [48]. These simulations are pivotal in optimizing the design and functionality of wood-based evaporators. By understanding the specific pathways and mechanisms through which light and water interact within the wood structure, engineers and researchers can fine-tune the evaporator design to enhance performance and stability. This tailored approach not only maximizes the evaporation rate but also ensures that the device can operate under a variety of conditions, making it a useful tool for addressing water scarcity in remote regions.

5.2 Microscale Simulation

Utilizing computational techniques such as the Finite Difference Time Domain (FDTD) method [49], Density Functional Theory (DFT) [50], Molecular Dynamics (MD) [7], and Monte Carlo (MC) simulations [51], microscale simulation offers a detailed computational and visual analysis of the interrelated structural and physicochemical properties of materials at the atomic scale. FDTD and DFT are employed to investigate potential enhancements in the photothermal properties of materials [52]. MD simulations elucidate the microscopic evolution of systems at the atomic level by modeling the motion of water molecules during evaporation. Additionally, MC simulations are instrumental in identifying the key factors influencing system behavior. The subsequent passage elaborates on the application of microscale numerical simulation in solar interfacial evaporation.

5.2.1 Finite Difference Time Domain

FDTD is a computational technique in electromagnetics, which employs recursive operations in the temporal domain to simulate wave propagation and ascertain the distribution of electromagnetic fields [53]. This method obviates the need for object segmentation or deriving specific equations, facilitating straightforward implementation in parallel computing environments. FDTD represents a fundamental and optimal approach for addressing problems in electromagnetic simulation, extensively applicable across a range of scenarios including the analysis of optical absorption properties in photothermal materials [54, 55] (Fig. 5.4). Photothermal conversion constitutes the initial phase in the operation of solar interfacial evaporators. The effectiveness of these devices in achieving evaporation depends on the material's light absorption and photothermal conversion capabilities, highlighting the importance of selecting suitable materials. Notably, materials such as carbon-based substances, semiconductors, plasmonic metals, and organic polymers have demonstrated superior photothermal conversion in the context of interfacial evaporation [56, 57]. By leveraging simulations, the underlying mechanisms of photothermal conversion at the atomic scale can be elucidated, thereby enhancing the photothermal performance of materials via structural modifications [5].

Tellurium (Te) nanoparticles, with their diverse size distributions in broadband absorbers, exhibit significant light absorption efficiency (85%) and surpass the photothermal conversion performance of previously discussed plasmonic nanoparticles. Employing the FDTD method, the scattering behavior and optical responses of Te nanoparticles were analyzed, revealing that the electric field enhancement values of Te nanoparticle oligomers increase substantially with wavelength, comparable to those of gold nanoparticle oligomers. Furthermore, as particle size escalates, Te nanoparticles self-assemble into oligomers, which results in a marked increase in scattering intensity, beneficial for augmenting light absorption. These nanoparticles exhibit optical duality, as evidenced by the shift in the real part of their permittivity from negative to positive across the UV-vis-NIR spectrum [49]. The pronounced field enhancement and substantial imaginary component of the permittivity are pivotal in determining their light absorption capacity, positioning Te nanoparticles as promising candidates for nanophotothermal applications.

Inspired by the microstructures on the wings of black butterflies, Ying et al. successfully synthesized transition metal carbide (MXene) nanosheets with parallel or wedge-like features by selectively etching the Al layer from Ti_3AlC_2. These nanostructures, mimicking the ridge arrays observed on butterfly wings, are highly effective in capturing light and reducing reflection. The interlayer spacing in the reduced graphene oxide (rGO) sublayer, adjusted for selective ion transport, is particularly suited for applications necessitating salt resistance. The integration of MXene material with a light-trapping structure on the rGO sublayer resulted in

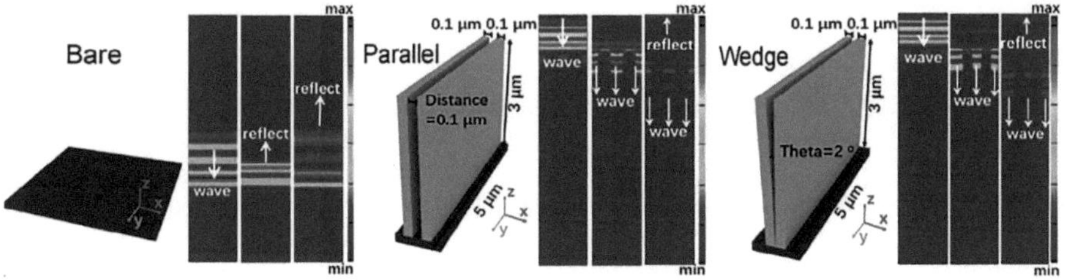

Fig. 5.4 FDTD simulation of electric fields during reflection of three different surface structures. (Reproduced with permission from [58] Copyright 2021 Elsevier)

the MXene@rGO film, which exhibited remarkable capabilities in light trapping and salt resistance during interfacial evaporation. FDTD simulations provided insights into the light-trapping effects of MXene nanostructures, demonstrating that while a significant amount of energy is reflected off a bare surface upon interaction with an incident plane wave, the presence of parallel or wedge-shaped nanostructures on the surface confines the energy within the structures, thereby reducing electric field reflection and enhancing the material's light absorption capacity (Fig. 5.4) [58].

5.2.2 Density Functional Theory

Density Functional Theory (DFT) is a computational quantum mechanical method used to study the electronic structure of atomic and molecular systems. DFT enables precise predictions of the configuration and properties of diverse systems, including molecules, crystals, surfaces, and interfaces, without explicitly considering interaction effects [59]. This method is pivotal for exploring a range of phenomena such as chemical reactions, surface adsorption, and the mechanical properties of materials. As a result, DFT serves as a powerful tool for elucidating the fundamental characteristics and behaviors of matter, which significantly aids in the design and optimization of materials and chemical processes (Fig. 5.5) [52, 60].

In semiconductor materials, DFT is crucial for analyzing energy band characteristics, which are essential for developing strategies to enhance photothermal properties. By doping semiconductor materials or modifying their dimensions, the bandgap structure can be effectively adjusted. This adjustment enables the selective absorption of near-infrared light, leading to expanded light absorption capabilities and

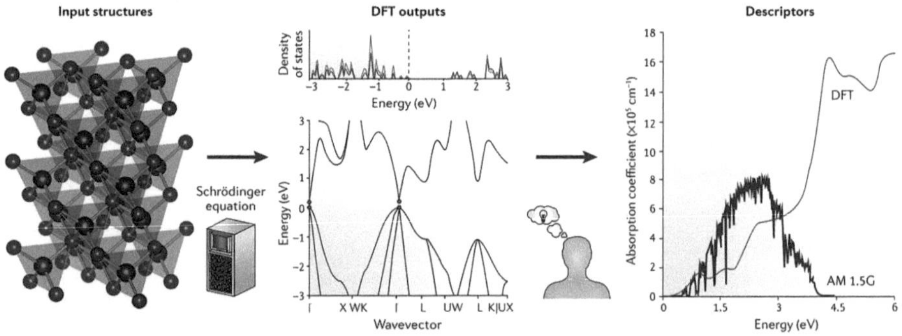

Fig. 5.5 Screening Material Properties via Density Functional Theory: This requires an initial input comprising a well-founded hypothesis of the crystallographic structure of the target material or molecule. Then, it involves selecting suitable approximations and establishing numerical convergence parameters. The outputs derived from the simulations include properties like electron charge density, total energy, optimized crystal structure, and band structure. (Reproduced with permission from [52] Copyright 2016 Nature Portfolio)

impressive photothermal conversion efficiencies [61, 62]. These properties make such materials highly suitable for applications like solar interfacial evaporation. Semiconductor materials demonstrate an absorption spectrum that is contingent upon their bandgap energy. Exposure to light with energy equivalent to the semiconductor's bandgap generates electron-hole pairs, which subsequently release energy through non-radiative relaxation processes, thus achieving photothermal conversion. For instance, $CuFeSe_2$ nanoparticles, a novel narrow bandgap semiconductor photothermal material, are employed for efficient interfacial photothermal conversion [63]. When applied onto porous, low-density, and highly hydrophilic wood surfaces, $CuFeSe_2$ nanoparticles form a narrow bandgap semiconductor film that achieves an evaporation efficiency of 86.2% at 5 kW/m^2. DFT calculations reveal that the bandgap of $CuFeSe_2$ is approximately 0.45 eV, allowing for complete spectral absorption across a comprehensive range.

Another example involves Ti_2O_3 nanoparticles, which, due to their small bandgap energy of 0.09 eV, can absorb light across the entire spectral range. Notably, reducing the size of Ti_2O_3 particles to the nanoscale enhances their capacity to absorb light, thereby achieving nearly 100% internal and 92% external solar thermal conversion efficiencies [64]. Advanced insights into the band characteristics of Ti_2O_3 nanoparticles were obtained through DFT calculations. The analysis of the projected and total electronic density of states (PDOS/DOS) reveals that the bandgap of Ti_2O_3 is approximately 0.09 eV, corroborating previous findings (Fig. 5.6c). With a significant presence of Ti3d orbitals spanning from −2 to 2 eV, Ti_2O_3 is capable of absorbing a wide range of energy, from the band edge at 0.09 eV to greater than 4 eV (less than 310 nm) [65]. This characteristic enables the absorption of the full spectrum of solar energy. Moreover, the majority of photons in solar

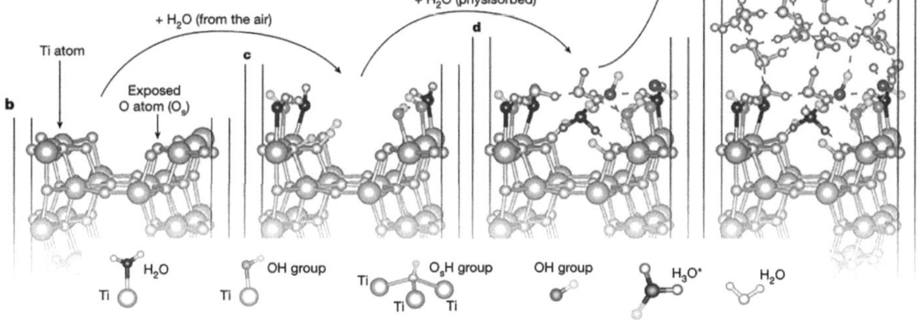

Fig. 5.6 MD for water molecule motion and evaporation, and the state of water molecules on the surface of λ-Ti_3O_5 (**a–e**). (Reproduced with permission from [7] Copyright 2023, Nature Portfolio)

radiation possess energy levels significantly higher than the bandgap of Ti_2O_3, facilitating the rapid conversion of solar energy into thermal energy. Besides, the light-absorbing materials Cu_2SnSe_3 (CTSe) and $Cu_2ZnSnSe_4$ (CZTSe) exhibit favorable internal electronic band structures, density of states, and absorption coefficients, allowing their absorption encompass the full solar spectrum [66]. Consequently, the innovative hydrophilic/hydrophobic nanoporous double-layer structure employing CTSe and CZTSe as photothermal materials delivers continuous and efficient water supply and evaporation, while offering resistance to salt deposition.

5.2.3 Molecular Dynamics

MD simulations represent a computational methodology for modeling atomic and molecular interactions to elucidate the dynamics and states of particles within a system, grounded in the principles of Newtonian mechanics. These simulations are instrumental in delineating the interaction mechanisms among particles, their reactive behaviors, and state transitions under variable environmental conditions such as temperature and pressure [67]. Consequently, MD simulations offer profound insights into the particulate behaviors of complex systems, finding extensive applications across materials science, biology, and medical research [68]. Specifically, utilizing MD simulations to model the transport dynamics of water molecules and salt ions during the evaporation process facilitates a detailed examination of factors influencing evaporation efficiency and enables predictions of kinetic properties at the nanoscale. This enhanced understanding of the evaporation mechanism can improve the performance of evaporation devices.

The phase transition of water is critical and directly impacts the evaporation rate. It is thus vital to analyze the dynamic states of water molecules during evaporation. A novel interfacial evaporator employing a piezoelectric $BaTiO_3$ ultrathin nanofilm has been developed, which harnesses wave-driven electrical energy to augment evaporation. The conversion of wave energy into electrical energy stimulates adjacent water molecules, thereby facilitating enhanced evaporation rates. MD simulations have been employed to assess the effects of an alternating electric field, generated by barium titanate nanoparticles, on nearby water molecules. The simulations reveal that the alternating electric field accelerates the reduction in water molecule count within the droplet, correlating well with the observed evaporation processes [69]. The facilitated evaporation evidenced by shorter droplet lifespans confirms the role of the electric field in promoting evaporation, supporting the efficiency of piezoelectric ultrathin nanofilms as solar absorbers.

Alterations in surface wettability states of the evaporator further augment evaporation rates. The creation of hydrophobic island-shaped patches on hydrophilic hydrogel evaporators introduces variations in surface wettability. By modifying surfaces with trichloro(octadecyl)silane (OTS), a hydrophobic island patch surface on a polyvinyl alcohol substrate was engineered. MD simulations confirmed the impact of surface wettability on evaporation rates, aligning with experimental observations

[70]. During evaporation, many water molecules remain in the hydrophilic region, migrating towards and evaporating at the hydrophobic boundaries. Increasing OTS coverage results in a thicker water film in the hydrophilic region, weakening water-surface interactions. This effect reduces the average total interaction energy, lowering the energy barrier for water molecule evaporation. At 30% OTS coverage, the evaporation rate peaks, influenced by film thickness, hydrophilic surface area, and boundary line length. The evaporator achieves an optimal hydrophilic surface area and water film thickness, ensuring efficient evaporation from the hydrophobic region and demonstrating superior performance.

A novel strategy for developing photothermal materials involves enhancing light absorption capacity by regulating the joint densities of states. Yang et al. suggested that a flat-band electronic structure could augment the joint density of states [7]. An evaporator incorporating λ-Ti_3O_5 into a porous hydrogel with a conical cavity was fabricated, achieving an impressive evaporation rate under solar irradiation. The elevated absorption rate of λ-Ti_3O_5 is attributed to a Ti-Ti dimer-induced flat-band electronic structure, which promotes the evaporation of water molecules as small clusters, effectively reducing water vaporization enthalpy and enhancing evaporation efficiency. MD simulations exploring the interaction between interfacial water molecules and λ-Ti_3O_5 surface particles revealed that the unique U-shaped groove structure of the λ-Ti_3O_5 surface facilitates water molecule adsorption and cluster formation, thereby reducing the energy required for hydrogen bond disruption and promoting water evaporation (Fig. 5.6).

A widely adopted method to enhance evaporation rates without additional energy inputs involves lowering the evaporation enthalpy. Molecular dynamics simulations have proven instrumental in examining changes in the enthalpy of evaporation. The application of polyaniline for light absorption and a hierarchically porous hydrogel (HPH) for brine transportation via evaporation has been facilitated through the complexation of opposite charged polyelectrolytes within an HPH structure [71]. This study reveals that HPH usage yields superior evaporation rates in brine compared to pure water. The interaction between the oppositely charged groups in the polyion complex (PIC) and the anions and cations in brine enhances PIC's hydration. Molecular dynamics data show an increase in average hydrogen bonds in PIC-water as salt concentration increases, peaking at 7.2 wt% NaCl solution. With higher salt concentrations, PIC–water interactions intensify, ion hydration capabilities augment, and average water–water hydrogen bonding weakens. Salt ion hydration generates highly active water molecules, which in turn boosts hydrokinetics and decreases the enthalpy of vaporization, thereby promoting evaporation.

J.J. Koh et al. developed bilayer evaporation structures incorporating hydratable plastic in the form of regenerated cellulose as a water conveyance medium, and commercial carbon black nanoparticles as a light-absorbing element embedded within the cellulose matrix. These structures exhibited high evaporation rates, flexibility, and resistance to salt accumulation [72]. Molecular dynamics simulations explored water molecule organization within cellulose, corroborating with Raman spectroscopy findings. Simulation results suggest that increased hydration leads to expanded water clusters and their migration into sub-nano channels, enhancing cellulose swelling and water molecule diffusion. A connected water network formed by sub-nano channels enhances water transport capacity, with water molecules displaying higher local order in hydrated cellulose compared to bulk water, resulting in longer and less linear hydrogen bonds in cellulose-water than in water-water, thus reducing the enthalpy of evaporation.

The use of hydrophobic nanoconfined water molecule channels (NCWMCs) has been effective in diminishing the evaporation enthalpy. The interstitial spaces among 1D-O-doped MoS_{2-x} nanosheet assemblies are critical in aiding water transport. The presence of nanochannels between ultrathin O-MoS_{2-x} nanosheets lowers the enthalpy of water evaporation by promoting cluster-based water molecule evaporation. Simulation results comparing the NCWMC system with bulk water reveal significant water cluster presence in the vapor phase of the NCWMC system, indicating enthalpy reduction and accelerated evaporation rates [73]. Liu et al. constructed hierarchical polyacrylonitrile@copper sulfide fabrics for solar interfacial evaporation, demonstrating a significant reduction in water evaporation enthalpy compared to pure water, achieving an evaporation rate of 2.27 kg m^{-2} h^{-1} [74]. Molecular dynamics simulations explored the disruption of water molecule hydrogen bonding by CuS nanosheets, facilitating enhanced evaporation on copper sulfide surfaces due to weaker interfacial constraints and disordered hydrogen bonding, effectively reducing the evaporation enthalpy.

Yu et al. synthesized $MoN_{1.2}$ nanosheets integrated with rGO nanosheets to form $MoN_{1.2}$-rGO heterostructures. MD simulations indicate that water molecules interact distinctively with the 2D $MoN_{1.2}$ and rGO within the heterostructure, leading to unbalanced water states that facilitate hydrogen bond breaking between water molecules, lowering the evaporation enthalpy [75]. Wood has been utilized for its cost-effectiveness and efficiency. Removing lignin from natural wood cell walls produces wood sponges (WS) with highly hydrated cellulose networks. Post-synthesis, porous carbon material is applied as a coating on WS surfaces, creating the C800-ZIF-8WS double-layer solar evaporator. The enhanced evaporation performance of C800-ZIF-8WS is attributed to the rapid water absorption and low thermal conductivity of WS. Molecular dynamics simulations elucidate the dynamics of water adsorption in different wood cell wall components, showing that pure cellulose, rich in hydroxyl groups, envelops a substantial quantity of water molecules, facilitating faster water transport compared to lignin-wrapped cellulose [76]. The elimination of lignin enhances evaporation processes, with the hydrated cellulose network in C800-ZIF-8WS providing ample water during evaporation. This finding is advantageous and desirable for other applications.

5.2.4 Monte Carlo

Monte Carlo simulation is powerful tools used to model and analyze complex systems, and its application in the field of solar interfacial evaporation is no exception [77]. Solar interfacial evaporation refers to the process where sunlight heats the interface between a liquid and air, causing the liquid to evaporate more efficiently. The first aspect of Monte Carlo simulation in this context involves the modeling of solar radiation. Solar radiation is inherently random and its intensity varies with weather conditions, time of day, and geographical location. Monte Carlo methods help in simulating these variations to predict how different patterns of sunlight impact the evaporation rate [5, 78]. This simulation can be crucial for optimizing the design of solar evaporation to maximize efficiency under varying environmental conditions. Solar evaporators often incorporate a top cover to enhance vapor condensation and simplify the collection of potable water. However, this configuration may induce energy and optical losses, with the transparency of the cover playing a pivotal role in influencing the efficiency of evaporation processes. As such, adopting a design that features a highly transparent top cover is a strategic approach to boost water production rates. In a study conducted by Yang et al., the development of a transparent radiative cooling film designed for freshwater harvesting was documented. The assessment of this film's solar transmissivity and radiative cooling properties was executed using precise computational techniques including Mie theory, MC simulation, and MC ray tracing methods [79]. Additionally, the analysis of solar transmissivity and emissivity of the sky window in polymer films embedded with silicon microspheres was undertaken, examining the effects of variables such as particle size, volume fraction, film thickness, and condensate droplets on the film's properties throughout the diurnal cycle. The results reveal a decrease in solar transmissivity coupled with an increase in sky window emissivity as film thickness increases. Variability in the sky window solar transmissivity and emissivity is also influenced by the silica volume fraction present in the films, exhibiting differences across various compositions. Films with a lower silica volume fraction and smaller particle size demonstrated enhanced daytime effective power. Optimal performance in terms of daytime effective power, solar transmissivity, and sky window emissivity was achieved using polydimethylsiloxane doped with SiO_2, with a particle diameter of 4 μm and a volume fraction of 2%. The employment of a super-hydrophilic surface further mitigates negative impacts on solar transmissivity and emissivity.

Secondly, Monte Carlo simulations are used to understand and optimize the thermal dynamics at the liquid-air interface [80]. In solar evaporation, the efficiency of heat transfer and the subsequent rate of evaporation depend on numerous factors, including the properties of the liquid, the surface area, and ambient conditions. By using Monte Carlo simulations, researchers can model these complex interactions by generating multiple scenarios and averaging the results to predict overall system performance. This helps in identifying the most effective materials and configurations for enhancing evaporation rates. The direct simulation Monte Carlo method simulated evaporation from nanopores into half-space, adhering to the true nanopore geometry with an axisymmetric computational domain [51]. The evaporation regime was divided into pinning and receding based on the liquid–vapor interface position. In the pinning regime, the interface stayed at the nanopore's top. Various Knudsen and Mach numbers were analyzed, showing that the Knudsen layer's structure and thickness increased from under 10 k at low Mach numbers to nearly 40 k at Ma = 0.785. The study also examined porosity's impact on evaporation mass

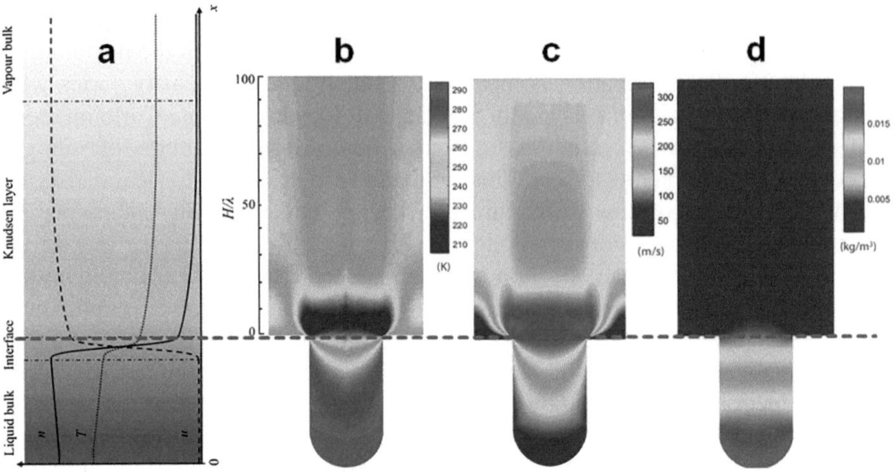

Fig. 5.7 (**a**) A molecular-scale interface separates a bulk liquid region from a bulk vapor region, with the non-equilibrium Knudsen layer adjacent to the vapor side. (Reproduced with permission from [80], Copyright 2024 Cambridge University Press). Contours of the macroscopic flow quantities through the nanopore outlet region, (**b**) Temperature. (**c**) Velocity. (**d**) Density. (Reproduced with permission from [51], Copyright 2022 Elsevier)

flux and investigated the effects of receding length in the receding regime. Downstream density decreased from 2.4×10^{-3} to 4.29×10^{-4} kg/m^3 as the receding ratio rose from 1 to 20. Additionally, an evaporation coefficient was established to evaluate its influence on gas flow (Fig. 5.7).

Lastly, Monte Carlo methods contribute to the scalability and sustainability assessments of solar evaporation systems. By simulating large-scale deployments and their potential impacts on local climates and ecosystems, these tools can inform policy decisions and guide the development of regulations. Additionally, they help in evaluating the long-term sustainability of these systems by assessing resource usage, environmental impact, and energy output, ensuring that solar interfacial evaporation technologies are both effective and ecologically responsible [81, 82]. Overall, Monte Carlo simulations provide a versatile framework for researching and developing more efficient and sustainable solar evaporation systems, addressing both the technical and environmental challenges associated with this promising technology.

5.3 AI Modeling

In solar desalination, considerable research has focused on developing smart multifunctional materials and systems, while less attention has been given to solar evaporation models and data frameworks [83]. Physical design parameters (e.g., material properties, structural design, and heat exchange area) and system performance are crucial for practical solar desalination. Parameters such as ambient temperature, relative humidity, wind speed, solar radiation, inlet water temperature, and total dissolved solids should not be neglected. Integrating Machine Learning (ML) [84] and Artificial Neural Networks (ANN) [85] into solar evaporation represents a significant advancement in optimizing and enhancing this sustainable technology [8]. ML and ANN can analyze and model complex datasets, predicting and simulating optimal operating conditions to maximize evaporation rates while minimizing energy consumption.

5.3.1 Machine Learning

Machine Learning (ML) is a pivotal branch of artificial intelligence that has seen extensive application across various sectors, including renewable energy technologies such as solar desalination [84]. In solar interface evaporation, ML algorithms play a crucial role in enhancing the efficiency and adaptability of evaporation systems. The efficiency of solar evaporation is influenced by numerous factors including environmental conditions (temperature, humidity, solar intensity), material properties of the evaporation surface, and system configurations. Traditional methods rely on static settings that do not adapt to changing environmental conditions, often leading to sub-optimal performance. ML algorithms, by processing historical and real-time data, enable dynamic adaptation of system settings to optimize performance under varying conditions. For instance, supervised ML regression algorithms can be trained using experimental data to predict the most effective configurations of a solar evaporation system. These algorithms include Stepwise Linear Regression (SLR), which simplifies model building by adding or removing predictive variables to find the best-fitting model.

Maddah et al. provide a compelling example of how ML can be applied in this field. They used theoretical heat transfer models combined with experimental data to assess the performance of an evaporator under different flow rates (Fig. 5.8). By employing SLR and extending datasets through randomization techniques, they developed robust training models that accurately predict the efficiency of the solar evaporator system (Fig. 5.8b, d) [86]. This approach not only correlates inputs and outputs effectively but also allows for dimensional analysis to enhance the understanding and prediction of system behaviors. Training ML models involves several stages, starting with data collection, where historical data on environmental conditions and system performance are gathered. This data is then used to train the ML model, enhancing its ability to predict outcomes based on input variables. The trained model can forecast optimal system settings for new environmental conditions, thus guiding adaptive control strategies that adjust the system in real-time to maintain or enhance efficiency. Moreover, ML's capability to process complex and multidimensional data allows for the exploration of optimal material structures and module parameters. By training AI models to predict the interactions between different materials and environmental factors, researchers and engineers can discover more effective material compositions and configurations for solar evaporation surfaces. ML enables systems to dynamically adapt to environmental changes,

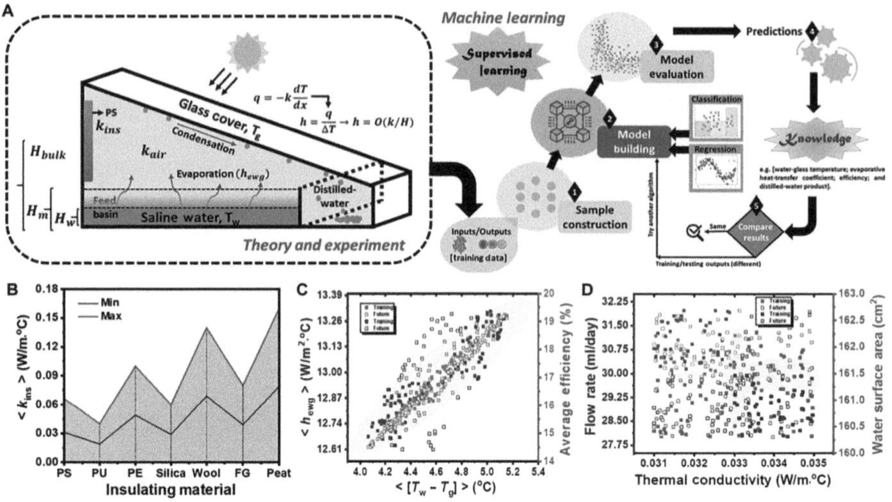

Fig. 5.8 (**a**) The heat transfer mechanisms and system parameters employed in ML, with theoretical and experimental results from a solar still forming the training datasets. A stepwise linear regression learner ML algorithm predicts the performance of different insulating materials. (**b**) Thermal conductivities of studied insulating materials. (**c**) Output datasets for PS training/future, correlating water-glass temperature difference with water evaporative coefficient and still efficiency. (**d**) Input datasets for PS training/future model calculations. (Reproduced with permission from [86], Copyright 2020 Elsevier)

significantly increasing the efficiency of solar desalination. This reduces energy consumption and increases the volume of clean water produced, which is crucial for regions with limited freshwater resources.

The integration of Machine Learning into solar evaporation represents a significant advancement. By the use of sophisticated algorithms to process and learn from data, ML facilitates the development of adaptive, efficient, and more intelligent solar evaporation systems. This improves individual system performance and supports the broader goal of sustainable and efficient renewable energy use. Additionally, the principles and models developed can be applied to other areas of renewable energy and environmental management. For instance, similar ML approaches can be used to optimize photovoltaic solar panels, wind turbines, and other renewable energy technologies. As research continues to advance in this area, it is expected that the applications of ML in solar energy technologies will become increasingly widespread, driving further improvements in efficiency and sustainability.

5.3.2 *Artificial Neural Network*

Artificial Neural Networks (ANNs) play a crucial role in transforming solar interfacial evaporation, especially for desalination and water purification [85]. Inspired by the neural structures of the human brain, these computational models excel at identifying complex patterns and anomalies in large datasets that traditional analytical methods might overlook. This capability is crucial for optimizing the performance, where environmental conditions can vary significantly. Theoretical Framework of ANNs in Solar Evaporation Systems: ANNs process information through interconnected nodes or neurons organized into layers: an input layer, one or more hidden layers, and an output layer. Neurons in one layer connect to those in the next, with connection strengths represented by weights. These weights are adjusted during training to minimize prediction error [87]. In solar interfacial evaporation, ANNs can be used to model and predict various system outputs based on inputs such as solar intensity, ambient temperature, humidity, and material properties of the evaporation interface. The multilayer perceptron (MLP), a type of ANN, is commonly employed due to its ability to approximate nonlinear functions effectively.

Case Studies include (1) Optimization of Process Parameters: Salem et al. utilized an ANN with the Adam optimizer to predict and optimize desalination efficiency based on solar energy inputs (Fig. 5.9) [9]. By tuning parameters like the activation function, learning rate, number of hidden layers, and neurons per layer, they developed an optimized MLP (OMLP). This model surpassed other intelligence-based models in predicting desalination productivity, as confirmed by statistical measures and error rates. In another innovative approach, Wang et al. used a combination of deep reinforcement learning (DRL) and convolutional neural networks (CNNs) to identify the most efficient graphene nanopores for seawater desalination [88]. Over a training period of one week, their AI framework evaluated 7999 different nanopores, demonstrating the AI capability to discover optimal materials for enhancing desalination performance. (3) Predictive Maintenance and Operational Efficiency: ANNs can forecast system failures and maintenance requirements, reducing downtime and enhancing operational efficiency. By continuously monitoring system performance and environmental conditions, ANNs can alert operators to potential issues before they escalate, ensuring consistent system performance.

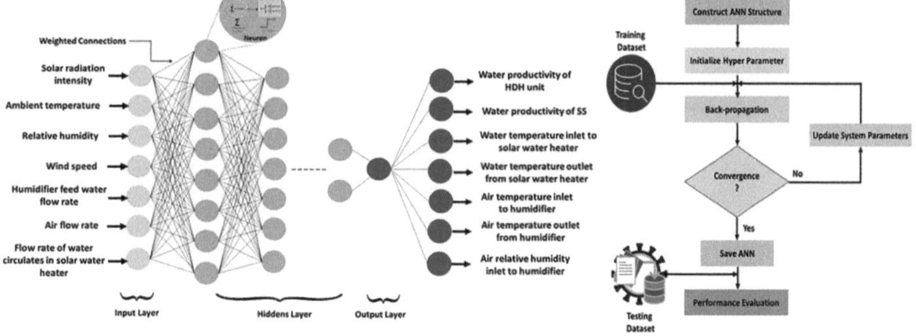

Fig. 5.9 ANN model used for the SS-HDH system performance prediction and Adam optimizer flowchart. (Reproduced with permission from [9], Copyright 2022 Elsevier)

The primary advantage of ANNs is their ability to learn and adapt to new data without explicit programming. This is particularly beneficial in dynamic environments with frequently changing operating conditions. Additionally, ANNs can process large volumes of data, making them ideal for applications where multiple variables impact system performance. Despite the advantages, the application of ANNs in solar-driven interfacial evaporation faces several challenges. One major issue is the need for large datasets to train the networks effectively, which can be difficult to obtain in specialized applications like solar desalination. Additionally, ANNs can be black boxes, meaning their decision-making process is not always transparent, which can be a barrier in applications requiring stringent validation and verification. Future research could focus on integrating newer AI techniques such as explainable AI to make ANNs more transparent and trustworthy. Additionally, combining ANNs with other computational methods, such as computational fluid dynamics, could enhance the predictive accuracy and reliability of these models.

The significance of utilizing ML and ANN in solar evaporation extends beyond mere efficiency improvements. These approaches promote the development of smarter, more autonomous solar evaporation systems that can dynamically adjust to optimize water purification processes. Consequently, this leads to more sustainable management of water resources, reduced operational costs, and increased accessibility to clean water, particularly in regions where water scarcity is a pressing issue [5]. This technology alignment thus holds the promise of contribution to global water sustainability, marking a pivotal step in the intersection of renewable energy technology and advanced computational methods.

References

1. L. Li, X. Sun, J. Miao, H. Wang, Y. Song, D. Tang, Gradient engineering in interfacial evaporation for water, energy, and mineral harvesting, ***Energy & Environmental Science***, (2025).
2. F. Liu, H. Pang, Y. Gu, Y. Ning, R. Bradley, B. Zhao, W. Wu, Electrically powered artificial black body for low-voltage high-speed interfacial evaporation, ***Journal of Materials Chemistry A***, 10 (2022) 22992-23000.
3. X. Sun, J. Miao, H. Wang, Y. Song, D. Tang, L. Li, Multiscale Heat and Mass Transport Optimization for Solar-Driven Interfacial Evaporation: Opportunities and Challenges for Industrial Applications, ***Small***, (2025) e2502639.
4. R. Fillet, V. Nicolas, V. Fierro, A. Celzard, Modelling heat and mass transfer in solar evaporation systems, ***International Journal of Heat and Mass Transfer***, 181 (2021) 121852.
5. Y. Wei, Y. Yang, Q. Zhao, Y. Ma, M. Qiang, L. Fu, Y. Liu, J. Zhang, Z. Qu, W. Que, Numerical Simulation Technologies in Solar-Driven Interfacial Evaporation Processes, ***Small***, 20 (2024) e2312241.
6. X. Ma, X. Jia, G. Yao, D. Wen, Umbrella evaporator for continuous solar vapor generation and salt harvesting from seawater, ***Cell Reports Physical Science***, 3 (2022) 100940.
7. B. Yang, Z. Zhang, P. Liu, X. Fu, J. Wang, Y. Cao, R. Tang, X. Du, W. Chen, S. Li, H. Yan, Z. Li, X. Zhao, G. Qin, X.Q. Chen, L. Zuo, Flatband lambda-Ti(3)O(5) towards extraordinary solar steam generation, ***Nature***, 622 (2023) 499-506.
8. M. Wang, Y. Wei, R. Li, X. Wang, C. Wang, N. Ren, S.H. Ho, Sustainable Seawater Desalination and Energy Management: Mechanisms, Strategies, and the Way Forward, ***Research***, 6 (2023) 0290.

9. H. Salem, A.E. Kabeel, E.M.S. El-Said, O.M. Elzeki, Predictive modelling for solar power-driven hybrid desalination system using artificial neural network regression with Adam optimization, ***Desalination***, 522 (2022) 115411.
10. Y. Lu, D. Fan, Y. Wang, H. Xu, C. Lu, X. Yang, Surface Patterning of Two-Dimensional Nanostructure-Embedded Photothermal Hydrogels for High-Yield Solar Steam Generation, ***ACS nano***, 15 (2021) 10366-10376.
11. Q. Wang, Z. Zhu, G. Wu, X. Zhang, H. Zheng, Energy analysis and experimental verification of a solar freshwater self-produced ecological film floating on the sea, ***Applied Energy***, 224 (2018) 510-526.
12. M. Tabatabaian, COMSOL for Engineers, ***Mercury Learning & Information***, ISBN 10:1938549538 (2014) 254.
13. L.S. Mayboudi, Comsol Heat Transfer Models ***Mercury Learning & Information***, ISBN 10:1683922115 (2019) 402.
14. H. Jin, G. Lin, A. Zeiny, L. Bai, D. Wen, Nanoparticle-based solar vapor generation: An experimental and numerical study, ***Energy***, 178 (2019) 447-459.
15. M.R. Karami, S. Saffar-Avval, M. Shokri, H. Aghakhani, Z. Mansoori, M. Saffar-Avval, Numerical and experimental investigation of floating wick solar still with a porous-media system, ***Solar Energy***, 282 (2024) 112958.
16. M.S. Zielinski, J.W. Choi, T. La Grange, M. Modestino, S.M. Hashemi, Y. Pu, S. Birkhold, J.A. Hubbell, D. Psaltis, Hollow Mesoporous Plasmonic Nanoshells for Enhanced Solar Vapor Generation, ***Nano letters***, 16 (2016) 2159-2167.
17. F. Wang, N. Xu, W. Zhao, L. Zhou, P. Zhu, X. Wang, B. Zhu, J. Zhu, A high-performing single-stage invert-structured solar water purifier through enhanced absorption and condensation, ***Joule***, 5 (2021) 1602-1612.
18. J. Choi, S. Jeon, Simultaneous evaporation and vapor condensation enhancement in 3D solar steam generators via rotational evaporation, ***Chemical Engineering Journal***, 505 (2025) 159372.
19. W. Han, J. Gao, J. Yu, R. Wang, Z. Xu, Efficient and low-cost solar desalination device with enhanced condensation on nail arrays, ***Desalination***, 544 (2022) 116132.
20. Y. Luo, B. Fu, Q. Shen, W. Hao, J. Xu, M. Min, Y. Liu, S. An, C. Song, P. Tao, J. Wu, W. Shang, T. Deng, Patterned Surfaces for Solar-Driven Interfacial Evaporation, ***ACS applied materials & interfaces***, 11 (2019) 7584-7590.
21. X. Liu, F. Chen, Y. Li, H. Jiang, D.D. Mishra, F. Yu, Z. Chen, C. Hu, Y. Chen, L. Qu, W. Zheng, 3D Hydrogel Evaporator with Vertical Radiant Vessels Breaking the Trade-Off between Thermal Localization and Salt Resistance for Solar Desalination of High-Salinity, ***Advanced materials***, 34 (2022) e2203137.
22. Y. Guo, L.S. de Vasconcelos, N. Manohar, J. Geng, K.P. Johnston, G. Yu, Highly Elastic Interconnected Porous Hydrogels through Self-Assembled Templating for Solar Water Purification, ***Angewandte Chemie***, 61 (2022) e202114074.
23. J. Ren, T. Gu, S. Ma, X. Li, Z. Zhou, D. Hao, K. Fang, S. Tian, Key factor in continuous salt harvesting via solar interfacial evaporation: Water supply to evaporation ratio, ***Desalination***, 607 (2025) 118800.
24. L. Zhang, X. Li, Y. Zhong, A. Leroy, Z. Xu, L. Zhao, E.N. Wang, Highly efficient and salt rejecting solar evaporation via a wick-free confined water layer, ***Nature communications***, 13 (2022) 849.
25. Y. Wang, X. Wu, P. Wu, H. Yu, J. Zhao, X. Yang, Q. Li, Z. Zhang, D. Zhang, G. Owens, H. Xu, Salt isolation from waste brine enabled by interfacial solar evaporation with zero liquid discharge, ***Journal of Materials Chemistry A***, 10 (2022) 14470-14478.
26. Y. Shao, J. Tang, N. Li, T. Sun, L. Yang, D. Chen, H. Zhi, D. Wang, H. Liu, G. Xue, Designing a bioinspired synthetic tree by unidirectional freezing for simultaneous solar steam generation and salt collection, ***EcoMat***, 2 (2020).
27. K. Xu, C. Wang, Z. Li, S. Wu, J. Wang, Salt Mitigation Strategies of Solar-Driven Interfacial Desalination, ***Advanced Functional Materials***, 31 (2020).

28. Y. Yang, H. Feng, W. Que, Y. Qiu, Y. Li, L. Guo, Q. Li, A Diode-like Scalable Asymmetric Solar Evaporator with Ultra-high Salt Resistance, ***Advanced Functional Materials***, 33 (2022).
29. J. Li, X. Wang, Z. Lin, N. Xu, X. Li, J. Liang, W. Zhao, R. Lin, B. Zhu, G. Liu, L. Zhou, S. Zhu, J. Zhu, Over 10 kg m−2 h−1 Evaporation Rate Enabled by a 3D Interconnected Porous Carbon Foam, ***Joule***, 4 (2020) 928-937.
30. A.K. Rao, A.J. Fix, Y.C. Yang, D.M. Warsinger, Thermodynamic limits of atmospheric water harvesting, ***Energy & Environmental Science***, 15 (2022) 4025-4037.
31. L. Wu, Z. Dong, Z. Cai, T. Ganapathy, N.X. Fang, C. Li, C. Yu, Y. Zhang, Y. Song, Highly efficient three-dimensional solar evaporator for high salinity desalination by localized crystallization, ***Nature communications***, 11 (2020) 521.
32. Y. Wang, X. Wu, T. Gao, Y. Lu, X. Yang, G.Y. Chen, G. Owens, H. Xu, Same materials, bigger output: A reversibly transformable 2D–3D photothermal evaporator for highly efficient solar steam generation, ***Nano Energy***, 79 (2021) 105477.
33. X. Li, J. Li, L. Zou, M. Chen, H. Wang, L. Yang, R. Yang, H. Zhang, Y. Zhou, Less is More: Reducing Evaporation Dead Zones to Tune Thermal Convection in Donut-Structure Evaporators for High-Efficiency Interfacial Solar Steam Generation, ***Advanced Functional Materials***, (2025).
34. X. Li, R. Yang, L. Zou, S. Zheng, M. Chen, J. Wen, H. Zhang, C. Wu, Y. Zhang, Y. Zhou, Reassessing the Role of Thermal Convection in Simultaneous Water Production and Pollutant Degradation in Interfacial Photothermal-Photocatalytic Systems, ***Advanced materials***, (2024) e2416283.
35. M. Morciano, M. Fasano, S.V. Boriskina, E. Chiavazzo, P. Asinari, Solar passive distiller with high productivity and Marangoni effect-driven salt rejection, ***Energy & Environmental Science***, 13 (2020) 3646-3655.
36. M. Zou, Y. Zhang, Z. Cai, C. Li, Z. Sun, C. Yu, Z. Dong, L. Wu, Y. Song, 3D Printing a Biomimetic Bridge-Arch Solar Evaporator for Eliminating Salt Accumulation with Desalination and Agricultural Applications, ***Advanced materials***, 33 (2021) e2102443.
37. I. ANSYS, Ansys Fluent Theory Guide, © ***2023 ANSYS, Inc.*** , (2023).
38. I. ANSYS, Ansys Fluent Tutorial Guide, © ***2023 ANSYS, Inc.***, (2023).
39. Y. Dong, Y. Tan, K. Wang, Y. Cai, J. Li, C. Sonne, C. Li, Reviewing wood-based solar-driven interfacial evaporators for desalination, ***Water research***, 223 (2022) 119011.
40. C. Jia, Y.J. Li, Z. Yang, G. Chen, Y.G. Yao, F. Jiang, Y.D. Kuang, G. Pastel, H. Xie, B. Yang, S. Das, L.B. Hu, Rich Mesostructures Derived from Natural Woods for Solar Steam Generation, ***Joule***, 1 (2017) 588-599.
41. C. Yan, J. Huang, Z. Liang, Z. Fang, D. Yang, Z. Li, Fabrication of a Highly Efficient Wood-Based Solar Interfacial Evaporator with Self-Desalting and Sterilization Performance, ***Langmuir : the ACS journal of surfaces and colloids***, 38 (2022) 12813-12821.
42. J. Tang, T. Zheng, Z. Song, Y. Shao, N. Li, K. Jia, Y. Tian, Q. Song, H. Liu, G. Xue, Realization of Low Latent Heat of a Solar Evaporator via Regulating the Water State in Wood Channels, ***ACS applied materials & interfaces***, 12 (2020) 18504-18511.
43. Y. Zhang, H. Zhang, T. Xiong, H. Qu, J.J. Koh, D.K. Nandakumar, J. Wang, S.C. Tan, Manipulating unidirectional fluid transportation to drive sustainable solar water extraction and brine-drenching induced energy generation, ***Energy & Environmental Science***, 13 (2020) 4891-4902.
44. Y. Xia, Q. Hou, H. Jubaer, Y. Li, Y. Kang, S. Yuan, H. Liu, M.W. Woo, L. Zhang, L. Gao, H. Wang, X. Zhang, Spatially isolating salt crystallisation from water evaporation for continuous solar steam generation and salt harvesting, ***Energy & Environmental Science***, 12 (2019) 1840-1847.
45. J.E. Matsson, An Introduction to SOLIDWORKS Flow Simulation 2019 ***SDC Publications***, ISBN 13: 9781630572396 (2019).
46. M. Zhu, Y. Li, G. Chen, F. Jiang, Z. Yang, X. Luo, Y. Wang, S.D. Lacey, J. Dai, C. Wang, C. Jia, J. Wan, Y. Yao, A. Gong, B. Yang, Z. Yu, S. Das, L. Hu, Tree-Inspired Design for High-Efficiency Water Extraction, ***Advanced materials***, 29 (2017).

47. Y. Pang, X. Chu, L. Song, L. Jin, C. Ma, Y. Wu, L. Li, Y. Peng, X. Zheng, F. Wang, S. Wu, Z. Shen, H. Chen, Laser-engraved wood-based evaporators: A sustainable approach for solar interfacial evaporation, ***Chemical Engineering Journal***, 479 (2024) 147891.
48. H. Liu, C. Chen, G. Chen, Y. Kuang, X. Zhao, J. Song, C. Jia, X. Xu, E. Hitz, H. Xie, S. Wang, F. Jiang, T. Li, Y. Li, A. Gong, R. Yang, S. Das, L. Hu, High-Performance Solar Steam Device with Layered Channels: Artificial Tree with a Reversed Design, ***Advanced Energy Materials***, 8 (2017).
49. C. Ma, J. Yan, Y. Huang, C. Wang, G. Yang, The optical duality of tellurium nanoparticles for broadband solar energy harvesting and efficient photothermal conversion, ***Science advances***, 4 (2018) eaas9894.
50. J. Kim, S.S. Baik, S.H. Ryu, Y. Sohn, S. Park, B.G. Park, J. Denlinger, Y. Yi, H.J. Choi, K.S. Kim, 2D MATERIALS. Observation of tunable band gap and anisotropic Dirac semimetal state in black phosphorus, ***Science***, 349 (2015) 723-726.
51. J. Wang, G. Xia, R. Li, Numerical analysis of evaporation from nanopores using the direct simulation Monte Carlo method, ***Journal of Molecular Liquids***, 347 (2022) 118348.
52. A. Jain, Y. Shin, K.A. Persson, Computational predictions of energy materials using density functional theory, ***Nature Reviews Materials***, 1 (2016).
53. D.M.S. Jennifer E. Houle, Electromagnetic Simulation Using the FDTD Method with Python, ***Wiley-IEEE Press***, ISBN 10:1119565804 (2020) 224.
54. R.A.M. Umran S. Inan, Numerical Electromagnetics The FDTD Method ***Cambridge University Press***, ISBN 13: 9780521190695 (2011) 406.
55. L. Zhou, Y. Tan, J. Wang, W. Xu, Y. Yuan, W. Cai, S. Zhu, J. Zhu, 3D self-assembly of aluminium nanoparticles for plasmon-enhanced solar desalination, ***Nature Photonics***, 10 (2016) 393-398.
56. J. Liu, S. Zhang, J. Wang, Q. Lan, Recent progress in solar-driven interfacial evaporation: Evaporators, condensers, applications and prospects, ***Desalination***, 597 (2025) 118356.
57. H.T. Kim, L. Philip, A. McDonagh, M. Johir, J. Ren, H.K. Shon, L.D. Tijing, Recent Advances in High-Rate Solar-Driven Interfacial Evaporation, ***Advanced science***, 11 (2024) e2401322.
58. P. Ying, B. Ai, W. Hu, Y. Geng, L. Li, K. Sun, S.C. Tan, W. Zhang, M. Li, A bio-inspired nanocomposite membrane with improved light-trapping and salt-rejecting performance for solar-driven interfacial evaporation applications, ***Nano Energy***, 89 (2021) 106443.
59. J.A.S. David Sholl, Density functional theory A practical introduction, ***Wiley-Interscience***, ISBN 10: 0470373172 (2009) 253.
60. F. Giustino, Materials Modelling using Density Functional Theory Properties and Predictions ***Oxford University Press***, ISBN 10: 0199662436 (2014) 304.
61. M. Gao, L. Zhu, C.K. Peh, G.W. Ho, Solar absorber material and system designs for photothermal water vaporization towards clean water and energy production, ***Energy & Environmental Science***, 12 (2019) 841-864.
62. G. Liu, Thermoplasmonics, ***Springer***, ISBN 978-981-97-8331-1 (2024).
63. H. Liu, C. Chen, H. Wen, R. Guo, N.A. Williams, B. Wang, F. Chen, L. Hu, Narrow bandgap semiconductor decorated wood membrane for high-efficiency solar-assisted water purification, ***Journal of Materials Chemistry A***, 6 (2018) 18839-18846.
64. W. Li, H. Jian, W. Wang, D. Yu, Highly efficient solar vapour generation via self-floating three-dimensional Ti2O3-based aerogels, ***Colloids and Surfaces A: Physicochemical and Engineering Aspects***, 634 (2022) 128031.
65. J. Wang, Y. Li, L. Deng, N. Wei, Y. Weng, S. Dong, D. Qi, J. Qiu, X. Chen, T. Wu, High-Performance Photothermal Conversion of Narrow-Bandgap Ti2O3 Nanoparticles, ***Advanced materials***, 29 (2017).
66. Y. Yang, H. Zhao, Z. Yin, J. Zhao, X. Yin, N. Li, D. Yin, Y. Li, B. Lei, Y. Du, W. Que, A general salt-resistant hydrophilic/hydrophobic nanoporous double layer design for efficient and stable solar water evaporation distillation, ***Materials horizons***, 5 (2018) 1143-1150.
67. R. Santamaria, Molecular Dynamics ***Springer International Publishing***, ISBN 10: 3031370422 (2023) 389.

68. J.H. Dominik Marx, Ab initio molecular dynamics basic theory and advanced methods ***Cambridge University Press***, ISBN 10: 0521898633 (2009) 579.
69. S. Meng, C.Y. Tang, J. Jia, J. Yang, M.B. Yang, W. Yang, A Wave-Driven Piezoelectric Solar Evaporator for Water Purification, ***Advanced Energy Materials***, 12 (2022).
70. Y. Guo, X. Zhao, F. Zhao, Z. Jiao, X. Zhou, G. Yu, Tailoring surface wetting states for ultrafast solar-driven water evaporation, ***Energy & Environmental Science***, 13 (2020) 2087-2095.
71. F. Zhu, L. Wang, B. Demir, M. An, Z.L. Wu, J. Yin, R. Xiao, Q. Zheng, J. Qian, Accelerating solar desalination in brine through ion activated hierarchically porous polyion complex hydrogels, ***Materials horizons***, 7 (2020) 3187-3195.
72. J.J. Koh, G.J.H. Lim, S. Chakraborty, Y. Zhang, S. Liu, X. Zhang, S.C. Tan, Z. Lyu, J. Ding, C. He, Robust, 3D-printed hydratable plastics for effective solar desalination, ***Nano Energy***, 79 (2021) 105436.
73. Q. Lu, W. Shi, H. Yang, X. Wang, Nanoconfined Water-Molecule Channels for High-Yield Solar Vapor Generation under Weaker Sunlight, ***Advanced materials***, 32 (2020) e2001544.
74. Z. Liu, Z. Zhou, N. Wu, R. Zhang, B. Zhu, H. Jin, Y. Zhang, M. Zhu, Z. Chen, Hierarchical Photothermal Fabrics with Low Evaporation Enthalpy as Heliotropic Evaporators for Efficient, Continuous, Salt-Free Desalination, ***ACS nano***, 15 (2021) 13007-13018.
75. H. Yu, D. Wang, H. Jin, P. Wu, X. Wu, D. Chu, Y. Lu, X. Yang, H. Xu, 2D MoN1.2-rGO Stacked Heterostructures Enabled Water State Modification for Highly Efficient Interfacial Solar Evaporation, ***Advanced Functional Materials***, 33 (2023).
76. T. Meng, Z. Li, Z. Wan, J. Zhang, L. Wang, K. Shi, X. Bu, S.M. Alshehri, Y. Bando, Y. Yamauchi, D. Li, X. Xu, MOF-Derived nanoarchitectured carbons in wood sponge enable solar-driven pumping for high-efficiency soil water extraction, ***Chemical Engineering Journal***, 452 (2023) 139193.
77. A. Stevens, Monte-Carlo Simulation An Introduction for Engineers and Scientists, ***CRC Press***, ISBN 10: 1032280778 (2022) 98.
78. B. Yang, C. Li, Z. Wang, Q. Dai, Thermoplasmonics in Solar Energy Conversion: Materials, Nanostructured Designs, and Applications, ***Advanced materials***, 34 (2022) e2107351.
79. R. Yang, D. Niu, J.H. Pu, G.H. Tang, X. Wang, M. Du, Passive all-day freshwater harvesting through a transparent radiative cooling film, ***Applied Energy***, 325 (2022) 119801.
80. S. Li, W. Su, B. Shan, Z. Li, L. Gibelli, Y. Zhang, Molecular kinetic modelling of non-equilibrium evaporative flows, ***Journal of Fluid Mechanics***, 994 (2024).
81. Y. Chen, L. Shen, Z. Qi, Z. Luo, X. Li, H. Bao, Large-scale implementation of solar interfacial desalination, ***Nature Sustainability***, (2025).
82. F. Nawaz, Y. Yang, S. Zhao, M. Sheng, C. Pan, W. Que, Innovative salt-blocking technologies of photothermal materials in solar-driven interfacial desalination, ***Journal of Materials Chemistry A***, 9 (2021) 16233-16254.
83. X. Ying, S.Y. Leng, H.F. Ma, Q. Nie, Y.C. Lai, W. Lin, Continuity Scaling: A Rigorous Framework for Detecting and Quantifying Causality Accurately, ***Research***, 2022 (2022) 9870149.
84. Z. Zhi-Hua, Machine Learning, ***Springer***, ISBN 10: 9811519668 (2021) 472.
85. C.C. Aggarwal, Neural Networks and Deep Learning A Textbook, 2nd Edition ***Springer***, ISBN 10: 3031296419 (2023) 553.
86. H.A. Maddah, M. Bassyouni, M.H. Abdel-Aziz, M.S. Zoromba, A.F. Al-Hossainy, Performance estimation of a mini-passive solar still via machine learning, ***Renewable Energy***, 162 (2020) 489-503.
87. H.H. Migaybil, B. Gopaluni, A performance neural network model for conventional solar stills via transfer learning, ***Applied Energy***, 375 (2024) 124118.
88. Y. Wang, Z. Cao, A. Barati Farimani, Efficient water desalination with graphene nanopores obtained using artificial intelligence, ***npj 2D Materials and Applications***, 5 (2021).

Chapter 6
Engineering Applications

6.1 Clean Water

6.1.1 Adsorption and Photodegradation

Water sources are frequently contaminated with volatile organic compounds from human activities and natural chemical substances. Due to their low boiling points, these compounds volatilize alongside water molecules, thus polluting the water supply [1, 2]. Wastewater treatment methodologies are classified into chemical, biological, and physical processes [3]. Chemical strategies encompass chemical precipitation, coagulation/ flocculation, and advanced oxidation processes, while biological approaches include biodegradation and bioadsorption [4, 5]. Physical methods are delineated into membrane technologies, flotation, and adsorption. These techniques basically operate on two principles: (1) transformation or decomposition of pollutants to attenuate toxicity by altering their chemical structure or properties, and (2) concentration and segregation of pollutants without a change in state, to enhance pollutant removal efficiency. A wastewater treatment apparatus utilizing solar evaporation technology comprises a photothermal layer and an adsorption layer (mass transfer layer) [6]. The photothermal layer absorbs solar energy, converting it into thermal energy, whereas the adsorption layer is tasked with pollutant absorption and vapor transport [7–9]. Some designs incorporate a support layer enabling the adsorbent to remain afloat on the wastewater surface [10]. The objective of this technology is to augment the heat and mass transfer capabilities of the evaporation system by leveraging solar energy, thereby enhancing adsorption effectiveness [11]. The operational principle of this technology is predicated on the thermal and concentration polarization effects, which facilitates an increase in the adsorption rate through elevated temperatures and enhanced evaporation rates, respectively.

G. Liu, T. Chen, *Solar Evaporation for Water-Energy-Resource Nexus*,
https://doi.org/10.1007/978-981-95-3815-7_6

Adsorption is considered as one of the most efficacious methods for water purification, surpassing other more costly and intricate techniques. This process capitalizes on porous solid adsorbents to extract pollutants from wastewater via surface adhesion, encompassing three principal phases: external diffusion, internal diffusion, and surface attachment [3, 12]. The mechanism of adsorption involves both physical and chemical interactions. Physical adsorption primarily occurs through electrostatic interactions, hydrogen bonding, and π-π stacking (Fig. 6.1a). Here, electrostatic interactions arise when the adsorbent and the pollutant bear opposite charges, facilitating the adsorption of the pollutant [13]. The zeta potential, which can be altered by modifying the pH of the solution, plays a crucial role in these interactions, as evidenced in the adsorption of Sr^{2+} by microplastics [14]. Hydrogen bonding, a significant type of intermolecular or intramolecular force, occurs between two electronegative atoms, one of which is hydrogen. These bonds are characterized by their saturation and directionality, providing robust molecular forces crucial in interactions, such as between ammonium groups and phosphate in wastewater [15]. Chemisorption, on the other hand, involves the formation of chemical bonds between the adsorbent and the pollutant, offering a more substantial capacity and stronger interactions compared to physical adsorption. Significant examples include the bonding of Pb^{2+} with micro/nano hydroxyl groups on Metal-Organic Frameworks (MOFs) [16] and the affinity of heavy metal ions for carboxyl/carbonyl groups in lignin [17].

Photocatalytic technology encounters challenges due to the widespread of pollutants. Integrating photocatalytic and photothermal evaporation offers a combined solution to improve freshwater availability [19–23]. Localized heating at catalytic sites can improve photodegradation efficiency. Inspired by plants' selective sunlight utilization, a spectrum-tailored solar harnessing aerogel has been developed. This aerogel comprises oxygen vacancy-rich semiconductor HNb_3O_8 nanosheets within a polyacrylamide framework, enhancing water evaporation by providing high kinetic energy to photogenerated electron-hole pairs and reactants [24]. This

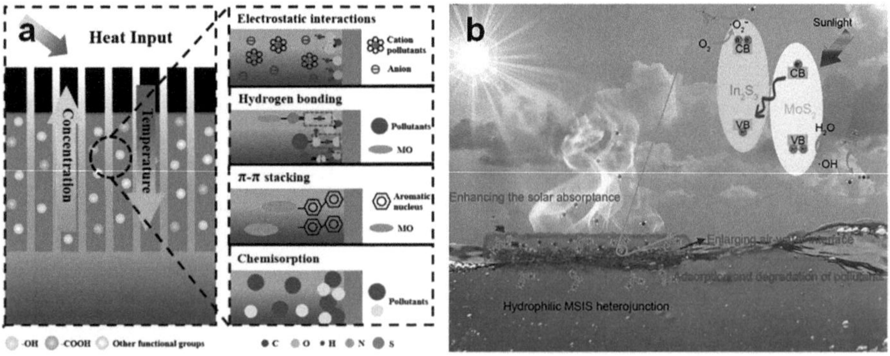

Fig. 6.1 (**a**) The adsorption coupled interface evaporation. (Reproduced with permission from [3], Copyright 2024 Elsevier). (**b**) Solar evaporation coupled with photocatalysis. (Reproduced with permission from [18], Copyright 2023 Elsevier)

configuration enhances charge carrier mobility, reduces recombination, and increases energetic interactions. Beyond superior water evaporation, this aerogel demonstrates enhanced photoactivity, fully oxidizing rhodamine B within 100 min. This is about 71% and 64% more efficient than pure water and polyacrylamide aerogel, respectively. Additionally, a novel technology combining solar distillation and photodegradation using a ZnO/gold-particle-decorated membrane has been developed. This technology achieves a steam generation rate of 8.70 kg m^{-2} h^{-1} and reduces the concentration of the organic pollutant rhodamine B in residual water by 70% within 2 h of solar irradiation [25]. β-In_2S_3 nanosheets decorated with 2H phase MoS_2 nanosheet arrays are developed for simultaneous solar water evaporation and photodegradation of organic pollutants (Fig. 6.1b) [18]. This hybrid nanosheets enhance solar water evaporation by adjusting the wettability of the absorber and improve photocatalytic degradation of organic pollutants by heterojunction interfaces [26]. The optimized absorber showed a 4.2-fold increase in the photocatalytic rate constant without compromising water evaporation efficiency. Additionally, incorporating photocatalysis into solar evaporation systems alleviates solar absorber fouling to a certain extent. Besides, dye pollution poses challenges due to the chemical stability and toxicity of synthetic dyes. MXene/carbon nanotubes and titanium dioxide nanorod-enhanced fabrics have shown potential in improving dye removal efficiencies and enhancing water purification processes, demonstrating the critical role of material science in addressing water pollution challenges [27, 28].

6.1.2 Sterilization

Solar steam sterilization represents a promising avenue for effective pathogen inactivation, particularly in regions where conventional energy sources are scarce or unavailable [29–31]. Vapor or steam sterilization, utilizing the large amount of latent heat released during vapor condensation, is acknowledged as one of the most effective methods for inactivating pathogens on medical devices and materials [32–34]. Traditional steam sterilization, however, predominantly relies on electricity or modern fuels-resources that are scarce in many developing regions inhabited by over one-quarter of the global population. In light of this, high-temperature/pressure solar vapor generation, leveraging the ubiquitous availability of sunlight, emerges as a highly promising solution in these settings [35, 36]. Effective sterilization requires precise control of temperature, pressure, and sufficient vapor contact time. Typically, complete sterilization needs saturated vapor conditions at temperatures above 120 °C for at least 15 min, or higher temperatures for shorter durations [34].

Photothermal sterilization is an economical and environmentally friendly method that deactivates bacteria and viruses in drinking water without producing harmful by-products, making it ideal for decentralized household applications [37–39]. Notable advancements include the development of evaporation design into a solar vacuum tube system, which generates steam above 120 °C and shows robust sterilization capabilities [40]. This integration enables the production of high-temperature steam under solar irradiation, effectively sterilizing Geobacillus stearothermophilus biological indicators and Escherichia coli bacteria. Zhu's group introduced an innovative solar autoclave, operating with a minimal carbon footprint, to achieve an impressive 99.999999% pathogen inactivation rate, as confirmed by tests on both bacterial vegetative cells and spores [34]. Additionally, a metallic MWCNTs and $HfTe_2$ van der Waals heterostructure was utilized for broadband solar harvesting, increasing the steam temperature to 132 °C in 20 min. A proof-of-concept sterilization demonstration showed a 99.04% reduction in E. coli bacteria [41].

To achieve high vapor temperatures, both the absorber and the system need meticulous design, often involving inorganic-based materials. Solar interfacial sterilizers offer advantages over traditional volumetric heating due to their reduced heat loss [31, 42]. However, most designs require an optical concentrator to achieve nucleate boiling, a process that demands a heat flux approximately seven times greater than the peak solar irradiance on Earth. This complexity increases both the cost and the complexity of the system. The sterilization systems that either do not require optical concentrators or only need minimal light-concentrating mechanisms are highly desirable [40, 43]. A novel approach utilizing thermal concentration has been reported, offering an efficient and practical solution (Fig. 6.2). Powered by an optimized, ultra-transparent silica aerogel, the device enhances local heat flux and temperature, eliminating the need for active optical concentrators. With an energy efficiency approximately twice that of previous devices (47%) tested at 100 °C, this device successfully demonstrated its sterilization capability during a field test in Mumbai, India [44]. It is crucial to acknowledge that the sterilization efficacy depends not only on the temperature and pressure but also on the volume of hot vapor produced, highlighting a trade-off between the quality and quantity of vapor for effective sterilization.

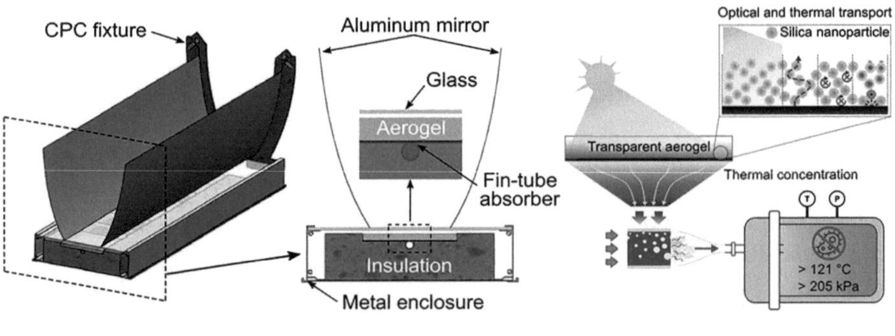

Fig. 6.2 A passive solar steam generator for medical sterilization. (Reproduced with permission [44] Copyright 2020 Cell Press)

6.1.3 Microplastic Removal

Microplastics defined as plastic particles smaller than 5 mm, are pervasive contaminants in various aquatic environments. Their ability to adsorb and concentrate harmful pollutants makes them significant ecological hazards [45–47]. Current techniques for extracting nano/microplastics from aquatic environments include flocculation, microbial ingestion, biofiltration, chlorination, UV oxidation, and membrane filtration [48]. However, flocculation introduces metal ions into the water, potentially causing unintended consequences and generating toxic byproducts. Microbial ingestion often results in suboptimal removal efficiencies and prolonged degradation periods. Oxidation technologies can increase toxicity and elevate hazardous contaminant levels due to changes in surface topography and chemical properties from incomplete oxidation processes [49]. While membrane filtration avoids secondary pollutant, it is hindered by surface fouling and pore clogging with plastic particles. Hereby, traditional water treatment methods often prove inadequate in eliminating microplastics due to their diminutive size and chemical robustness. This insufficiency highlights the urgent need for alternative, more effective, and sustainable technologies for microplastic removal.

Solar evaporation emerges as a promising low-energy alternative that utilizes renewable resources to address this pollution challenge [50, 51]. This method not only aims to purify water but also enhances the efficiency of microplastic recovery, leveraging solar thermal conversion to drive the process. The process of removing microplastics via solar evaporation involves multiple mechanisms that contribute to its efficacy [52]. One primary method is surface adsorption, where microplastics adhere to the photothermal materials used during the evaporation process [53, 54]. This method has shown promise in recovering microplastics from marine settings. Innovations in this area include the development of adsorbents that integrate with solar evaporation platforms, enhancing water purification while simultaneously targeting microplastic removal. This integrated approach not only mitigates the secondary pollution risk but also improves the removal efficiency, with studies showing a significant increase in removal ratios compared to traditional adsorbents [55]. Another mechanism for nano/microplastics removal is plasmonic heating, which involves processes such as photothermal decomposition and degradation, enhanced filtration, pollutant detoxification, aggregation, and sedimentation. Techniques like using light-driven microbubbles and localized heating zones to create large-scale convection demonstrated low energy consumption and reduced operational costs (Fig. 6.3) [48, 56].

However, challenges remain, particularly in the collection of negatively charged microplastics due to electrostatic repulsion, and the limitations in treating real seawater or wastewater contaminated with other pollutants [52, 57]. The effectiveness of solar evaporation systems significantly depends on the materials used and the design of the evaporation devices. Materials like graphene, carbon nanotubes, and plasmonic metals are explored for their superior photothermal properties, enhancing the solar thermal conversion efficiency crucial for effective microplastic

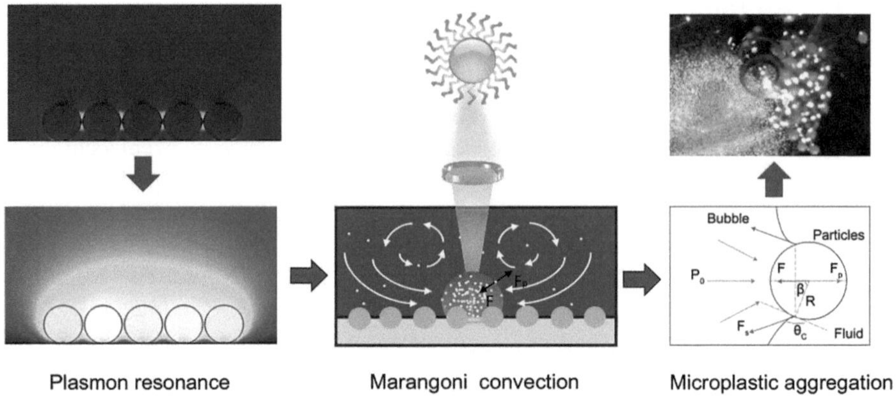

Fig. 6.3 Solar thermoplasmonic bubbles for sustainable removal of microplastics

exclusion [50, 53, 56]. The design elements of these devices often focus on maximizing surface area and thermal absorption, incorporating structures like floating arrays or multi-layered composites to optimize performance. Despite the advantages, the adoption of solar evaporation faces challenges such as scalability, durability of photothermal materials, and management of concentrated microplastic residues [53]. Future research directions include improving scalability, integrating the systems with existing water treatment infrastructures, and conducting comprehensive lifecycle assessments to ensure environmental and economic sustainability.

6.1.4 Seawater Desalination

Solar evaporation represents a promising advancement in desalination technology. Leveraging the abundant and high energy density of solar radiation, this method efficiently purifies seawater by eliminating salts and other contaminants, yielding potable water [58, 59]. This approach is particularly beneficial in remote or arid locales where conventional desalination infrastructures are untenable. Utilizing solar energy, which is essentially free and environmentally benign, solar seawater desalination stands out as an ecologically sustainable method for freshwater production [60–62]. Integrating condensation system with the solar evaporation enables direct application in seawater desalination. In this process, vapor generated through a photothermal system is condensed and collected, leading to reduced salinity levels in the produced water.

Innovative developments in this field have seen the use of nanoporous anodic aluminum oxide membranes, which facilitate water transport and continuous steam flow, enhancing the desalination process [63]. The plasmon-enhanced solar desalination utilizing broadband-absorbing aluminum nanoparticles has been demonstrated to effectively reduce the salinity of various seawater samples to below the thresholds set by the World Health Organization. The development of solar absorber films made from vanadium-doped molybdenum oxide nanospheres has been developed to achieve a high evaporation rate of 2.01 kg m^{-2} h^{-1} and a high evaporation efficiency of 93.44% [64]. This rapid evaporation rate is attributed to enhanced light harvesting properties facilitated by vanadium doping. The synergy between effective thermal management and robust light absorption capabilities of these nanospheres significantly boosts photothermal conversion efficiency.

Integrating hollow titanium oxynitride spheres into polyacrylamide-based films, combined with a two-dimensional water-wicking material, has shown substantial promise in solar seawater desalination under simulated sunlight. This setup has demonstrated a water evaporation rate of 1.49 kg m^{-2} h^{-1} and an evaporation efficiency of 89.1% [65, 66]. The process not only maintains high evaporation efficiency but also ensures that the freshwater generated meets WHO standards for potability. Moreover, the salts produced during solar evaporation can be automatically cleaned for continuous desalination [67, 68]. Salt management in solar seawater desalination is vital for maintaining the desalination performance with long-term stability. Innovative design solutions, such as the use of fabric wicks in floating solar evaporative structures and 3D umbrella architectures [69–71], have been developed to manage salt deposition effectively, ensuring continuous operation and long-term system stability.

Solar interfacial desalination remains at the experimental stage, with challenges persisting in scaling up for commercial application [72–74]. Concerns include potential decreases in water evaporation efficiency with larger system sizes and the impact of contaminants on system efficacy. Recently, a scaled-up solar interfacial system covering over 150 m^2 achieved a water production rate of up to 4.5 kg m^{-2} d^{-1} and a total fresh-water output of 300 kg/d over four months (Fig. 6.4) [75]. This system's scale necessitated significant changes in material selection and a substantial increase in thermal mass, altering water yield. Cost analysis indicated a heavy capital investment, contrasting with the low-cost operation typical of indoor solar interfacial desalination. The study also highlighted the importance of addressing fouling issues, such as sedimentation and green algae. By combining large-scale implementation with a solar desalination analysis, this work provides insights into its sustainable potential for remote areas.

6.1.5 Soil Water Harvesting

The terrestrial soil layer is estimated to encompass 165,000 cubic kilometers of water, a volume roughly eightfold greater than that contained within the world's rivers. Terrestrial hydric resources, primarily soil water, constitute a substantial yet underexploited reservoir of freshwater globally [76]. The extraction of water from soil, contingent upon regional conditions, presents a viable approach for mitigating potable water deficiencies in isolated areas (Fig. 6.5a) [77]. Fang et al. engineered a

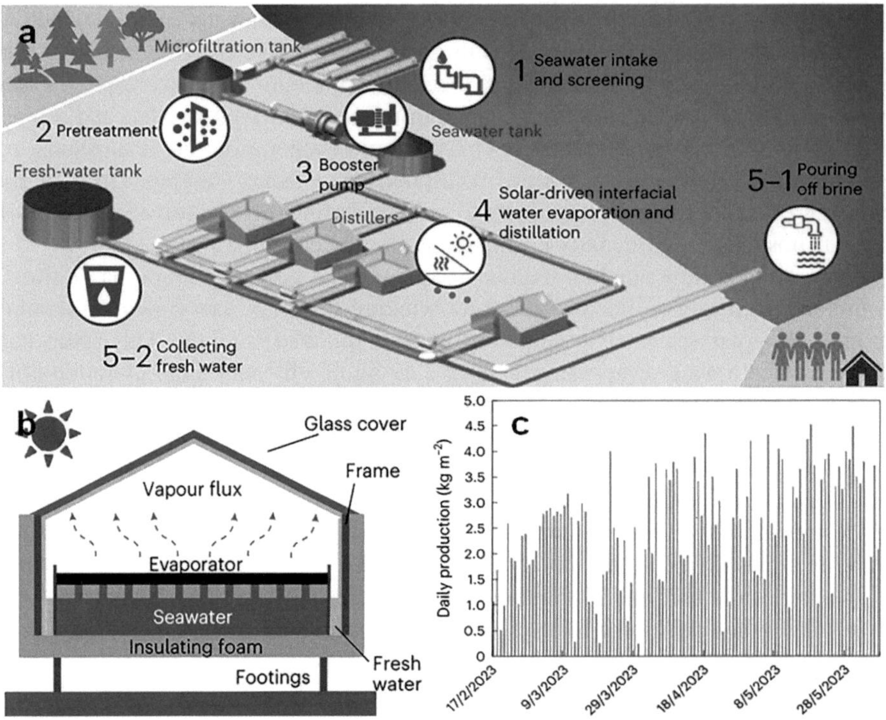

Fig. 6.4 Large-scale implementation of solar interfacial desalination. (**a**) A plant. (**b**) Schematic of one device. (**c**) Daily water production. (Reproduced with permission [75], Copyright 2025 Nature Portfolio)

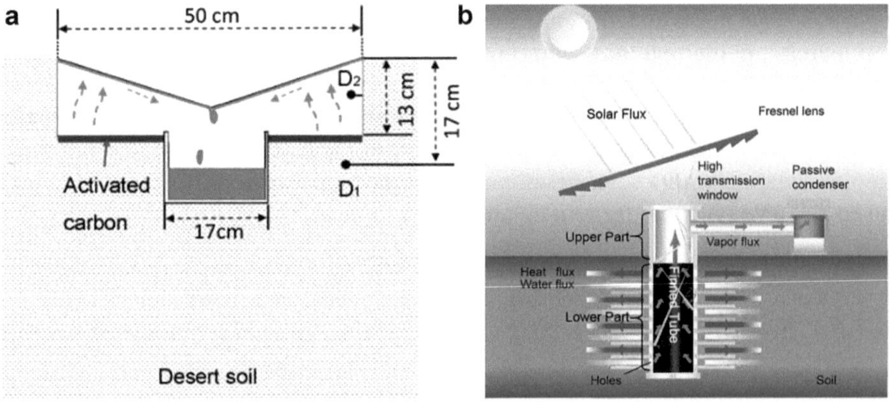

Fig. 6.5 (**a**) Water harvesting from desert soil. (Reproduced with permission from [77] Copyright 2022 Elsevier). (**b**) Water harvesting from soils by solar evaporation and capillary water migration. (Reproduced with permission from [80] Copyright 2020 Elsevier)

novel, entirely biomass-based solar evaporation apparatus capable of generating potable water from sandy and soil substrates [78]. This system utilized carbonized rice straw integrated with bacterial cellulose as an effective light-absorbing material, while the lower segments of rice stalks were adapted to facilitate water transport. The unique capillary structures and multi-tiered geometrical configurations of the rice stalks enabled seamless water extraction from moist and sandy soils, achieving impressive potable water production rates of 6.4 to 7.9 kg per square meter during sunny days.

Inspired by the natural root in terrestrial flora, Xiao et al. devised a biomimetic arboreal structure featuring photothermal foliage and paper-based root analogs with modifiable transpiration rates to enhance water evaporation efficiency. In a controlled experiment, this synthetic tree was embedded in sandy terrain within an enclosed chamber, yielding a water collection rate of 6.2 kg per square meter per hour under 10 h of irradiation [79]. Nevertheless, the efficiency of soil water evaporation can be significantly impaired by obstructions within the water transport channels, often caused by sand or other particulates. Such blockages lead to a marked decrease in water transport efficiency, thus impacting the sustainability of the evaporation process. It is imperative to note that the studies mainly took place in controlled environments or under conditions of saturated soil moisture, which starkly contrasts with the typically lower moisture levels observed in natural soil environments, where water content generally spans from 2% to 40% per kilogram of soil [80]. Therefore, it is crucial to conduct extensive research on soils with diverse moisture contents to fully assess the feasibility of solar evaporation for water extraction from soil.

Addressing the challenge of extracting water from soils with minimal moisture content, Li et al. introduced a cutting-edge solar-driven apparatus designed specifically to harvest freshwater from shallow subsurface layers. This device comprised a vertically embedded metallic tube with fins and a strategically placed Fresnel lens to concentrate solar energy through a transparent top onto the tube (Fig. 6.5b) [80]. The internal solar radiation was effectively reflected, and a cover equipped with numerous minute pores was employed to enhance vapor ingress. Capillary action facilitated the movement of soil moisture towards the heated tube, where vapor was condensed, and freshwater collected after passing through a passive condenser. This configuration not only achieved a daily clean water collection of 900 g over a 12-h period but also exhibited remarkable portability and scalability, rendering it a promising and sustainable option for securing freshwater in arid and economically disadvantaged regions.

6.1.6 Atmospheric Water Harvesting

An often-neglected yet sustainable water source is the moisture prevalent in atmospheric air. The Earth's atmosphere consistently contains an astonishing volume of approximately 37 million billion gallons of water [81, 82]. The challenge of extracting water from atmospheric humidity with minimal energy expenditure has catalyzed significant scientific inquiry [83, 84]. The World Economic Forum, in partnership with Scientific American, has recently ranked this domain as the second most transformative technology, with potential widespread impact on both everyday life and industrial processes [85, 86]. Among various Atmospheric Water Harvesting (AWH) technologies, the three predominant methods are fog harvesting, dew harvesting through water vapor condensation, and sorption-based AWH [87].

Fog harvesting, which employs biomimetic strategies, proves effective in environments with high ambient humidity, typically with a relative humidity (RH) exceeding 65% [88]. For example, a fog collector inspired by the Namib Desert beetle utilizes a fabric engineered with superhydrophobic and superhydrophilic properties, enhanced by copper particles, achieving water collection rates of 1.4 g/h/cm^2 [89]. Conversely, condensation technologies often rely on electrical energy to cool and condense water vapor using methods such as vapor compression cycles, thermoelectric cooling, and adsorption cooling. These active dewing-based AWH systems, particularly that using vapor compression refrigeration, tend to be energy-intensive and less efficient in arid regions [90].

In light of factors like onsite energy availability, infrastructure, and specific geographical and hydrological conditions, the most feasible technological solutions have been identified as Solar-Driven Clean Water Harvesting from Humid Air [91–93]. This approach emphasizes energy efficiency, semi-passive operation, and sustainability. Solar-driven sorption-based AWH typically involves two sequential processes: capturing water vapor from the ambient air and releasing this vapor through solar heating, subsequently condensing it into pure water [94–96]. The initial phase necessitates precise engineering of the sorbent's chemical and physical properties, whereas the latter phase's design closely parallels processes used in the purification of seawater, brackish water, and industrial wastewater. A novel metal-organic framework (MOF) has demonstrated the capability to harvest water from atmospheric humidity even in arid conditions with RH below 20%, producing up to 2.8 liters of water per kilogram of MOF utilized (Fig. 6.6) [97].

However, the application of adsorption techniques faces several challenges. These include the limited research conducted in environments with RH above 30%, the slow kinetics of adsorption and/or desorption of most materials, the potential structural degradation of hygroscopic materials during these processes, and the requirement for external power during desorption [98–100]. Despite the high average relative humidity levels (85% ~ 90%) over ocean surfaces and the increasing trend of atmospheric moisture over these areas, the lack of materials that can efficiently exploit this vast resource hinders its practical use [101, 102]. Innovations such as incorporating photothermal materials into adsorbents have the potential to obviate the need for external power sources. Nonetheless, the development of efficient, scalable devices may be more critical for practical application than the creation of high-performance materials alone [103, 104]. Current research suggests that continuous operation devices significantly outperform intermittent systems [87, 105], highlighting the importance of advancing both the materials and the device designs for effective and scalable implementation of AWH technologies.

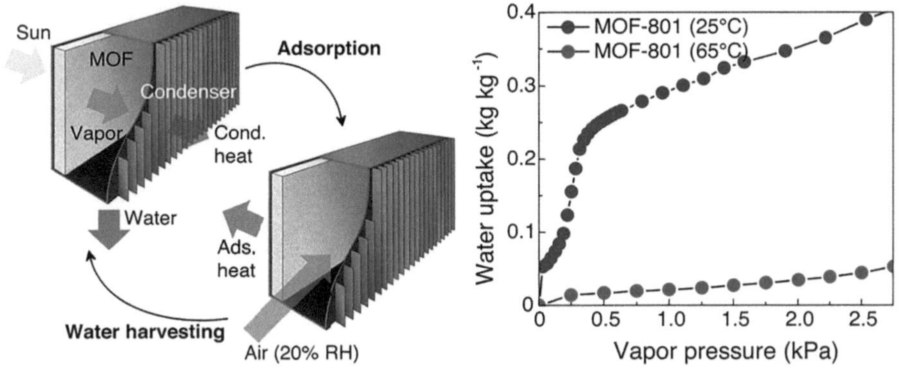

Fig. 6.6 Water harvesting from air with metal-organic frameworks upon natural sunlight. (Reproduced with permission from [97] Copyright 2017 Science AAAS)

6.2 Sustainable Energy

6.2.1 Evaporation Cooling

Cooling is indispensable in numerous facets of everyday life. The necessity for cooling, particularly for space cooling, is anticipated to escalate dramatically due to the growing frequency, duration, and intensity of extreme heatwaves, coupled with continuous improvements in living standards [106–108]. In this context, the development of cost-effective cooling technology that functions without electricity is imperative and has the potential to improve the life quality. Evaporative cooling emerges as a paradigmatic passive cooling technology, achieving temperature reduction through the intrinsic energy dissipation during water evaporation processes [108–111]. This technology is extensively applied across various domains of daily and industrial activities, from ambient cooling to dust control in settings like yards and roadways, and is especially prevalent in the cooling systems of condensers at thermal power plants. The introduction of interfacial solar steam/vapor generation technologies has markedly increased evaporation rates, fostering significant progress in evaporative cooling techniques [76, 112].

In solar evaporation, the resultant cooling effect is capable of reducing the temperature of the evaporation surface below the ambient temperature, thus facilitating the extraction of energy from the surroundings and bolstering the effectiveness of passive cooling approaches [113]. Tang et al. have devised a conical photothermal evaporator tailored for room cooling via water evaporation [109]. The conical configuration and the dynamic transition in water states along the transport direction ensure that the apex temperature of the device remains substantially lower than the surrounding air temperature, thereby maintaining a consistent energy transfer from the environment to the evaporator (Fig. 6.7a). Outdoor testing scenarios revealed a sharp temperature differential of 13.7 °C between the internally cooled space and the external environment, robustly validating the potential of solar-driven

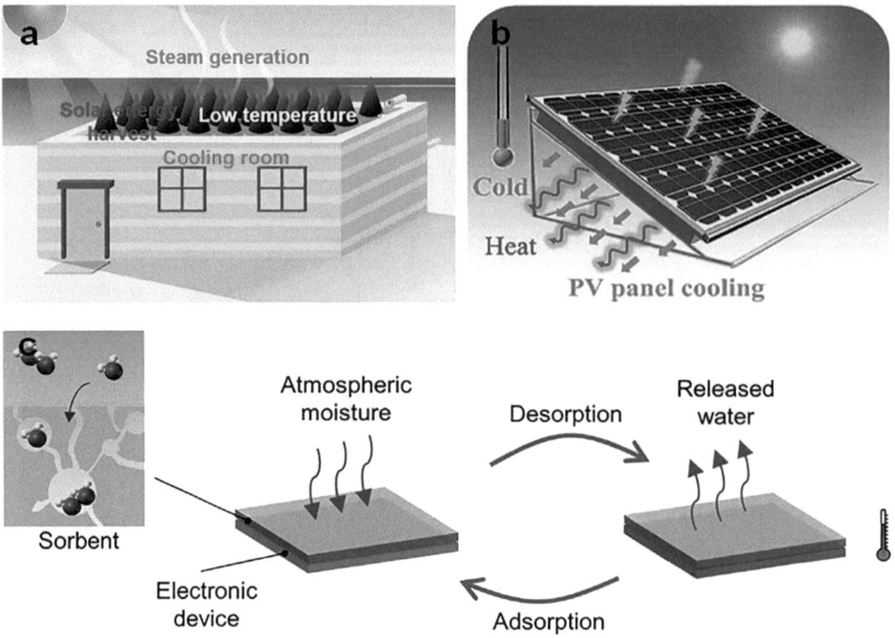

Fig. 6.7 Evaporation cooling. (**a**) A conical aerogel evaporative cooling system for household cooling. (Reproduced with permission from [109], Copyright 2021 Elsevier). (**b**) The PV working with a PAM-CNT-$CaCl_2$ hydrogel. (Reproduced with permission from [114], Copyright 2020 Nature). (**c**) The strategy of the adsorption and desorption cycles for electronic cooling. (Reproduced with permission from [116], Copyright 2020 Cell Press)

evaporative cooling to enhance environmental temperature control. Furthermore, the integration of evaporative cooling effects with commercial energy systems, such as solar panels and batteries, can significantly enhance heat dissipation, thus improving both the energy efficiency and operational reliability of these systems (Fig. 6.7b) [114]. For example, during the operation, solar panels can undergo temperature increases up to 40 °C above ambient levels, which can diminish both power output and device longevity. To mitigate these challenges, Wang et al. introduced a multi-stage distillation unit at the rear of a solar panel, which utilizes the excess heat generated during operation to drive water evaporation, effectively lowering panel temperature and consequently boosting electricity production by 8% [115].

Moreover, evaporative cooling has shown considerable promise in the thermal management of electronic devices [106, 116, 117]. Wang et al. suggested an innovative sweat cooling utilizing a sorption-desorption mechanism to regulate device temperatures (Fig. 6.7c). This study rigorously assessed the performance of various sorbents under different relative humidity conditions through comprehensive simulations, identifying MIL-101(Cr) as an ideal sorbent at an RH of 40%. The findings revealed that applying 0.288 g of MIL-101(Cr) powder to the device surface resulted in a maximum temperature reduction of 8.6 °C over a period of 25 min at a heating power of 1.5 W [116]. The researchers also outlined strategic directions for future improvements, advocating for the development of climate-specific sorbents, the selection of sorbents demonstrating linear water uptake across varying RH levels, and the exploration of kinetic mechanisms to guide the design of sorbents with optimal geometric and morphological properties. Additionally, the subambient cooling in energy conservation and the preservation of food and pharmaceuticals was studied by Aili et al. by a comparison of evaporative and radiative cooling via theoretical analysis. The results suggest that evaporative cooling is more efficacious under conditions of high temperature and low humidity, for instance, achieving sub-ambient temperatures of −15.0 °C at an RH of 13% and an ambient temperature of 26 °C, compared to −13.5 °C achieved by radiative cooling [118]. These findings highlight the superior performance of evaporative cooling in specific environmental conditions due to its enhanced responsiveness to temperature and humidity variables, in contrast to the relatively modest dependency exhibited by radiative cooling.

6.2.2 Electricity Generation

Water is indispensable for all life forms and harbors substantial energy reserves, manifesting as mechanical, latent heat, and chemical energies. A significant fraction of this energy, particularly from evaporation, remains untapped [119–121]. In solar evaporation, the evaporators create conditions like accelerated water evaporation, enhanced water flow, and established salinity gradients. These systems also maintain high humidity and a temperature gradient across the evaporation interface and the bulk water. By leveraging these conditions, the system not only yields potable water but also generates supplemental electrical power, making solar evaporation technologies advantageous for decentralized water and power supplies [122–124]. Recent research has increasingly concentrated on energy harvesting from water's phase transitions, propelled by significant advancements in nanomaterials and solar evaporation technologies [119, 123, 125]. This section explores the latest innovations in electricity generation via solar interfacial evaporation, examining mechanisms such as salinity gradients, hydrovoltaics, and temperature differentials, etc.

6.2.2.1 Hydrovoltaic Effect

Xue et al. pioneered the development of an electrical generator that utilizes water evaporation over nanostructured carbon materials to generate sustained voltages of up to 1 V under ambient conditions [76, 126]. Figure 6.8 shows the electrical generator activated by evaporation. This generator consists of a porous carbon black film, equipped with carbon nanotube electrodes at both ends. One extremity of the apparatus is submerged in water. The evaporation at the gas-liquid boundary creates a capillary force within the porous carbon structure, which drives water molecules near the evaporated area to move and replace the lost water, thus maintaining a

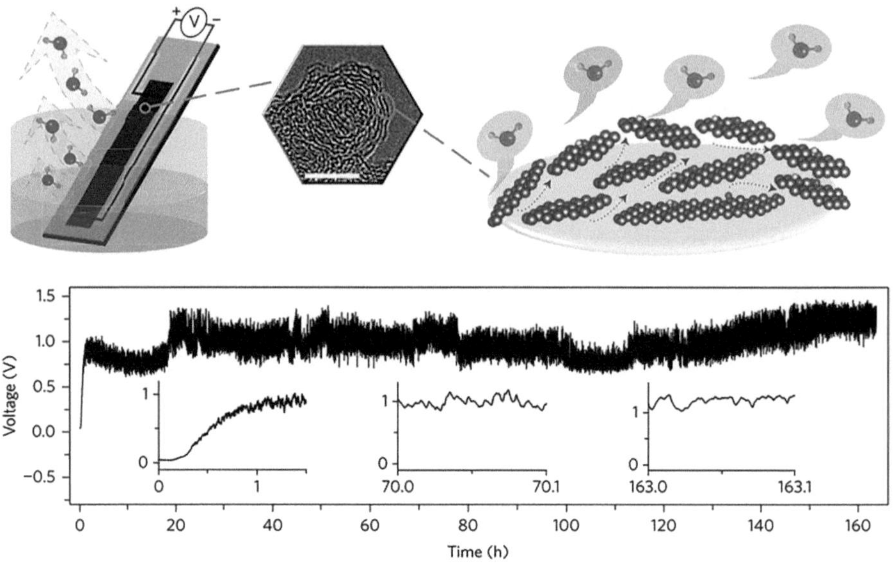

Fig. 6.8 Evaporation-induced electricity in carbon black film. (Reproduced with permission from [126], Copyright 2017 Nature)

steady flow within. This flow triggers the movement of ionic solutions, leading to the generation of an electric potential difference across the narrow channels or porous structures. The interaction between mass and heat transfer, enhanced by nanostructured materials, enables the efficient conversion of thermal energy from water evaporation into electrical energy, sparking significant interest for further investigation.

Recent advancements enhanced our understanding of hydroelectric mechanisms, primarily attributed to the Streaming potential [127]. When the charged solid surfaces come into contact with water, an Electric Double Layer (EDL) spontaneously forms at the interface due to Coulomb interactions. This layer consists of a Stern layer and a diffusion layer. Within the Stern layer, polar ions like cations and anions adhere electrostatically to the solid surface. Coulomb forces then attract counter ions from the diffusion layer toward the Stern layer, effectively shielding its electrostatic potential. The shear plane between these layers defines the ζ-potential, a critical parameter in the dynamics of water-solid interactions [119, 128]. The imbalance of anions and cations within the diffusion layers, influenced by the Stern layer's Coulomb force, prompts the solution to revert to its bulk phase beyond the diffusion layer. The Debye length (λ_D), ranging from a few nanometers to several hundred nanometers, describes the extent of electrostatic interactions within the EDL and provides insights into the spatial charge distribution and ion behavior near solid-liquid interfaces [129–131]. When external forces act on the liquid, relative motion between the diffusion and Stern layers generates various electrokinetic phenomena, including electroosmosis, electrophoresis, and streaming potential [132].

Specifically, the streaming potential efficiently converts kinetic energy into electrical energy, harvesting water energy through interfacial interactions between water and solids. When an electrolyte is confined in a charged nanochannel, comparable in size to λ_D, the confined electrolyte becomes dominated by counterions due to EDL overlap. A pressure gradient induces hydrodynamic flow through the channel, directing counter-ion transport and establishing a constant potential difference across the channel, known as the streaming potential. The streaming potential (V_{str}) and current (I_{str}) are expressed as functions of

$$V_{str} = \frac{\varepsilon_0 \varepsilon_r \zeta}{4\pi\eta\sigma} \Delta P \quad (6.1)$$

$$I_{str} = \frac{\pi a^2 \varepsilon_0 \varepsilon_r}{\eta} \left(\frac{\Delta P}{L} \right) \zeta \quad (6.2)$$

the relative permittivity (ε_r), dynamic viscosity (η), specific conductivity (σ) of the liquid, vacuum permittivity (ε_0), nanochannel radius (a), length (L), ζ-potential (ζ), and the pressure difference (ΔP) across the nanochannel [132]. Enhancing the ζ-potential and increasing the external pressure difference, as well as reducing the nanochannel diameter, can augment both the streaming potential and current.

Diverse materials have been explored for constructing hydrovoltaic devices. Initial focus was on carbon materials such as graphene, carbon black and carbon nanotubes [126, 133–135]. Subsequently, materials including metal oxides, hydroxides, biomass, hydrogels, metal-organic frameworks, and composites have gained attention for their potential in evaporative hydrovoltaic generation [105, 136, 137]. For instance, Xu et al. reported using polyvinyl alcohol sponges with poly(3,4-ethylenedioxythiophene), poly(styrene sulfonate), and carbon nanotubes as 3D evaporators for simultaneous desalination and power generation [138]. These sponges, characterized by hierarchical porous structures and strong light absorption, achieved an impressive evaporation rate of 6.8 kg m^{-2} h^{-1} and an evaporation efficiency of 94.9% under one sun. The high-flux directional flow of seawater through the sponges' negatively charged ionic channels generated a stable potential of 117.8 mV. Given the high zeta potential exhibited by metal oxides due to the adsorption of hydroxyl complexes from the amphiprotic dissociation of surface hydroxyl groups or hydrolysis products, these materials are promising for energy harvesting. A notable example is a two-dimensional monolayer of titanium oxide ($Ti_{1-\delta}O_2$) nanosheet with specific atomic vacancies, used for efficient electricity generation induced by water evaporation [139]. By controlling the concentration of Ti atomic vacancies, the device with optimized 2D $Ti_{0.87}O_2$ nanosheets maintained a stable open-circuit voltage of 1.32 V for over 250 h, surpassing the performance of commercial TiO_2 nanosheets. MOFs, recognized for their high specific surface area and variable functionality, have been identified as potential candidates in various energy-related applications. A flexible Cu-CAT-1 MOF nanorod membrane, exhibited a unique hierarchical structure, excellent solar absorption, efficient photothermal conversion, high Zeta potential, and outstanding hydrophilicity. This membrane demonstrated exceptional desalination performance and all-weather electricity output [140]. Significantly, the electrical output of the solar evaporation power generation device increased under irradiation compared to dark conditions. This confirms that solar evaporation accelerates both water evaporation and the flow rate within the MOF membrane, leading to higher power output. As many materials exhibit both photothermal and hydrovoltaic effects [120, 141–143], integrating evaporative power generation with simultaneous clean water production presents a promising technology for distributed water and power supply.

6.2.2.2 Photovoltaic Effect

Photovoltaics (PVs) are widely used to transform solar energy into electrical power, yet they suffer from low efficiency. Only photons with energies above the bandgap of silicon are absorbed for electricity generation, leaving ~70% of solar energy unused. Integrating photovoltaic technology with solar evaporation provides an innovative approach to enhance solar energy utilization [123]. This integration overcomes the limitations of traditional PV systems by offering dual benefits: electricity and freshwater production. The technology harnesses solar radiation to generate heat, facilitating interfacial evaporation that converts water into steam, thereby maximizing solar energy utilization [1, 76, 123, 144]. This process converts solar energy into both usable heat and electricity, significantly enhancing overall energy conversion efficiency. Additionally, SES utilizes wastewater as the operational medium, using interfacial evaporation for clean water production. This aspect is beneficial in water-scarce regions, offering a sustainable solution for managing water resources.

There has been an increasing integration between PV and solar evaporation systems. Cheng et al. developed photovoltaic dual-responsive nanocapsules through a rapid oxidative chemical polymerization process, using industrial paraffin wax (IPW) as the core phase change material, with crosslinked polystyrene (CLPS) and polypyrrole (PPy) as cross-linking agents [145]. These IPW@CLPS@PPy capsules exhibit exceptional properties, including excellent shape stability, a heat of phase change of 114.2 J/g, superior thermal stability, and impressive cycling performance. This innovation has produced shape-stabilized phase change nanocapsules with controllable photothermal and electrothermal response properties. When exposed to light, the nanocapsules absorb light energy, generate heat, and induce a phase change in the encapsulated material. Additionally, applying an electric field to the nanocapsules also triggers a phase change in the core material. This dual-responsive behavior allows the nanocapsules to effectively modulate and convert both photothermal and electrothermal energy. However, the mechanism of photovoltaic power generation requires further refinement.

Xu et al. proposed a coupled PV system with interfacial evaporation to utilize the full solar spectrum [146]. As shown in Fig. 6.9a, the solar cell was placed on top to absorb above-bandgap photons, while below-bandgap photons were absorbed and

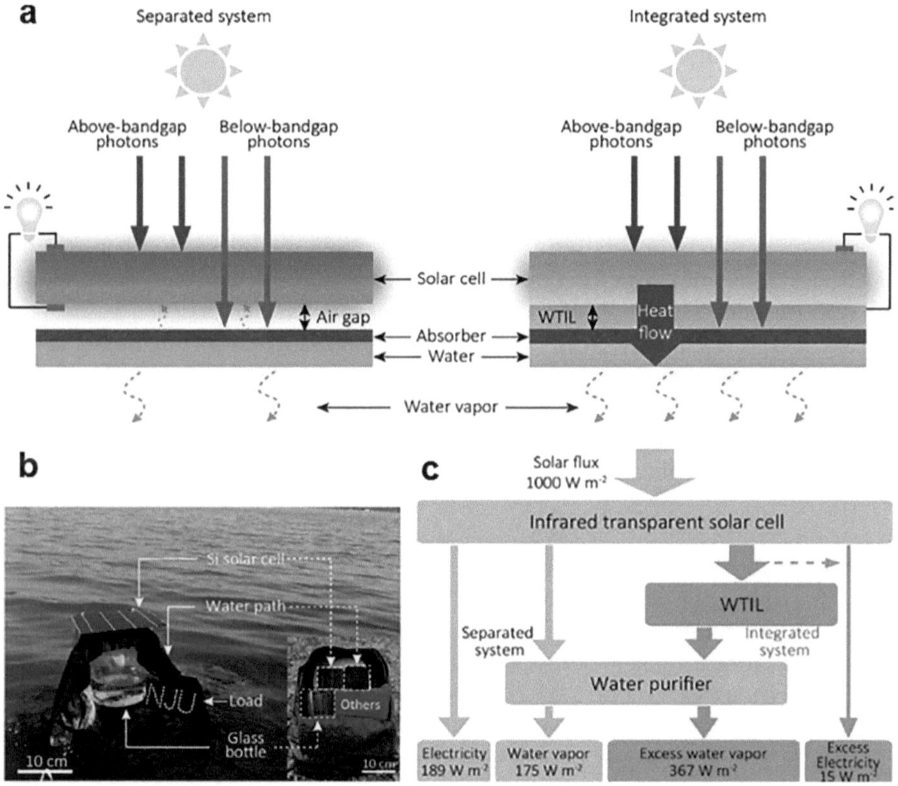

Fig. 6.9 Coupling cells for simultaneous water and power generation: (**a**) Diagrams comparing separate and integrated systems of photovoltaic cells with interfacial evaporation. (**b**) Photo of the setup for outdoor use. (**c**) Energy flow analysis. (Adapted from [146], Copyright 2019 Cell Press)

converted into heat by a reduced graphene oxide fabric. To ensure synergy between these components, a water-proof thermal interconnecting layer (WTIL) was incorporated. This coupled device demonstrated significant capacity for generating both electricity and freshwater (Fig. 6.9b). In this configuration, approximately 74.6% of the solar energy was effectively harnessed, compared to only 36.4% in the separated system (Fig. 6.9c). The PV and solar evaporator's synergistic effect maintained the solar cell at a lower temperature through water cooling, while simultaneously using the absorbed heat to generate additional steam. This lower temperature environment is beneficial for increased electricity generation [144]. Consequently, the coupled system showed remarkable synergy in producing additional electricity and fresh water resources, with the electrical output power sufficient to meet daily life's demands.

Ji et al. developed a solar-hydroelectric generator system by integrating a semi-transparent photovoltaic (ST-PV) with a solar interfacial evaporator [147]. This system leverages the unique inverted device structure of the ST-PV, optimized by precisely controlling light transmission and thermal properties through specialized windows. The ST-PV/SDIE hybrid system achieves exceptional performance, with a high power output of 122 W/m^2 and a consistent water evaporation rate of 1.30 kg k m^{-2} h^{-1}. This work marks the first integration of a solar evaporator with the emerging field of organic photovoltaic cells. However, evaluating the device's sustainable use requires understanding the solar evaporator's impact on aquatic environments and organisms. Zhang et al. investigated the application of semitransparent photovoltaic polymer solar cells in intelligent buildings, emphasizing efficiency and aesthetic integration [148]. By optimizing the interaction between the electron acceptor, donor, and polymer device, they developed three high-performance ST-PSC variants. These variants achieved an optimal balance between power generation and transparency, with specific transmission windows regulating water evaporation rates and algae growth. This development highlights that the roles of solar evaporator and water-electricity cogeneration extend beyond mere radiation absorption and electricity generation. The innovative features of ST-PSCs pave the way for practical applications and interdisciplinary collaborations.

6.2.2.3 Salinity Gradient Effect

During the operation of solar desalination, a persistent evaporation of seawater results in a salinity gradient across porous photothermal materials, offering a potential for electricity generation [76]. It has been documented that the salinity variance between seawater and freshwater harbors substantial amounts of renewable energy, extractable during their mixing and separation, a phenomenon termed blue energy [149–151]. Theoretically, the maximal energy output from such mixing is quantifiable via the Gibbs free energy difference (ΔG_{mix}), calculated as follows:

$$\Delta G_{mix} = RT\left[c_m \ln(c_m) - \chi c_s \ln(c_s) - (1-\chi) c_r \ln(c_r)\right] \tag{6.3}$$

The concentration of the mixed solution is represented by c_m, seawater concentration by c_s, river water concentration by c_r, and the seawater volume fraction by χ. In practical applications, the usable energy per unit volume is less than the theoretical Gibbs free energy due to unavoidable entropy production [76]. Current methods for harnessing energy from salinity gradients include reverse electrodialysis [152], pressure retarded osmosis [153], and electric double-layer capacitors [154]. Among these, reverse electrodialysis is the most promising technology for salinity gradient power generation, as it exploits the chemical potential gradient to facilitate directional charge transport across ion exchange membranes, converting chemical potential into electricity.

A salinity gradient power system consists of two solutions with different salinities, separated by an ion-selective membrane featuring charged nanopores [155]. The disparity in salt concentration across the membrane allows ions to migrate through the nanopores from the higher to the lower concentration solution. Assuming a negatively charged membrane, cations predominantly diffuse through the nanopores in a directed fashion, creating a net ionic charge flow. This movement generates an electrical potential difference across the membrane, commonly referred to as the osmotic electrostatic potential, which can be approximated by the Nernst equation [152]:

$$\Delta E_{osm} = \frac{k_B T}{ze} S ln \frac{c_h}{c_l} \tag{6.4}$$

Here, $S = (D_+ - D_-)/(D_+ + D_-)$ represents the ion selectivity, $D_\pm$ denotes the effective diffusion constants of cations and anions, and $k_B T/e$ approximates 25.7 mV at 25 °C, with c_h and c_l being the salt concentrations on the high and low salinity sides. Ion-selective pores that preferentially allow the flux of one type of charged ion over another can generate a non-zero short-circuit potential difference, converting some Gibbs free energy from mixing into electricity [156]. The maximum power density in salinity gradient power systems is determined by pore conductance and selectivity, which are influenced by intrinsic properties such as functional groups on the pore walls, their degree of dissociation, and the physical dimensions of the pores. Yang et al. synthesized a covalent organic framework monolayer with uniform thickness, ordered nano-channels, high ion conductivity, and superb ion selectivity through the Schiff base condensation of amine-terminated tetraphenyl porphyrin zinc (ZnTPP) monomer and 2,5-dihydroxy-benzaldehyde [157]. This ZnTPP-COF monolayer was then utilized to construct a high-output osmotic power generator delivering a power density of 135.8 W/m^2.

During evaporation, rapid water evaporation under solar irradiation increases the salt ion concentration near the evaporation surface, creating a salinity gradient relative to the bulk seawater. Consequently, solar evaporation not only produces clean water but also opens new avenues for harvesting salinity gradient energy. Yang et al. designed a coupling system combining carbon nanotube-modified filter paper and a commercial Nafion membrane for concurrent solar desalination and salinity gradient power generation (Fig. 6.10a) [158]. Carbon-based materials convert solar energy into heat, accelerating water evaporation and enhancing the concentration disparity between the evaporation surface and bulk seawater, thus generating a

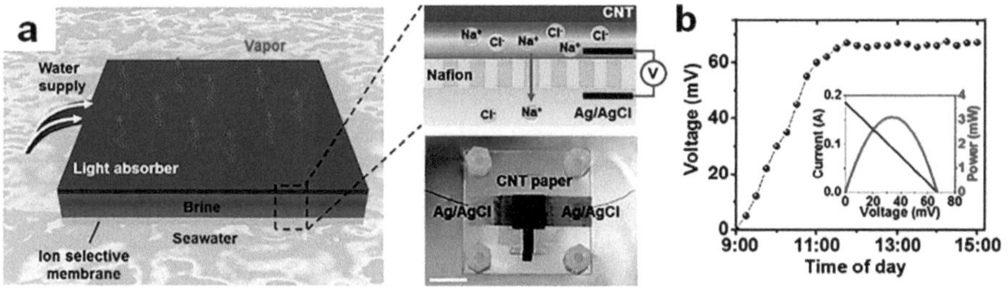

Fig. 6.10 (**a**) Water and electricity co-generation via Salinity Gradient Effect. (**b**) The V_{oc} of device, the inset is a typical IV curve, and the output power. (Reproduced with permission from [158], Copyright 2017 Royal Society of Chemistry)

potential difference across the Nafion membrane. As shown in Fig. 6.10c, this setup achieved an open circuit potential of 66 mV, a steady-state short-circuit current of approximately 180 μA, and an optimized power output of 3 mW during steady-state solar evaporation. Wang et al. introduced a trinity-integrated system for solar evaporation, seawater desalination, power generation, and cultivation, capable of harvesting oceanic osmotic energy through the reverse electrodialysis process [159]. In such device, a Janus evaporator with a hydrophobic top layer was employed to maximize solar absorption for water evaporation, while the hydrophilic beneath layer ensured a continuous water supply. During seawater evaporation, salt concentration linearly increased from 3.50 wt% to 5.49 wt% over 4 h, leading to an increase in maximum power from approximately 0.106 to about 0.25 W/m^2. This innovative method of using solar evaporation to produce high-salinity seawater, combined with sustainable reverse electrodialysis for power generation, has proven both feasible and promising.

6.2.2.4 Temperature Gradient Effect

During solar evaporation, the photothermal effect typically raises the temperature of the evaporation surface above that of the bulk water and the surrounding environment. This temperature difference results in the dissipation of thermal energy through radiation, convection, and conduction [123, 160]. To reduce energy loss, integrating thermoelectric devices into photothermal evaporators can utilize the temperature gradient between the evaporator and bulk water to generate electrical energy [161]. This approach provides the evaporator with dual functionality, improving the energy efficiency of solar evaporation while enabling electricity generation. To date, three main thermal energy harvesting mechanisms have been extensively explored: the thermoelectric effect [162], the thermogalvanic effect, and the pyroelectric effect [163].

A standard thermoelectric generator (TEG) module consists of n-type and p-type thermoelectric materials connected by conductive tabs. When a temperature gradient is applied across its junctions, the TEG utilizes the Seebeck effect to convert thermal energy into electrical energy [164]. Recent studies demonstrated that TEGs were integrated with solar evaporator to transform waste heat into electricity (Fig. 6.11a) [165–167]. Furthermore, thermoelectric modules can also function as insulative materials, reducing heat loss from the absorber surface to the bulk water. Cui et al. developed a donor-acceptor-type small organic molecule solar absorber material (DDPA-PDN) with extensive optical absorption in solid state and superior photothermal performance [168]. By attaching DDPA-PDN-loaded cellulose paper to a thermoelectric module, a multifunctional device was constructed that achieved a water evaporation rate of 0.89 kg m^{-2} h^{-1} under one sun irradiation and generated a voltage of up to 43 mV (Fig. 6.11b). This innovation paves the way for practical outdoor solar desalination and electricity generation.

The thermogalvanic cell operates as a electrochemical cell where two electrodes, connected by an electrolyte, experience different redox reactions and potential

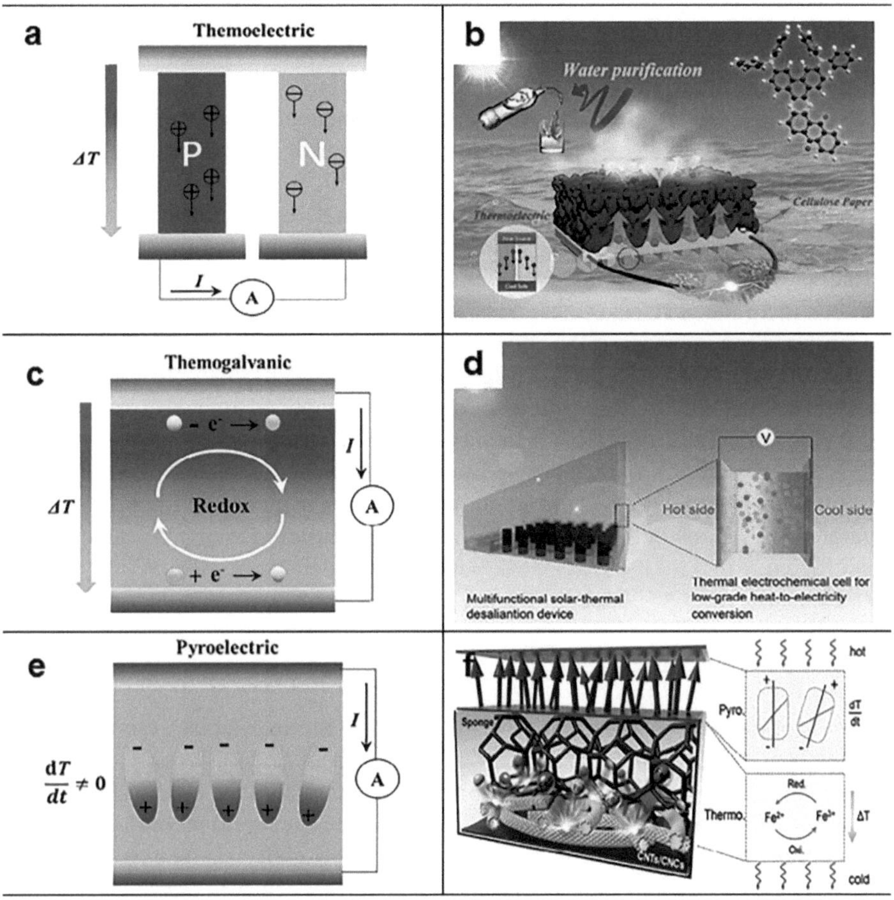

Fig. 6.11 Typical thermal-electric conversion based on (**a** and **b**) thermoelectric effect, (**c** and **d**) thermogalvanic effect, and (**e** and **f**) pyroelectric effect. (Adapted with permission from [76], Copyright 2024 Wiley-VCH)

differences under varying temperatures (Fig. 6.11c) [169, 170]. When connected to an external circuit, these devices deliver current and power, effectively converting thermal energy into electrical energy [171]. Thermogalvanic cells, known for their high Seebeck coefficients and low-cost, flexible, and scalable properties, present a viable alternative to TEGs for harvesting low-grade heat energy. Duan et al. introduced the chaotropic cations and highly soluble amide derivatives into aqueous ferric/ferrocyanide electrolytes, significantly enhancing the Seebeck coefficient and power output of thermogalvanic cells [172, 173]. The prototype module generated a high open circuit voltage of 3.4 V at a minor temperature difference of 18 K. Xu's group designed thermogalvanic cells using cellulose-based porous aerogel, liquid electrolyte, rGO, and carbon nanotube electrodes, integrating the devices into a solar evaporation allows harnessing latent heat from vapor condensation for power generation (Fig. 6.11d) [174]. The low thermal conductivity of cellulose porous aerogel effectively maintains the temperature differences between the electrodes, optimizing thermal energy recycling within the system.

Pyroelectric materials, which are ferroelectric and exhibit spontaneous polarization without any applied electric field, convert waste heat into power energy through their unique thermal response (Fig. 6.11e, f) [163, 175]. When a pyroelectric material is heated, thermal vibrations cause its internal dipoles to lose orientation, decreasing spontaneous polarization and reducing the number of free charges on its surfaces. Under short circuit conditions, this results in an electric current flow between the polar surfaces. Conversely, when the material cools, the dipoles regain their orientation, enhancing the spontaneous polarization and generating a reverse electric current flow [176, 177]. Differing from other cells, Pyroelectric materials can convert temperature fluctuations into pulsed voltage. Jiang et al. created a self-regulating temperature interface system by integrating tungsten-doped VO_2 nanoparticles with PVDF film, facilitating pyroelectric power generation during solar evaporation [178]. Under solar irradiation, the system achieved self-regulating temperature oscillations with a maximum temperature difference of 7 °C, an evaporation efficiency of 71.43%, and a peak output power density of 104 $\mu W/m^2$.

6.2.2.5 Piezoelectric Effects

Solar evaporation piezoelectricity (SEP) integrates the interfacial evaporation and piezoelectric effects, positioning itself at the forefront of solar energy conversion [123, 179]. This technology has shown immense potential in the clean energy sector by harnessing solar energy. The process begins when the liquid medium undergoes evaporation. The vapor exerts pressure on the piezoelectric material, causing mechanical deformation. This deformation, via the piezoelectric effect, is converted into electrical energy [151, 180]. This is conversion mechanism involves key components such as a solar interfacial evaporator, a liquid medium, piezoelectric materials, and necessary electrodes and circuits. As vapor pressure persists, the piezoelectric material undergoes additional mechanical deformation, enhancing energy conversion [181]. The generation of an electric potential difference from the charge separation within the piezoelectric material enables the production of electricity, which can then be utilized to power external devices or for the storage of electrical energy.

SEP technology offers numerous advantages, particularly in its operational efficacy under varying environmental conditions, irrespective of light intensity or angle. This feature underscores the technology's adaptability and robustness in diverse settings. Moreover, the approach is sustainable and eco-friendly, presenting a significant advantage over traditional methods that rely on fossil fuels. The technology to convert pressure from vapor into electricity has been explored by researchers like Zhu et al., who have applied it for electricity generation from waste heat or vapor pressure. The integration of a specially designed honeycomb carbon sponge have paved the way for the combination of thermoelectric and piezoelectric modalities within the SEP framework [182]. Wang et al. have made significant

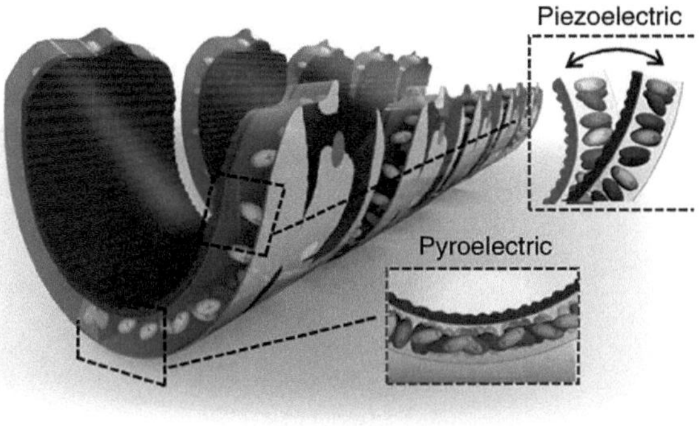

Fig. 6.12 A bimorph actuator based on thermo-mechano-electrical conversion. (Reproduced with permission from [181], Copyright 2018 Nature Publishing Group)

advancements in developing soft-matter devices aimed at recovering and utilizing low-grade waste heat (Fig. 6.12) [181]. These devices autonomously and automatically adjust to temperature fluctuations and mechanical movements, aiding in the creation of flexible, self-powered systems. Furthermore, Meng et al. fabricated thin piezoelectric films that successfully incorporated into solar evaporation systems [183]. These systems not only generate electrical energy but also enhance water evaporation in adjacent areas, exhibiting a remarkable efficiency increase of over 20% compared to other non-piezoelectric evaporators and demonstrating durability through 3000 operational cycles. Despite challenges such as material optimization, interface design, and efficiency enhancement, ongoing scientific and technological progress is expected to enhance the practical application of these electricity generation techniques, leading to breakthroughs in solar energy conversion.

6.2.2.6 Triboelectric Effects

Triboelectric effects in droplet-based systems have shown potential in the field of energy conversion [144, 184]. A notable advancement is the development of a triboelectric nanogenerator (TENG), which leverages the combined effects of triboelectrification and electrostatic induction. This innovative device effectively converts low-frequency mechanical energy into electrical energy, featuring high power density, cost-effectiveness, and environmental sustainability [185]. An interesting application is the development of a cost-effective 3D plasmonic solar absorber gel. This gel combines optimal optical, thermal, and wetting properties to enable efficient solar vaporization. Its high performance is attributed to broadband absorption and strong near-field intertip enhancement from sparsely distributed gold nanoflowers, which enhance light-to-heat conversion [184]. The thermal insulation properties of the silica gel are critical in retaining and directing plasmonic heat to the water

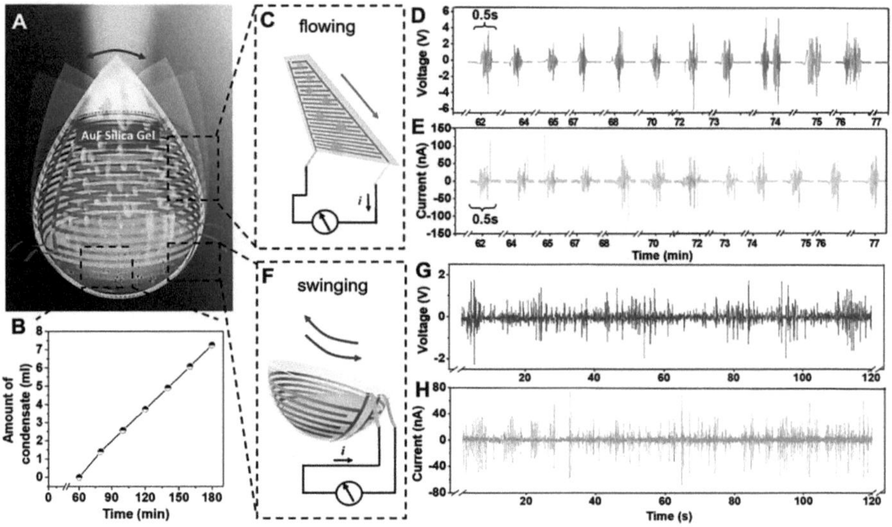

Fig. 6.13 The hybrid device for fresh water and triboelectricity. (Reproduced with permission from [184], Copyright 2018 Wiley-VCH)

pathways within its porous structure, achieving an 85% vaporization efficiency under standard solar irradiation. The gel's structure provides high mechanical stability and antifouling properties, making it suitable for evaporating polluted or turbid water. Additionally, it enables the harvesting of water condensation-induced triboelectricity, allowing for the simultaneous production of clean water and power (Fig. 6.13).

The integration of TENG into wave-driven systems presents another fascinating application. By embedding TENGs into the sides of an evaporator designed to float in water bodies such as seas or lakes, it becomes possible to capture and convert the mechanical energy of water waves into electricity. This approach not only aids in fresh water production but also enhances the utility of the device by enabling self-powered water wave detection. A notable implementation of this concept is seen in development of a 3D tumbler-shaped evaporator [186]. This generator is equipped with self-righting properties and TENGs on its sides to harvest wave energy. Demonstrating an impressive water evaporation rate and responsive output currents across various wave frequencies, this design introduces a novel approach to engineering self-powered water wave detection devices. Although the electrical output power from these systems may be limited, they hold significant potential for applications in detection sensors and wearable electronics.

The hybrid power system utilizing TENG technology represents a stride towards optimizing energy utilization from multiple sources. The proposed generator, which combines a water-wave TENG with an evaporation generator, exemplifies a strategic approach to enhancing the overall efficiency and energy output [187]. This hybrid system addresses the challenge of low energy output from wave TENGs in calm water conditions by optimizing the peak power output and integrating a capacitor that can be rapidly charged by TENG in actual water waves. Additionally, the evaporation generator is based on a wood block, consistently generates direct currents during the natural process of water evaporation. Achieving commendable solar-vapor conversion efficiency, this generator also delivers high current and power density under specific conditions. Its performance is further improved by using a multi-stage boost circuit with pulse charging, which significantly increases voltage output and reduces charging time. This hybrid approach not only demonstrates the capability of TENGs to collect energy from both water waves and steam evaporation efficiently but also broadens the scope of potential applications for triboelectric energy harvesting technologies.

6.2.3 Energy Storage

The integration of heat storage into desalination systems becomes increasingly significant due to the intermittent nature of solar irradiation. These systems enable continuous operation regardless of weather conditions (Fig. 6.14a) [188–190]. One effective method of storing energy is through the evaporation of a working fluid, which allows the energy to be captured as latent heat. This stored energy can then be released on demand to either generate electricity or provide thermal energy. The clean water production is dependent on the effective utilization or storage of solar energy [191, 192]. In this context, phase change materials (PCMs) have been proposed to enhance the process of solar evaporation [193–195]. An innovation is the PCM-integrated evaporative system as demonstrated by Gong et al. [196]. This system utilizes a network of nano/microchannels to capture solar energy, which is then used to heat water. The surplus heat is transferred to the PCM, allowing for effective energy management and enhanced evaporation rates. Such system has shown a substantial increase in clean water production, and exhibited excellent long-term resistance to salt, further proving its efficacy and sustainability. Further advancements in PCM technology include the use of organic PCMs to maintain continuous solar vapor generation throughout the day. Innovations such as polyimide/MXene hybrid aerogels and polyethylene glycol were specifically tailored for solar evaporation, showing high evaporation rates and stability under solar exposure [197]. Another notable development is the combination of PCM octadecane with carbonized ultralong PPy nanotube aerogels, which together with a photothermal conversion layer, achieves high solar absorption and efficient all-weather evaporation [198].

Domestic heating in buildings has rapidly advanced with the widespread adoption of solar-thermal technologies. Solar-thermal water heaters with vacuum tube technology are commonly installed on rooftops to produce hot water. Phase change materials (PCMs) like paraffin wax and hydrate salts are integrated into building materials and installed in roofs and walls to passively regulate indoor temperatures [189]. High-temperature molten salts offer greater thermal storage capacity than water and low-temperature PCMs, allowing them to provide more robust heating

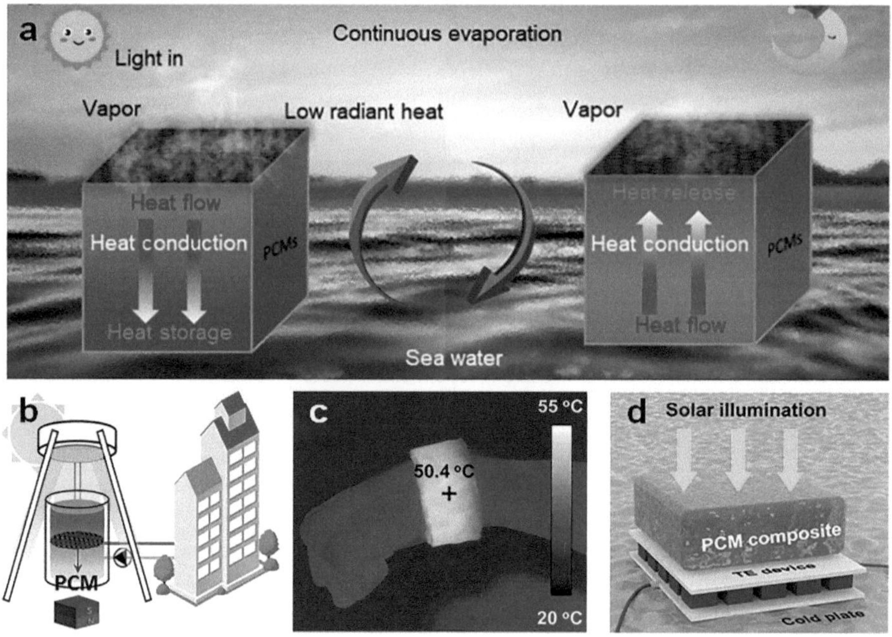

Fig. 6.14 (**a**) A heat storage/release process for sustainable evaporation. (Reproduced with permission from [190], Copyright 2024 Elsevier). (**b**) Molten salts charging by magnetically movable meshes. (Reproduced with permission from [199], Copyright 2019 Royal Society of Chemistry). (**c**) Human thermal managemen on a human wrist. (Reproduced with permission from [204], Copyright 2020 Royal Society of Chemistry). (**d**) Power generation through connecting metal foam-thermal composites with a thermoelectric generator. (Reproduced with permission from [209], Copyright 2020 Elsevier)

suitable for district heating systems, not just individual buildings. Deng et al. demonstrated that magnetically driven directional movement of solar thermal meshes can significantly enhance solar thermal energy storage in molten salts [199, 200]. This mobile charging system can be combined with heat exchangers, retaining its fast charge/discharge advantages to extract large-capacity, high-temperature heat for residential heating (Fig. 6.14b). The output temperature of the heat transfer oil can be regulated by adjusting flow rates. Connecting multiple storage units ensures a stable output temperature within a narrow range. The system can be powered by a cost-effective beam-down solar concentrator, avoiding the extensive infrastructure and capital investments usually required for concentrated power plants.

Beyond domestic and industrial heating, developing flexible, leakage-proof PCM composites can expand the applications of solar-thermal technologies [201–203]. In a proof-of-concept demonstration, Chang et al. showed that flexible paraffin wax composites encased in rGO-modified polyurethane sponges could be used as wearable heat packs for thermotherapy [204]. Unlike conventional heat packs that rely on electrical controllers or chemical reactions, these PCM heat packs can be solar-charged and consistently release substantial heat within a comfortable temperature range dictated by the solid-liquid phase transition [205]. The charged PCM roll can contain melting liquids and be applied to human arms, providing warmth without restricting movement.When used, the surface temperature of the wrist immediately rose to 50 °C and maintained a therapeutic temperature of approximately 40 °C for 15 min. Recently, they advanced heat packs using supercooled sugar alcohol PCM composites, enabling room-temperature, non-leakage storage and controlled heat release (Fig. 6.14c) [206]. These PCM composites can be reused as heat packs to relieve body pain and fatigue and potentially integrated into textiles for personal thermal management.

Thermal storage provides a distinct method for harnessing solar energy. On an industrial level, the heat retained in high-temperature molten salts can drive steam turbines to sustain electricity generation [207]. Unlike solar power plants, thermoelectric conversion provides a simple way to generate electricity from temperature differences [208]. Due to their high latent heat storage capacity and stable heat release temperatures, charged PCM composites can be attached to the hot side of a thermoelectric generator to maintain a consistent temperature differential. For example, carbon-modified copper foam can capture solar photons and retain melted PCMs, allowing for efficient direct solar thermal energy storage and effective release of stored energy across interconnected networks [209]. The output voltage can be regulated by choosing PCMs with different phase change temperatures, which affect the temperature difference between the hot and cold sides. Medium-temperature sugar alcohol composites can discharge heat to produce an output voltage of around 3 V, enough to power electronic devices like LEDs and electric fans (Fig. 6.14d). Additionally, when the solid-liquid phase change temperature is close to room temperature, PCM composites impregnated in foams with high thermal effusivity are reported to enable the thermoelectric conversion of ambient temperature fluctuations into electricity [210]. In sum, energy storage systems benefit solar desalination by balancing supply-demand mismatches and recovering waste heat, but their practical application requires more rigorous efficiency evaluations.

6.2.4 Photothermal Motor

Photothermal actuators and robots are created by embedding plasmonic particles, organic photothermal agents and carbon nanomaterials into various matrices, including shape-memory polymers, hydrogels, elastomers, liquid crystals, and biopolymers [58, 211]. Upon exposure to light, these materials undergo a photothermal conversion, leading to mechanical deformations that encompass bending, twisting, rotating, and jumping actions. This transformative capability enables the devices to function in a variety of applications, serving as grippers, mills, swimmers, and syringes [212–215]. The incorporated light-absorbing materials not only capture light but also act as thermally responsive elements within the actuators. To enhance the performance, several critical parameters are meticulously considered during the design phase. These parameters encompass generated stress and strain, Young's modulus, durability, response time, and energy efficiency [216]. For example, incorporating two types of gold nanocrystals—nanospheres and nanorods—into thermoplastic polyurethane shape-memory polymer films illustrates this approach [213]. These films are specifically engineered to respond at a transition temperature of 55 °C, with the nanocrystal sizes tailored to resonate with light-emitting diode wavelengths of 530 nm and 860 nm. This selective wavelength responsiveness facilitates controlled, sequential activation of the devices.

Advancements extend to the development of desynchronized liquid crystalline network (LCN) actuators, which are capable of reversing their deformation [217]. These actuators are designed to trigger deformation at different temperatures on each side by employing asymmetrical cross-linking and stretching techniques. Additionally, layers of polydopamine act as photothermal agents on these cross-linked LCNs, allowing them to bend and straighten in response to light, emulating natural biological movements. A more significant innovation is the creation of an anisotropic hydrogel that mimics the peristaltic crawling of an earthworm within a glass capillary [218]. This hydrogel incorporates gold nanoparticles for efficient photothermal conversion, a thermoresponsive polymer network to alter electrical permittivity, and magnetically aligned titanate nanosheets to synchronize changes in anisotropic electrostatic repulsion. This design enables reversible expansion and contraction of the hydrogel, effectively reducing friction with glass during movement. Plasmonic motors use localized heating from nanostructures upon light absorption. The mechanisms for motion generation include: [219, 220] (1) Thermally induced variations in surface tension along fluid interfaces, known as the Marangoni effect, drive microscale or nanoscale devices through specially designed thermoplasmonic structures [221, 222]. (2) Photons exhibit mechanical properties, applying momentum transfer or optical forces on objects through scattering or absorption mechanisms [223]. (3) Nanomotors containing plasmonic photocatalysts convert energy from light and chemical fuels into mechanical motion [224, 225]. (4) Plasmonic heating can also induce phase transitions, such as liquid to gas, creating bubbles whose rapid expansion or collapse generates propulsive forces [226–229]. A typical design includes Janus SiO_2 nanomotors with a 10 nm gold layer sputtered on one hemisphere. A near-infrared laser heats this gold segment locally, creating a thermal gradient that induces opto-thermophoretic propulsion opposite to the Au-coated side, reaching speeds up to 950 body lengths per second at a laser intensity of 703 kW/m^2 [230]. Photon momentum generates rotational forces [231–234]. By changing the light wavelength to excite different plasmonic modes, these motors can rotate silica microdisks that are larger in volume by four orders of magnitude [223]. Simultaneously, photocatalytic activities on the particle surfaces may generate tiny gas bubbles; the directed expulsion of these bubbles from the particle adheres to momentum conservation principles, propelling the particle in the opposite direction [224]. Moreover, photothermal activation has enabled nanobubble generation by using wavelength-specific illumination aimed at the localized surface plasmon resonance peaks of particles, utilizing both continuous and pulsed light

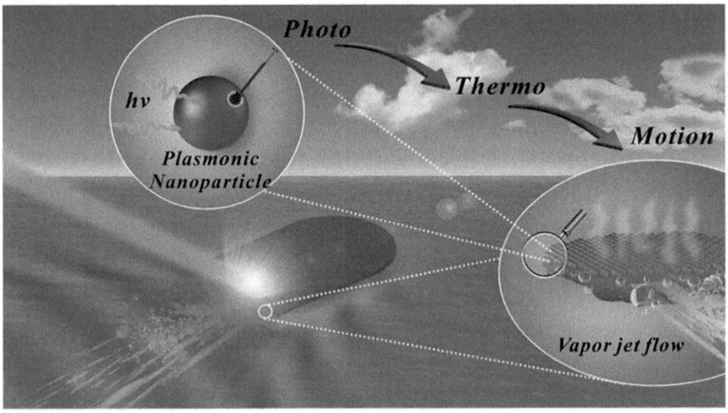

Fig. 6.15 Plasmonic motor at liquid/air interface. (Reproduced with permission from [226], Copyright 2017 American Chemical Society)

[235–237]. This approach allows small boat-like structures with plasmon-active surfaces to move across liquids, propelled by the expulsion of hot vapor jets, providing innovative methods for controlling the movement of photothermal motors (Fig. 6.15) [226]. The key advantage of thermoplasmonic motors is their ability to operate wirelessly with light, facilitating remote activation and control [220].

The control mechanisms for photothermal motors extend to manipulating the wavelength, intensity, and polarization of the light sources [58, 238]. This control method allows for non-contact manipulation, leading to faster response times, improved energy conversion efficiency, and enhanced mechanical durability. The ability to precisely control at micro- and nanoscale makes these systems well-suited for a wide range of applications, such as targeted drug delivery, microfluidics, sensing, and the assembly of nanostructures [239]. Advancements in manufacturing methods, such as electron beam lithography and colloidal synthesis, are crucial for creating the complex nanostructures needed by these motors [240, 241]. These breakthroughs mark significant strides in the development of smart autonomous systems and underscore the vast potential of photothermal technology in scientific and industrial applications.

6.2.5 Fuel Production

Solar evaporation can be used to concentrate solar energy to drive chemical reactions that produces hydrogen or other biofuels, offering a pathway to sustainable energy production.

6.2.5.1 Water-Electricity-Hydrogen Cogeneration

Water, energy, and chemical fuels are vital for sustainable development and economic growth [242, 243]. As the global focus on carbon neutrality intensifies, industries are adopting policies to cut carbon emissions. Water shortages, fossil fuel depletion, and energy demands have worsened greenhouse effects and caused ecological harm [76, 244]. In support of carbon neutrality, a range of sustainable, carbon-free technologies has been developed, including solar desalination, photocatalysis, and solar brine treatment [72, 144, 245]. Integrating the energy-water-fuel balance system with a clean water-green electricity-hydrogen energy system through solar water evaporation can greatly improve the efficiency of solar interfacial evaporation [150, 246–248]. The system operates simply: it generates electricity through the thermoelectric effect during evaporation and uses it to electrolyze clean water, producing hydrogen.

For instance, a combined solar interfacial evaporation with photovoltaic electrolysis can be used to simultaneously produce clean water and hydrogen. Lu et al. designed an electrocatalyst of perovskite oxide and 2D MXene on nickel foam for a system combining interfacial solar evaporation with photovoltaic-assisted water splitting [249, 250]. In this system, some solar irradiation is used for photovoltaic electrolysis to produce hydrogen, while the rest is harnessed for solar vapor generation. This results in a hydrogen production rate of 150 mL m^{-2} h^{-1} under natural conditions with an average solar flux of 0.8 kW/m^2 and a water evaporation rate of 1.41 kg m^{-2} h^{-1} under one sun. Ho's team has developed a solar-driven hybrid system that enables both water-electricity coproduction and water-electricity-hydrogen cogeneration by managing uneven water distribution and evaporation (Fig. 6.16) [243]. The system employs a specially designed hydrogel to enhance evaporative cooling and solar absorption heating, alongside a Bi_2Te_3-based thermoelectric generator to efficiently convert waste heat into electrical energy and drive electrochemical water splitting for hydrogen production, regardless of sunny or cloudy conditions. This integrated approach achieved a water evaporation rate of 1.42 kg m^{-2} h^{-1}, a

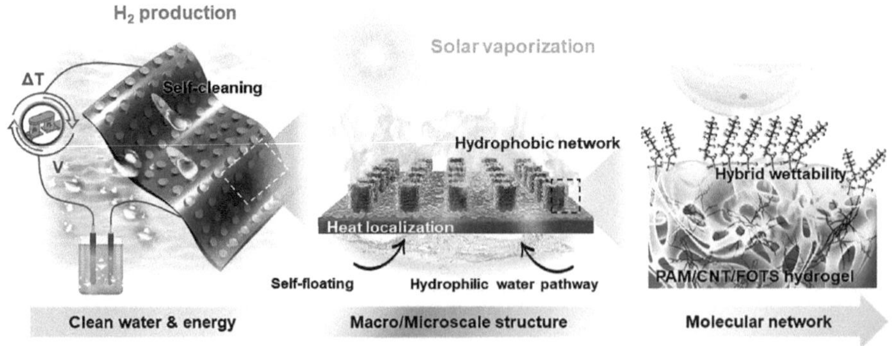

Fig. 6.16 Wter-electricity-hydrogen co-roduction. (Reproduced with permission from [243] Copyright 2020, Elsevier)

power output of 4.8 W/m^2, and hydrogen production of 0.3 mmol/h. This innovative design not only advances the development of highly integrated hybrid systems but also holds significant potential for providing the vital resources of water, energy, and fuel crucial for sustainable development.

6.2.5.2 Steam-Hydrogen Coupling System

Vapor, a key product of solar evaporation, is an excellent reactant for hydrogen production [144, 242]. Enhancements in photocatalytic hydrogen generation can be achieved by the water vapor splitting. Photolysis, electrolysis, and photochemical water splitting involve two half-reactions: oxygen evolution at the anode and hydrogen evolution at the cathode. These techniques have been widely studied for hydrogen production. However, there has been limited research on enhancing the efficiency of these processes, especially using seawater, by integrating solar steam and catalytic technologies. The interdisciplinary of photothermal-photocatalytic coupling encompasses the creation of novel evaporators and photocatalysts, alongside comprehending their synergistic interactions. This process involves incorporating photocatalysts into structural evaporators, employing photoabsorbers to capture solar energy for water evaporation, and selectively converting the resulting water vapor into hydrogen [251].

For example, Guo et al. developed an integrated system that combines photothermal and photocatalytic technologies, employing carbonized wood-coated photocatalysts to convert water vapor into hydrogen [249, 252]. This dual-phase interface system, which reduces both catalyst corrosion and hydrogen transport resistance, significantly surpasses traditional three-phase systems, achieving a hydrogen production rate of 220.74 μmol cm^{-2} h^{-1} and a water evaporation rate of 0.748 kg m^{-2} h^{-1}. By combining solar absorbers with photocatalysis, Gao et al. created solar thermal collector nanocomposites (SiO_2/Ag@TiO_2 core-shell) with effective photothermal properties, enhancing targeted interfacial phase transitions that improve seawater catalysis and desalination [253]. Plasmonic nanoparticles generate localized heating, promoting surface-dominated catalysis and steam production with minimal heat loss and reduced thermal mass. These durable nanocomposites efficiently convert seawater into hydrogen and fresh water under sunlight. Additionally, a photothermal catalytic absorber using TiO_2/Ag nanofiber-based photocatalysts combined with hydrophilic chitosan polymers was demonstrated, enabling simultaneous water purification and hydrogen production [254]. Ag nanoparticles extend the solar absorption spectrum of TiO_2 via localized surface plasmon resonance, thereby enhancing charge transfer and hydrogen production. However, the hydrogen production rate may be limited by the trapping of hydrogen bubbles within the gel, which obstructs the active sites of the photocatalyst.

Recently, Lee et al. developed a floatable nanocomposite consisting of porous elastomer-hydrogel composites embedded with Cu/TiO_2 photocatalysts for solar hydrogen production (Fig. 6.17). This design keeps the photocatalyst above the water surface, enhancing sunlight absorption and hydrogen separation, and

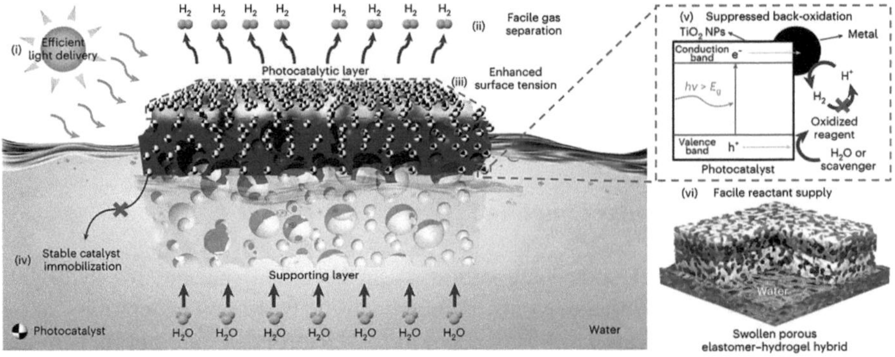

Fig. 6.17 A floatable nanocomposite for water splitting. (Adapted with permission from [256], Copyright 2023 Nature)

achieving a hydrogen production rate of 163 mmol m^{-2} h^{-1}. This system demonstrates significant potential for large-scale hydrogen production [247]. Reisner et al. introduced an integrated device that combines photocatalysts on a carbon-based solar evaporator for both water purification and water vapor splitting [255]. The device consistently supplies vapor to the photocatalysts at a rate of 0.95 kg m^{-2} h^{-1}, resulting in a hydrogen production rate of 19.8 mmol m^{-2} h^{-1}, with the remaining vapor being condensed and collected. The hydrophobic-treated evaporator floats on water, isolating the photocatalysts from contaminants and preventing their dissolution. This system has demonstrated stability in seawater and various aqueous waste streams for over 154 h. These findings highlight the potential of combining solar evaporation with photocatalytic and photoelectrochemical water splitting as effective solutions for off-grid or small-scale desalination and energy harvesting from seawater and other saline sources.

6.2.5.3 Other Chemical Fuels

Solar evaporation has been used for the production of various chemical fuels including ethanol, methane, and a range of hydrocarbon fuels [1, 257, 258]. Aiming to satisfy energy needs, particularly at off-grid locations, Halas et al. showcased a method for solar-driven ethanol production from cellulosic feedstock as illustrated in Fig. 6.18a [259]. This approach produces high-yield steam by irradiating light-absorbing nanoparticles. The steam generated from solar energy is then used to hydrolyze feedstock into sugars and purify ethanol during the final processing stage. In a similar vein, Grimes and his team showcased the conversion of carbon dioxide and water vapor into methane and other hydrocarbons using nitrogen-doped titania nanotube arrays [257]. These titania arrays, with a modified bandgap, capture and utilize the visible spectrum of sunlight, while dispersed cocatalysts facilitate the redox reactions of the reactants. A hydrocarbon rate of 111 ppm cm^2 per hour, or approximately 160 μL g^{-1} h^{-1} under standard solar conditions, was achieved. These

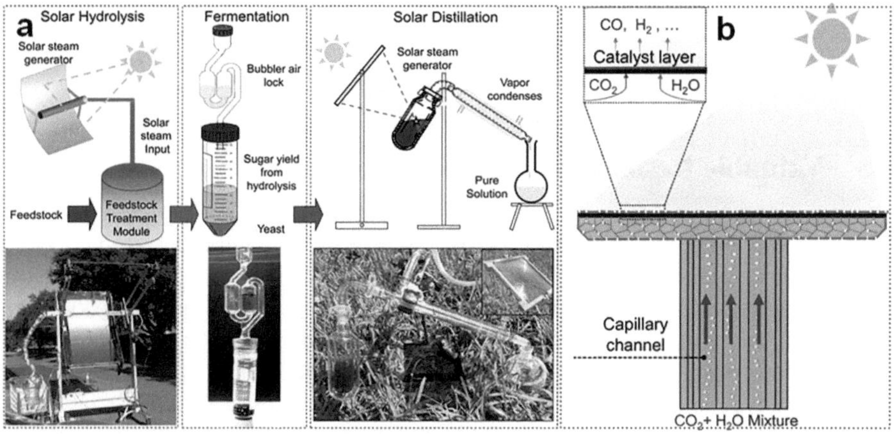

Fig. 6.18 (**a**) solar biofuel process. (Reproduced with permission from [259], Copyright 2017 American Chemical Society). (**b**) A CO_2 photo-thermal reduction system. (Reproduced with permission from [260], Copyright 2025 Elsevier)

innovations not only provide alternative methods for fuel production but also pave the way for carbon recycling using renewable sources.

Xuan et al. recently introduced a lotus-inspired triphase continuous photothermal reduction system to improve CO_2 conversion, as illustrated in Fig. 6.18b [260]. This innovative system goes beyond traditional liquid-solid and gas-solid reactions by leveraging capillarity and evaporation to optimize the delivery paths of reaction gases and liquids, thereby adjusting their supply ratios near the catalyst. The researchers used Bi and Ti co-doped g-C_3N_4 as the catalyst and systematically assessed its performance across three reaction types: those with sacrificial agents, those without sacrificial agents, and coupled oxidation-reduction reactions. The findings reveal that the system significantly boosts the efficiency of all tested reactions, automatically maintaining an optimal reactant supply ratio near the catalyst and allowing for easy replenishment of the auxiliary solution. In the coupled reaction system, the highest triphase efficiency reached was 0.0458%, nearly ten-fold higher than that of the liquid-solid phase reaction. This innovative system not only enhances the catalytic activity across various reaction conditions but also supports the potential for large-scale application of photocatalysis in the future.

Enhanced by steam production, solar fuel generation presents a more sustainable way to meet energy demands. Various photochemical processes have been shown to effectively use natural sunlight for producing commodity chemicals such as terebic acid, benzyl bromide, and benzaldehyde [261, 262]. Unlike the challenges faced by solar fuel generation, where conversion efficiencies are too low for industrial investments, the production of fine chemicals on a small scale and in decentralized settings effectively utilizes natural sunlight. This is especially pertinent in industries like flavor, fragrance, and vitamins [1, 263]. Compared to traditional methods, solar-driven production of fuels and chemicals, enhanced by high temperatures and steam in solar evaporation [20], represents a cost-effective and sustainable alternative.

6.3 Valuable Resources

As the global energy landscape shifts towards sustainability, there is an increasing demand for vital minerals such as lithium and uranium. However, the uneven distribution of these minerals across the globe presents significant supply risks, with certain countries struggling to meet their domestic needs [264, 265]. In contrast, the ocean harbors vast, untapped reserves of these minerals, which could be extracted through innovative techniques. Solar evaporation plays a pivotal role in addressing the mineral harvesting and water scarcity by providing freshwater for agricultural use [266, 267]. This is valuable in coastal areas where resources are limited, and traditional desalination methods are costly and energy-intensive. By integrating solar desalination with agricultural practices, these systems not only make water desalination more cost-effective but also enable soil remediation by washing saline or contaminated soils, thereby increasing the availability of arable land. This section explores the potential applications of solar evaporation in two critical areas: mineral harvesting and sea agriculture for food production.

6.3.1 Salt Harvesting

A critical byproduct of solar evaporation is the progressive concentration of dissolved salts, which presents both challenges and opportunities. While salt accumulation on evaporative surfaces can impede light absorption and reduce evaporation rates, recent advancements in system design have transformed this phenomenon into a sustainable strategy for salt harvesting and resource recovery [1, 267, 268]. This dual functionality addresses environmental concerns related to brine discharge while creating value-added products for industrial applications ranging from food processing to chemical manufacturing. A key innovation in salt management is the implementation of zero liquid discharge (ZLD), where salt crystallization is spatially separated from active evaporation zones through deliberate structural designs [59, 269–271]. Wang et al. demonstrated this principle using a cup-shaped photothermal evaporator that achieved continuous ZLD desalination of high-salinity brines (5–15 wt%) [272, 273]. By leveraging humidity gradients, salt precipitation preferentially occurred on the outer evaporator surface, preserving the inner light-absorbing region's functionality [274]. This configuration maintained stable evaporation rates (no significant decline over 72 h) despite salt crust formation due to the loose packing of surface deposits. They further advanced this concept using a 3D evaporator with thermally conductive aluminum partitions, spatially isolating light

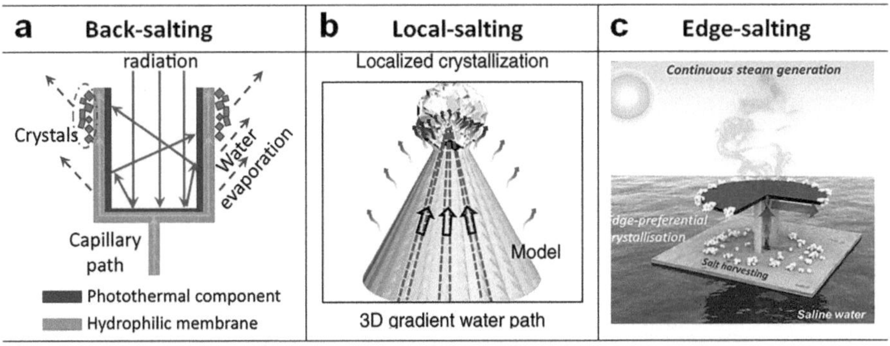

Fig. 6.19 Salt harvesting (**a**) solar crystallizer by back-salting. (Reproduced with permission from [272], Copyright 2021 Nature). (**b**) Solar-driven localized crystallization. (Reproduced with permission from [277], Copyright 2020 Nature). (**c**) The salt crystalizes at the evaporator edge. (Reproduced with permission [289], Copyright 2019 Royal Society of Chemistry)

absorption (inner surfaces) from evaporation/salt deposition zones (outer surfaces) (Fig. 6.19a), attaining 1.61 kg m^{-2} h^{-1}vapor rates in 24 wt% NaCl brine [272].

To enhance long-term stability in hypersaline environments (>20 wt%), researchers developed antifouling absorbers with physical separation mechanisms [275, 276]. Zhu's group created a biomimetic nanostructured absorber inspired by water lily surfaces, featuring hydrophobic upper layers that repel brine while allowing subsurface salt crystallization [275]. This design achieved 80% solar-to-vapor efficiency in 30 wt% heavy metal solutions, simultaneously recovering valuable salts like $CuSO_4·5H_2O$ and $HAuCl_4·4H_2O$. To sustain high evaporation rates in high-concentration brine, they further introduced a nanostructured solar absorber that supports salt crystallization beneath the absorber surface [276]. This absorber is designed with a hydrophobic surface to prevent water ingress and promote salt deposition beneath it. Its nanoscale light-trapping and controlled water supply reduce losses in light absorption and heat conduction. As a result, the absorber attains around 80% solar-to-vapor efficiency and maintains long-term stability in high-concentration brine (10 wt%) and wastewater with heavy metals (30 wt%). The system effectively separates water from heavy-metal salts, enabling the recovery of the salts. Another strategy to maintain high cvaporation rates is to localize salt crystallization to specific areas [59, 269, 277, 278]. For instance, Song et al. designed a 3D conical absorber with asymmetric capillary ratchets mimicking avian beak fluid transport [277]. The gradient microcavity arrays created position-dependent water film thickness and thermal profiles, concentrating crystallization at the apex (Fig. 6.19b). This achieved record evaporation rates of 2.63 kg m^{-2} h^{-1} (96% efficiency) in 25 wt% NaCl while enabling simple mechanical salt harvesting.

Further advancements in directional salt crystallization control were achieved through hydraulic management [279–283]. The emerging systems now synergize water production with mineral extraction [284–288]. Zhang's horizontal evaporation dish with vertical uptake threads demonstrated simultaneous freshwater generation and NaCl recovery (0.42 kg m^{-2} day^{-1}) from 10 wt% brine over 600 h (Fig. 6.19c) [289]. Such paper-based evaporators exploiting lateral concentration gradients have extended this capability to $MgCl_2$ and heavy metal recovery from industrial wastewater, outperforming traditional solar farms in both productivity and process simplicity. Besides, Xu et al. engineered an inclined evaporator with gravity-assisted flow patterns, directing salt deposition to vertical surfaces while maintaining evaporation performance on the angled light-absorbing region [290]. Wu et al. designed an asymmetric evaporator that establishes varying water supply and loss ratios across its surface, leading to directional salt concentration gradients [284]. This mechanism causes solid salt to precipitate and gather on the side opposite the seawater supply center, simplifying the collection process. By integrating positively charged polyacrylamide gel, the movement of cations is slowed, creating a significant potential difference between areas close to and far from the seawater supply center, facilitating continuous electrical energy production. As a result, this asymmetric evaporation system simultaneously derives freshwater, salt, and electrical energy from seawater. These technological innovations not only tackle the historical challenge of salt fouling but also transform solar evaporators into multifunctional platforms for sustainable resource management.

6.3.2 Lithium Extraction

The growing demand for lithium-ion batteries in renewable energy storage has intensified efforts to develop sustainable lithium extraction technologies. Solar evaporation systems offer a promising route to concentrate lithium-rich brines while simultaneously addressing energy and environmental constraints of conventional evaporation ponds [249, 266, 291]. This section examines recent advancements and challenges in selective lithium recovery using solar absorbers, focusing on crystallization control, ion selectivity, and system scalability. Initial attempts to integrate lithium extraction with solar evaporation systems revealed fundamental thermodynamic limitations [292–296]. Santoro et al. demonstrated a tungsten disulfide (WS_2) nanofiller-based photothermal membrane capable of inducing heterogeneous nucleation of lithium chloride (LiCl) from supersaturated brines (18 M LiCl) [297]. While interfacial solar evaporation promoted LiCl crystallization, the system exhibited an exceptionally low evaporation rate (0.067 kg m^{-2} h^{-1} under one sun) due to two synergistic factors: (1) strong lithium hydration energy ($\Delta H_{hyd} = -522$ kJ/mol) that traps water molecules via Li^+-H_2O coordination, reducing vapor pressure, and (2) optical interference from densely packed LiCl crystals (reflectivity >80%) that obstructed light absorption and vapor diffusion pathways. These findings highlight the inherent conflict between lithium recovery efficiency and evaporation performance in concentrated brines, necessitating alternative strategies to decouple crystallization from photon management.

Real-world lithium extraction faces the challenge of isolating Li^+ from complex ionic matrices (e.g., Na^+, K^+, Mg^{2+}) in natural brines and seawater [298–301]. Chen et al. addressed this through a cellulose fiber-based absorber exploiting differential salt solubility gradients [302]. The twisted fiber architecture (height/radius ratio > 15:1) created vertically stratified crystallization zones: sodium salts (NaCl, solubility ~357 g/L) precipitated at lower regions, while lithium salts (LiCl, solubility ~832 g/L) migrated upward due to slower supersaturation kinetics. After 60 h operation, the Li/Na weight ratio increased from 2% to 77% in brine and from 0.0015% to 1.02% in seawater. Radial separation further enhanced selectivity via advection-dominated transport, forming a Na-rich peripheral shell (thickness ~ 300 μm) removable by mechanical stripping. The system achieved record evaporation rates (9.8 kg m^{-2} h^{-1} under wind-assisted 1 sun), theoretically reducing traditional pond evaporation times from 18 months to <20 days. Limitations in single-stage selectivity (Li purity <80%) could be mitigated through multistage cascading or ion-selective membrane integration.

Recently, bio-inspired systems have been designed to combine lithium harvesting with in situ storage. Song et al. developed a solar transpiration–powered lithium extraction and storage device featuring: (i) A hierarchical evaporator generating 2.3 kPa transpiration pressure gradients. Lithium-selective membranes (e.g., $LiMn_2O_4$-coated PVDF) with >90% Li^+/Na^+ selectivity. (ii) Vascular hydrogel reservoirs for continuous Li^+ accumulation (capacity: 1.2 mmol/g per cycle). (iii) Long-term testing (1000 h) under natural sunlight demonstrated stable lithium recovery from hypersaline brines (Li^+: 0.8 g/L) with <5% efficiency decay. The system's modular design enabled scalability from lab-scale (0.1 m^2) to prototype arrays (5 m^2), achieving lithium yields equivalent to 8.7 kg Li_2CO_3 per hectare daily - a 40-fold improvement over conventional solar ponds (Fig. 6.20) [303]. This biomimetic approach not only bypasses energy-intensive post-processing but also enables direct integration with battery manufacturing supply chains.

Although the current solar-driven lithium extraction systems show promise in addressing the water-energy-resource nexus, key challenges remain: (1) optimizing photon-to-lithium conversion efficiency in multicomponent brines, (2) developing

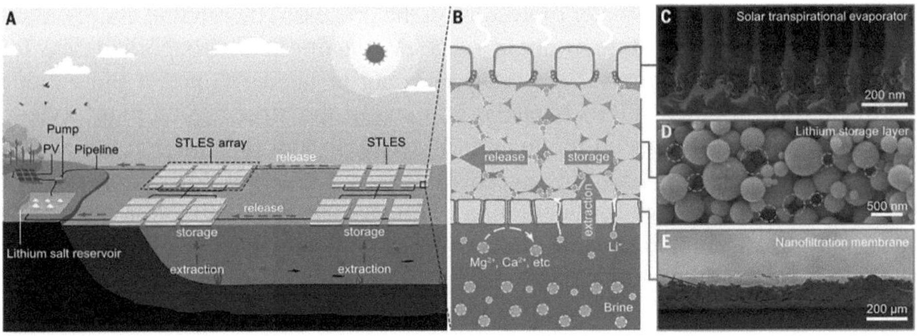

Fig. 6.20 Solar transpiration-enabled lithium extraction. (Reproduced with permission from [303], Copyright 2024 Science AAAS)

antifouling membranes with molecular-scale ion selectivity, and (3) scaling systems for large-scale production demands [300, 304, 305]. The convergence of photothermal nanomaterials, crystallization engineering, and circular design principles positions solar-driven lithium recovery as a cornerstone technology for sustainable mineral harvesting.

6.3.3 Uranium Extraction

Uranium, a critical resource for nuclear energy, is abundantly present in seawater, albeit in ultralow concentrations (approximately 3 ppb) [144, 249, 266]. The extraction of uranium from seawater has long been seen as a promising approach to supplement terrestrial mining, providing a vast, renewable supply of this resource [306–308]. However, the process faces significant challenges due to the complex ionic composition of seawater, which is dominated by sodium (Na^{+}), calcium (Ca^{2+}), and various competing metal ions. Conventional extraction methods rely on energy-intensive processes, such as pumping and filtration, which drive up costs to over $200 per kilogram of uranium oxide (U_3O_8). Furthermore, these techniques suffer from low selectivity, meaning that they are inefficient in isolating uranium from other ions in seawater [12, 309]. As a result, researchers have sought alternative approaches, such as solar interfacial evaporation, which combine the dual functions of seawater desalination and uranium recovery [310–313]. Such systems capitalize on two key phenomena: the concentration of brine through localized evaporation, which increases uranium availability, and photothermal effects that enhance the chemisorption kinetics of uranium.

The development of functionalized absorbers, which selectively coordinate with uranyl ions (UO_2^{2+}), while maintaining high evaporation rates, offers a promising pathway for addressing energy security concerns. Amidoxime-functionalized absorbers gain attention due to their ability to selectively bind to uranyl ions [314, 315]. The amidoxime functional group forms a bidentate coordination with the uranyl ion, utilizing its oxygen and nitrogen atoms to bind to the equatorial plane of the uranium atom, displacing water molecules in the process. The binding energy associated with this coordination is about −1.8 eV, making it a strong candidate for efficient uranium extraction. For instance, Wang et al. developed a biomimetic hydrogel with oriented microchannels, mimicking the structure of plant xylem, which significantly enhanced the material's ability to absorb seawater and adsorb uranium (Fig. 6.21a) [310]. This hydrogel achieved uranium adsorption rates of 6.42 mg/g over 7 days, while maintaining a high evaporation rate of 2.86 kg m^{-2} h^{-1} under one sun. The photothermal effect in these systems raised local temperatures by 38 °C, thereby accelerating the adsorption kinetics, with the adsorption rate increasing from 0.09 h^{-1} in the dark to 0.24 h^{-1} under light. Despite these advancements, challenges remain in terms of selectivity, as ions such as vanadium, which has a similar ionic radius and charge density as UO_2^{2+}, can compete for binding sites on the absorber material. This competition can reduce the uranium-to-vanadium

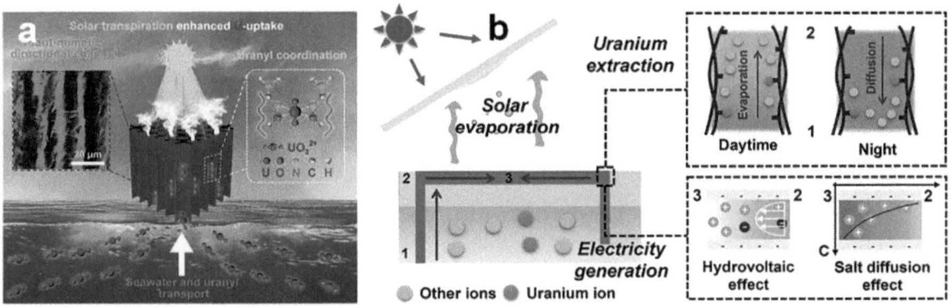

Fig. 6.21 (**a**) The enhanced uranium extraction from seawater. (Adapted from [310], Copyright 2021 Wiley-VCH). (**b**) Ion pump-evaporation for uranium extraction. (Reproduced with permission from [322], Copyright 2024 Elsevier)

mass ratio to below 3:1, leading to additional costs for post-processing, which may undermine the overall economic feasibility of these systems.

Recent advances in molecular engineering focused on overcoming selectivity and biofouling challenges [316, 317]. Selectivity improvements have been achieved by integrating alternative functional groups into the absorbers. For example, Li et al. fabricated a glyphosine-modified hydrogel, in which phosphate groups preferentially bind to uranyl ions via stronger Lewis acid-base interactions. The modified hydrogel demonstrated a uranium uptake of 9.18 mg/g over 14 days from seawater, with significantly improved uranium-to-vanadium mass ratios (>8:1) compared to amidoxime-based materials [318]. The glyphosine-modified hydrogel also maintained a stable evaporation rate of approximately 2.5 kg m^{-2} h^{-1} and preserved its hydrophilic properties, as indicated by a water contact angle of 28°. In addition to selectivity, biofouling resistance has emerged as another major concern, as biofilm formation on the absorber material can significantly reduce uranium adsorption capacity by 40–60% within a month [309, 319, 320]. To address this, researchers have incorporated quaternary ammonium-functionalized silica nanoparticles into the hydrogel matrix. These nanoparticles effectively inhibit bacterial growth, achieving 99.9% inhibition against key marine pathogens such as Pseudomonas aeruginosa and Bacillus subtilis, thus extending the operational stability of the material to over 60 days without significant performance decay.

Emerging research in uranium extraction has also explored the development of advanced membrane structures that combine multiple functionalities, including photothermal effects, selective ion sieving, and rapid water transport [318, 321–324]. One promising design involves Janus-structured membranes that incorporate different layers with distinct functionalities. For example, Fu et al. developed a Janus fiber from polyamidoxime and MXene, enhancing uranium extraction to 6.05 mg/g under one sun, a 30.1% increase over dark conditions [325]. Liu et al. created a biomimetic interfacial evaporation inspired by ion pumps, integrating seawater desalination, electricity generation, and uranium extraction (Fig. 6.21b) [322]. In long-term tests upon one sun, this platform maintained an evaporation rate of 1.80 kg m^{-2} h^{-1} and achieved over 80% uranium adsorption within 48 h in actual seawater. Qiu et al. developed a covalent organic framework hydrogel that performs exceptionally well in both seawater desalination and uranium extraction [324]. In natural seawater, the device demonstrated a notable evaporation rate of 0.744 kg m^{-2} h^{-1} and a uranium extraction capacity of 4.15 mg/g. This performance is attributed to its hydrophilic hydrogel network and capillary microporous channels, which enhance water transport, uranyl ion mass transfer, and the availability of binding sites. Despite these promising developments, several challenges remain in scaling up these systems for large-scale use. Future directions will focus on improving selectivity, durability, scalability, and environmental sustainability.

6.3.4 Food Production

Solar evaporation can improve agricultural water management, thereby enhancing food production [76, 267, 326, 327]. By desalting and purifying irrigation water, it ensures a reliable supply of clean water for crop cultivation, ultimately boosting agricultural productivity and strengthening food security [31, 328].

6.3.4.1 Sea Farm

Earth's oceans cover about 71% of the planet's surface, they present a unique dual opportunity: they offer vast reserves of desalinated seawater and serve as a potential site for marine agriculture. Seawater desalination could draw from a volume of 1.4×10^{21} liters, while marine agriculture offers an avenue to ease pressure on terrestrial resources, particularly in coastal areas [76, 329]. Solar-powered sea farming systems, which integrate interfacial solar steam generation with hydroponic cultivation, are emerging as promising closed-loop platforms to address freshwater shortages [330, 331]. These systems have the potential to revitalize degraded coastal lands (targeting 1.5 million km^2), offering an alternative to the challenges posed by soil salinity, which is prevalent in arid coastal zones with salt content exceeding 20%.

The prototypes of solar sea farming laid the groundwork for the concept in terms of scalability. One example is the desalination-cultivation platform developed by Zou et al., which demonstrated the viability of autonomous operation. This platform achieved a freshwater yield of 1.42 kg m^{-2} h^{-1} while simultaneously supporting plant growth [328]. However, a major limitation of this system was the modular segregation of seawater intake, evaporation, and irrigation units, which resulted in a large spatial footprint. This space requirement hampered the feasibility of deploying such systems on a large scale. In response, Tan's group introduced advancements by embedding floating photothermal evaporators within cultivation chambers [326, 332]. This innovation reduced the spatial occupancy by 60%, while maintaining evaporation rates of 2.1 kg m^{-2} h^{-1}. Despite this improvement, the system still faced significant issues, particularly with localized temperature spikes, which required manual thermal management and disrupted the system's continuous operation, further complicating scalability.

Third-generation sea farming systems made progress in overcoming the spatial and thermal challenges. One promising development is Xu's vertical double-layered sea farm, which separates the photothermal evaporation in the lower chamber from the plant growth process in the upper chamber [333]. This design utilizes hydrophilic cellulose belts to facilitate continuous freshwater transfer between chambers, achieving an impressive water flux of 0.67 kg m^{-2} h^{-1}. As a result, the system benefits from passive thermal regulation, with temperature variations in the growth chamber maintained below 5 °C throughout the day. These innovations led to stable crop yields, such as Brassica chinensis biomass production at a rate of 0.87 kg/m^2 per week. Field trials have proven successful, with a 98% survival rate for marine-deployed units over 120 days-outperforming terrestrial saline agriculture, which typically has a survival rate of less than 40%. Furthermore, Ho et al. developed a trinity system that integrates osmotic energy harvesting with desalination [159]. This hybrid system harnesses osmotic power through membrane-based processes during water condensation, utilizing the salinity gradient between concentrated seawater and surface water via solar desalination. The purified water is employed for irrigation and agricultural purposes (Fig. 6.22). Demonstrating a peak power output of ~0.25 W/cm^2 and a consistent water production rate of 1.42 kg m^{-2} h^{-1}, this integrated approach provides an efficient solution for coastal agriculture. Together, these developments represent significant advancements toward achieving energy-efficient and scalable sea farming systems.

As we transition from pilot-scale systems to large-scale implementations (>1 km^2), several key innovations are required to fully realize the potential of solar sea farming. First, the durability of materials must be improved to withstand the harsh marine environment. Biofouling-resistant photothermal coatings, such as

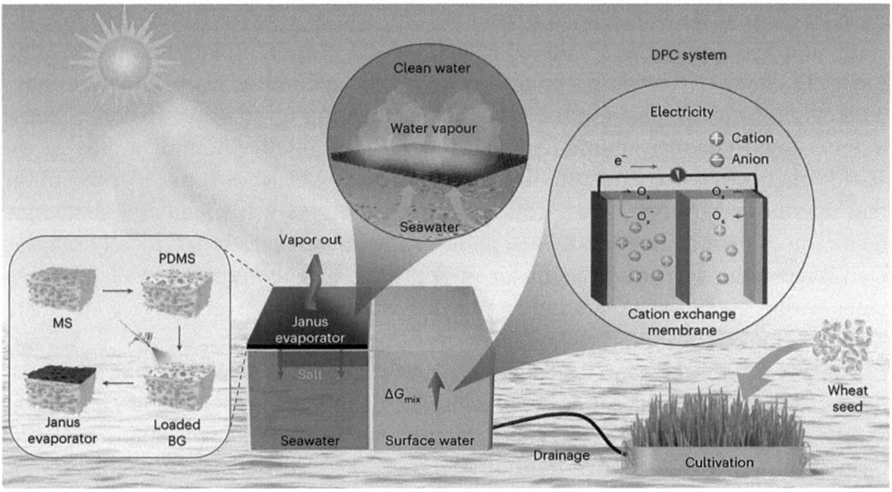

Fig. 6.22 An integrated system based on Janus evaporator and the reverse electrodialysis for crop irrigation. (Reproduced with permission from [159] Copyright 2023 Nature)

zwitterionic polymer membranes, could maintain evaporation efficiency above 90% over a five-year period of marine exposure [110, 334]. Furthermore, to ensure continuous operation, hybrid systems that combine solar steam generation with offshore photovoltaics, with efficiencies exceeding 23%, could provide a reliable 24/7 power supply [267, 335]. Another crucial area of development involves ecological integration. Submerged nutrient recovery modules, such as algal bioreactors, could be deployed to sequester excess nitrogen from agricultural runoff, potentially achieving an 85% reduction in nitrate concentrations, which would improve water quality and promote sustainable agricultural practices in coastal zones [267]. Additionally, the economic viability of these systems must be addressed. By utilizing recycled polymer composites, the cost of floating platforms could be reduced, making these systems more economically feasible for large-scale deployment [72, 336]. Computational models suggest that scaling current prototypes to cover 10,000 km^2 of coastal waters could produce up to 9.3×10^9 m^3 of freshwater annually, along with 1.7×10^8 tons of crops. This scale of production could sustain 200 million people while simultaneously rehabilitating 500,000 km^2 of degraded land [76]. These developments position solar sea farming as a pivotal technology for achieving the sustainable development goals, particularly those related to hunger, clean water, and marine conservation, by enabling the circular use of marine resources in agricultural production.

6.3.4.2 Water-Saving Greenhouse

Greenhouse agriculture exemplifies advanced controlled-environment cultivation yet grapples with inefficiencies in water resource management. Although greenhouses employ advanced irrigation techniques such as drip irrigation (with a water use efficiency of 80–90%), aeroponics, and ebb-and-flow system, they still lose the water supplied due to the combined effects of soil evaporation and crop transpiration [337, 338]. Urban horticultural innovations, including plant factories, have made slight improvements in the efficiency, achieving up to 40%. However, these systems tend to increase humidity levels to over 90% relative humidity, thus necessitating energy-intensive dehumidification processes which consume 30–50% of the total operational energy [339]. This scenario highlights a fundamental challenge: the thermodynamic linkage between photosynthetic transpiration, typically ranging from 3–10 mm/day per square meter of leaf area, and the regulation of microclimates within greenhouses, creating a paradox where efforts to conserve water inadvertently lead to substantial atmospheric moisture loss.

The advent of Sorption-Atmospheric Water Harvesters/SAWHs offers a promising dual-functional approach for both dehumidification and irrigation in greenhouse settings [100, 340]. These systems utilize hygroscopic materials to capture moisture during high nocturnal relative humidity (RH > 80%) and release this moisture via solar-driven desorption when temperature changes exceed 40 °C under solar exposure. Initial implementations in greenhouses involved materials like silica gel, with a water uptake rate of 0.3 g/g, and Metal-Organic Frameworks with 1.2 g/g.

However, these materials faced regeneration challenges, requiring high temperatures and large energy inputs [341–343]. Additionally, the deployment of SAWHs in low-canopy, leafy vegetable greenhouses revealed spatial inefficiencies, with 70% of vertical space underutilized. The integration of auxiliary components such as heat pumps [341] (with a Coefficient of Performance, COP, between 2.5 and 3.0) and photovoltaic panels (15–20% efficiency) further increased the system's footprint by 40% and reduced the photosynthetically active radiation by 25% due to shading effects [344], underscoring the need for more localized and energy-passive SAWH architectures.

A breakthrough in SAWH technology was achieved by Wang et al., who developed a multicomponent hygroscopic gel (TCP-Li) tailored for greenhouse-specific conditions (Fig. 6.23) [337]. This innovative composite combines Titanium nitride particles for plasmonic photothermal effects, achieving 92% solar absorption and a surface temperature of 68 °C under solar exposure; Curdlan, a thermoresponsive polysaccharide that forms hierarchical macropores facilitating rapid vapor transport; Sodium polyacrylate, a superabsorbent network capable of retaining 3.38 g/g sorbent at 95% RH; and LiCl, a deliquescent salt that enhances moisture capture kinetics. Incorporated into the Transpiration and Evaporation Adsorption Device (TEAD), this system absorbs moisture at elevated relative humidity levels overnight and discharges water for daytime irrigation through the utilization of natural sunlight. Field experiments conducted in Lactuca sativa greenhouses validated TEAD's effectiveness in supporting standard plant growth and delivering supplementary irrigation water (87.1 g per plant and 1890.6 g/m^2). This resulted in an average water conservation rate of 44.9%. The advancement of the TEAD system paves the way for the future of smart greenhouses, which autonomously regulate water cycles.

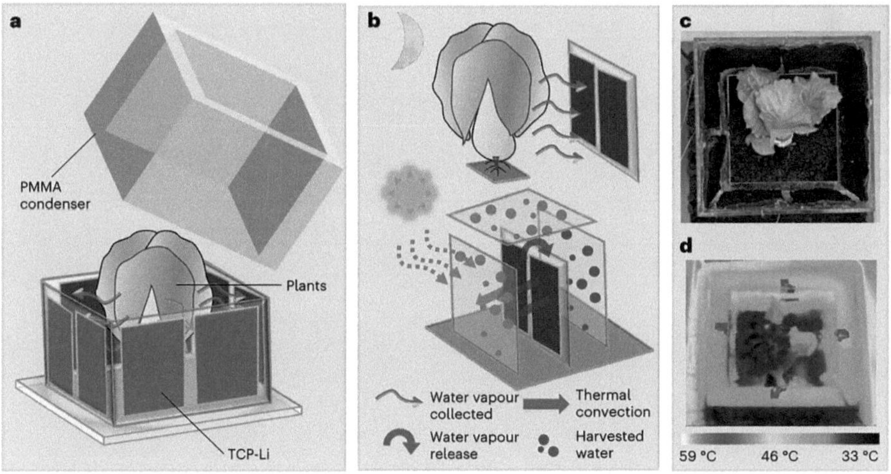

Fig. 6.23 Solar-hygroscopic-gel for recycling water. (**a**) The design system. (**b**) Heat and mass transfer analysis. (**c**) Photograph of the prototype and (**d**) Infrared temperature mapping during desorption. (Reproduced with permission from [337], Copyright 2024 Nature)

6.3.4.3 Soil Remediation

Solar evaporation facilitates the remediation of contaminated soils via artificial phytoremediation [345–351], soil washing [352], and biodegradable soil enrichment [15, 267]. Following thorough safety assessments, these treated soils can be repurposed for agricultural production.

Artificial phytoremediation presents a significant advancement over traditional plant-based decontamination methods by integrating solar evaporators with selective ion-binding mechanisms (Fig. 6.24a). This innovative approach employs photothermal membranes placed directly on contaminated soils, utilizing interfacial heating to achieve a temperature increase under solar exposure [345, 348, 351]. This enhancement facilitates heavy metal mobilization through thermally induced capillary transport, essentially mimicking and accelerating the natural transpiration process of plants [345, 353]. Key components include ion-selective hydrogels, functionalized with thiol groups for effective chelation of lead ions (Pb^{2+}). The system promotes directional vapor gradients that drive a metal-laden water flux from the soil to the evaporator. This setup not only increases the rate of contaminant extraction by two to three times compared to conventional methods but also enhances lead sequestration efficiency in the evaporator. Field trials in lead-contaminated soils have shown a 40% reduction in bioavailable lead within just 14 days, substantially outperforming traditional ryegrass phytoremediation methods [347].

The application of solar evaporation for soil washing represents a self-sustaining method to address soil salinity and contamination through closed-loop brine cycling

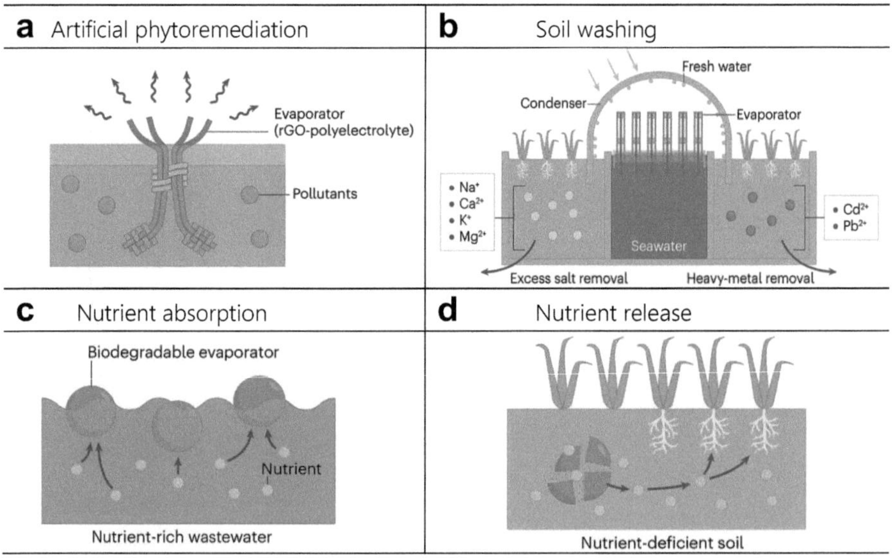

Fig. 6.24 Soil remediation via solar evaporation. (**a**) Artificial phytoremediation. (**b**) Soil washing. (**c**) Nutrient absorption. (**d**) Nutrient release. (Adapted with permission from [267], Copyright 2025 Nature Publication Group)

(Fig. 6.24b) [352]. This system incorporates a photothermal evaporator, featuring a cellulose membrane coated with reduced graphene oxide, to treat seawater for use in washing saline soils and providing agricultural irrigation. In outdoor trials, the evaporator achieved a remarkable freshwater production rate of 10.95 kg/m^2 per day and demonstrated soil washing efficiency three times higher than conventional distillation methods. Subsequent plant experiments revealed that soils with initially high salinity could be converted into productive agricultural land, with 65% germination rates observed after soil washing. Consequently, solar evaporation consistently supplies irrigation water to support continuous plant growth. This fully automated system operates exclusively on solar energy and seawater, effectively addressing both saline soil remediation and irrigation requirements [352]. However, one of the challenges noted is the substantial leaching of native nutrients such as nitrate (NO_3^-) and potassium (K^+), which necessitates a subsequent nutrient restoration process to maintain soil fertility [354].

Innovative biodegradable systems utilizing polysaccharide-based photothermal aerogels are designed to address nutrient loss through a dual-phase remediation process (Fig. 6.24c, d) [15, 267, 355]. Initially, chitosan-alginate beads with high porosity capture key nutrients such as nitrate and phosphate from the washate. Following this, an enzymatic hydrolysis process gradually releases these nutrients back into the soil over a period of 30 to 60 days, maintaining a favorable pH range of 6.5 to 7.2. The inclusion of Fe_3O_4 and TiO_2 nanoparticles not only facilitates solar-activated contaminant oxidation but also allows for the magnetic recovery of the beads for repeated use, with minimal efficiency loss over multiple cycles [356]. Additionally, the systems contribute to the regeneration of microbial communities, significantly enhancing the abundance of beneficial microbes such as Pseudomonadaceae. Field applications of the systems have demonstrated not only a high phosphate adsorption capacity but also a notable increase in crop yields, approximately 2.3 times higher than those achieved with chemical fertilization [15, 267]. Despite these advancements, challenges remain, particularly in the development of cost-effective, field-deployable contaminant sensors and the establishment of stringent regulatory frameworks for the management of nanomaterial residues.

References

1. T. Ding, Y. Zhou, W.L. Ong, G.W. Ho, Hybrid solar-driven interfacial evaporation systems: Beyond water production towards high solar energy utilization, ***Materials Today***, 42 (2021) 178-191.
2. D. Tong, B. Song, A high-efficient and ultra-strong interfacial solar evaporator based on carbon-fiber fabric for seawater and wastewater purification, ***Desalination***, 527 (2022) 115586.
3. J. Li, Y. Cao, K. Ding, J. Ye, F. Li, C. Ma, P. Lv, Y. Xu, L. Shi, Research progress of industrial wastewater treatment technology based on solar interfacial adsorption coupled evaporation process, ***The Science of the total environment***, 931 (2024) 172887.
4. X. Li, L. Liu, S. Li, Y. Wan, J.X. Chen, S. Tian, Z. Huang, Y.F. Xiao, X. Cui, C. Xiang, Q. Tan, X.H. Zhang, W. Guo, X.J. Liang, C.S. Lee, Biodegradable pi-Conjugated Oligomer

Nanoparticles with High Photothermal Conversion Efficiency for Cancer Theranostics, ***ACS nano***, 13 (2019) 12901-12911.
5. S. Fang, X. Lyu, T. Tong, A.I. Lim, T. Li, J. Bao, Y.H. Hu, Turning dead leaves into an active multifunctional material as evaporator, photocatalyst, and bioplastic, ***Nature communications***, 14 (2023) 1203.
6. T. Kumar, B. Zeng, H. Hamza, H. Zhao, X. Zhang, Root-based solar interfacial evaporation setup for efficient volumetric reduction of concentrated slurry waste, ***Desalination***, 585 (2024) 117686.
7. X. Zhou, F. Zhao, Y. Guo, B. Rosenberger, G. Yu, Architecting highly hydratable polymer networks to tune the water state for solar water purification, ***Science advances***, 5 (2019) eaaw5484.
8. S. Yan, H. Song, Y. Li, J. Yang, X. Jia, S. Wang, X. Yang, Integrated reduced graphene oxide/polypyrrole hybrid aerogels for simultaneous photocatalytic decontamination and water evaporation, ***Applied Catalysis B: Environmental***, 301 (2022) 120820.
9. X. Li, M. Wang, H. Tao, B. Ge, S. Liu, J. Liu, G. Ren, Z. Zhang, Constructing of efficient interface solar evaporator: In-situ colloid foaming strategy for solar desalination and visible light response sewage purification, ***Journal of colloid and interface science***, 649 (2023) 107-117.
10. X. Zhou, Y. Guo, F. Zhao, W. Shi, G. Yu, Topology-Controlled Hydration of Polymer Network in Hydrogels for Solar-Driven Wastewater Treatment, ***Advanced materials***, 32 (2020) e2007012.
11. L. Cui, P. Wang, H. Che, J. Chen, B. Liu, Y. Ao, Solar-driven interfacial water evaporation for wastewater purification: Recent advances and challenges, ***Chemical Engineering Journal***, 477 (2023) 147158.
12. Q. Sun, B. Aguila, J. Perman, A.S. Ivanov, V.S. Bryantsev, L.D. Earl, C.W. Abney, L. Wojtas, S. Ma, Bio-inspired nano-traps for uranium extraction from seawater and recovery from nuclear waste, ***Nature communications***, 9 (2018) 1644.
13. G. Lin, B. Zeng, J. Li, Z. Wang, S. Wang, T. Hu, L. Zhang, A systematic review of metal organic frameworks materials for heavy metal removal: Synthesis, applications and mechanism, ***Chemical Engineering Journal***, 460 (2023) 141710.
14. X. Guo, Y. Liu, J. Wang, Equilibrium, kinetics and molecular dynamic modeling of Sr(2+) sorption onto microplastics, ***Journal of hazardous materials***, 400 (2020) 123324.
15. J. Li, N. Li, X. Wu, S. Wang, S. Li, C. Guo, L. Yu, Z. Wang, P. Murto, X. Xu, Photothermal Aerogel Beads Based on Polysaccharides: Controlled Fabrication and Hybrid Applications in Solar-Powered Interfacial Evaporation, Water Remediation, and Soil Enrichment, ***ACS applied materials & interfaces***, (2022).
16. G.-R. Xu, Z.-H. An, K. Xu, Q. Liu, R. Das, H.-L. Zhao, Metal organic framework (MOF)-based micro/nanoscaled materials for heavy metal ions removal: The cutting-edge study on designs, synthesis, and applications, ***Coordination Chemistry Reviews***, 427 (2021) 213554.
17. Q. Hou, H. Zhou, W. Zhang, Q. Chang, J. Yang, C. Xue, S. Hu, Boosting adsorption of heavy metal ions in wastewater through solar-driven interfacial evaporation of chemically-treated carbonized wood, ***The Science of the total environment***, 759 (2021) 144317.
18. B. Yuan, Y. Sui, J. Dong, X. Lv, M. Guo, Y. Xiao, L. Yang, H. Yang, L. Bai, W. Wang, D. Wei, Y. Liang, H. Chen, A high-efficiency solar water evaporation-photocatalysis system achieved by manipulating surface wettability and constructing heterojunction, ***Applied Surface Science***, 611 (2023) 155678.
19. B. Peng, Y. Gao, Q. Lyu, Z. Xie, M. Li, L. Zhang, J. Zhu, Cationic Photothermal Hydrogels with Bacteria-Inhibiting Capability for Freshwater Production via Solar-Driven Steam Generation, ***ACS applied materials & interfaces***, 13 (2021) 37724-37733.
20. C. Chen, Y. Kuang, L. Hu, Challenges and Opportunities for Solar Evaporation, ***Joule***, 3 (2019) 683-718.
21. T. Weng, H. Du, H. Luo, M. Jiang, Z. Chen, J. Wang, F. Jiang, H. Chen, Synergistic photothermal conversion on Mn3CeOx/Coal-based carbon material membrane for simultaneous inter-

face evaporation and organic pollutant removal, ***Separation and Purification Technology***, 355 (2025) 129573.

22. S. Tian, Z. Zhou, X. Li, F. Wang, Y. Zhao, L. Tijing, H.K. Shon, B. Xu, J. Ren, Applications of solar-driven interfacial evaporation-coupled photocatalysis in water treatment: A mini review, ***Desalination***, 592 (2024) 118159.
23. Y. Lu, Z. Li, J. Liang, X. Xu, G. Zhang, H. Min, Vermiculite-derived MgFe nanosheet-enhanced hydrogel with dual photothermal and photocatalytic Fenton functions for solar water purification, ***Desalination***, 592 (2024) 118196.
24. M.Q. Yang, C.F. Tan, W. Lu, K. Zeng, G.W. Ho, Spectrum Tailored Defective 2D Semiconductor Nanosheets Aerogel for Full-Spectrum-Driven Photothermal Water Evaporation and Photochemical Degradation, ***Advanced Functional Materials***, 30 (2020).
25. X. Wang, Y. He, X. Liu, Synchronous steam generation and photodegradation for clean water generation based on localized solar energy harvesting, ***Energy Conversion and Management***, 173 (2018) 158-166.
26. J. Zhou, Z. Cui, X. Liu, Q. Wang, B. Chen, S. Zeng, J. Si, Multifunctional nanofiber membrane with antifouling properties for efficient seawater desalination and wastewater purification, ***Desalination***, 586 (2024) 117812.
27. A. Gnanasekaran, K. Rajaram, Rational design of different interfacial evaporators for solar steam generation: Recent development, fabrication, challenges and applications, ***Renewable and Sustainable Energy Reviews***, 192 (2024) 114202.
28. C. Niu, L. Yang, H. Sun, Z. Zhu, W. Liang, J. Li, A. Li, Three-dimensional macroporous foamy MXene membrane with high efficiency on interfacial evaporation and wastewater purification, ***Chemical Engineering Journal***, 476 (2023) 146522.
29. N. Xu, J. Li, C. Finnerty, Y. Song, L. Zhou, B. Zhu, P. Wang, B. Mi, J. Zhu, Going beyond efficiency for solar evaporation, ***Nature Water***, 1 (2023) 494-501.
30. M.P. Dion, W., Steam sterilization principles, ***Pharmaceutical engineering***, (2013) 1-8.
31. S.M. Parsa, F. Norozpour, S. Momeni, S. Shoeibi, X. Zeng, Z. Said, W. Guo, H.H. Ngo, B.-J. Ni, Advanced nanostructured materials in solar interfacial steam generation and desalination against pathogens: combatting microbial-contaminants in water – a critical review, ***Journal of Materials Chemistry A***, 11 (2023) 18046-18080.
32. O. Neumann, C. Feronti, A.D. Neumann, A. Dong, K. Schell, B. Lu, E. Kim, M. Quinn, S. Thompson, N. Grady, P. Nordlander, M. Oden, N.J. Halas, Compact solar autoclave based on steam generation using broadband light-harvesting nanoparticles, ***Proceedings of the National Academy of Sciences of the United States of America***, 110 (2013) 11677-11681.
33. L. Noureen, S. Zaman, W. Ali Shah, Q. Wang, M. Humayun, Q. Xu, X. Wang, Bifunctional photothermal membrane for high-temperature interfacial solar steam generation and off-grid sterilization, ***Chemical Engineering Journal***, 473 (2023) 145122.
34. J. Li, M. Du, G. Lv, L. Zhou, X. Li, L. Bertoluzzi, C. Liu, S. Zhu, J. Zhu, Interfacial Solar Steam Generation Enables Fast-Responsive, Energy-Efficient, and Low-Cost Off-Grid Sterilization, ***Advanced materials***, 30 (2018) e1805159.
35. D. Krueger, A. Graves, Y. Chu, J. Pan, S.E. Lee, Y. Park, Integrated Plasmonic Gold Nanoparticle Dimer Array for Sustainable Solar Water Disinfection, ***ACS Applied Nano Materials***, 6 (2023) 5568-5577.
36. X. Zhou, F. Zhao, P. Zhang, G. Yu, Solar Water Evaporation Toward Water Purification and Beyond, ***ACS Materials Letters***, 3 (2021) 1112-1129.
37. L. Wang, Y. Feng, K. Wang, G. Liu, Solar water sterilization enabled by photothermal nanomaterials, ***Nano Energy***, 87 (2021) 106158.
38. S. Loeb, C. Li, J.H. Kim, Solar Photothermal Disinfection using Broadband-Light Absorbing Gold Nanoparticles and Carbon Black, ***Environmental science & technology***, 52 (2018) 205-213.
39. S.K. Loeb, J. Kim, C. Jiang, L.S. Early, H. Wei, Q. Li, J.H. Kim, Nanoparticle Enhanced Interfacial Solar Photothermal Water Disinfection Demonstrated in 3-D Printed Flow-Through Reactors, ***Environmental science & technology***, 53 (2019) 7621-7631.

40. C. Chang, P. Tao, J. Xu, B. Fu, C. Song, J. Wu, W. Shang, T. Deng, High-Efficiency Superheated Steam Generation for Portable Sterilization under Ambient Pressure and Low Solar Flux, ***ACS applied materials & interfaces***, 11 (2019) 18466-18474.
41. A.M. Saleque, S. Ahmed, M.N.A.S. Ivan, M.I. Hossain, W. Qarony, P.K. Cheng, J. Qiao, Z.L. Guo, L. Zeng, Y.H. Tsang, High-temperature solar steam generation by MWCNT-HfTe2 van der Waals heterostructure for low-cost sterilization, ***Nano Energy***, 94 (2022) 106916.
42. Y. Zhang, D. Zhao, F. Yu, C. Yang, J. Lou, Y. Liu, Y. Chen, Z. Wang, P. Tao, W. Shang, J. Wu, C. Song, T. Deng, Floating rGO-based black membranes for solar driven sterilization, ***Nanoscale***, 9 (2017) 19384-19389.
43. L. Hao, N. Liu, H. Bai, P. He, R. Niu, J. Gong, High-performance solar-driven interfacial evaporation through molecular design of antibacterial, biomass-derived hydrogels, ***Journal of colloid and interface science***, 608 (2022) 840-852.
44. L. Zhao, B. Bhatia, L. Zhang, E. Strobach, A. Leroy, M.K. Yadav, S. Yang, T.A. Cooper, L.A. Weinstein, A. Modi, S.B. Kedare, G. Chen, E.N. Wang, A Passive High-Temperature High-Pressure Solar Steam Generator for Medical Sterilization, ***Joule***, 4 (2020) 2733-2745.
45. V. Nava, S. Chandra, J. Aherne, M.B. Alfonso, A.M. Antao-Geraldes, K. Attermeyer, R. Bao, M. Bartrons, S.A. Berger, M. Biernaczyk, R. Bissen, J.D. Brookes, D. Brown, M. Canedo-Arguelles, M. Canle, C. Capelli, R. Carballeira, J.L. Cereijo, S. Chawchai, S.T. Christensen, K.S. Christoffersen, E. de Eyto, J. Delgado, T.N. Dornan, J.P. Doubek, J. Dusaucy, O. Erina, Z. Ersoy, H. Feuchtmayr, M.L. Frezzotti, S. Galafassi, D. Gateuille, V. Goncalves, H.P. Grossart, D.P. Hamilton, T.D. Harris, K. Kangur, G.B. Kankilic, R. Kessler, C. Kiel, E.M. Krynak, A. Leiva-Presa, F. Lepori, M.G. Matias, S.S. Matsuzaki, Y. McElarney, B. Messyasz, M. Mitchell, M.C. Mlambo, S.N. Motitsoe, S. Nandini, V. Orlandi, C. Owens, D. Ozkundakci, S. Pinnow, A. Pociecha, P.M. Raposeiro, E.I. Room, F. Rotta, N. Salmaso, S.S.S. Sarma, D. Sartirana, F. Scordo, C. Sibomana, D. Siewert, K. Stepanowska, U.N. Tavsanoglu, M. Tereshina, J. Thompson, M. Tolotti, A. Valois, P. Verburg, B. Welsh, B. Wesolek, G.A. Weyhenmeyer, N. Wu, E. Zawisza, L. Zink, B. Leoni, Plastic debris in lakes and reservoirs, ***Nature***, 619 (2023) 317-322.
46. A. Stubbins, K.L. Law, S.E. Munoz, T.S. Bianchi, L. Zhu, Plastics in the Earth system, ***Science***, 373 (2021) 51-55.
47. M. MacLeo, H.P.H. Arp, M.B. Tekman, A. Jahnke, The global threat from plastic pollution, ***Science***, 373 (2021) 61-65.
48. P. Wang, Z. Huang, S. Chen, M. Jing, Z. Ge, J. Chen, S. Yang, J. Chen, Y. Fang, Sustainable removal of nano/microplastics in water by solar energy, ***Chemical Engineering Journal***, 428 (2022) 131196.
49. Y. Zhang, B. Sun, C. Cai, T. Wang, Y. Gao, D. Ma, Photothermocatalytic Wet Reforming of Waste Plastics to Syngas, ***Journal of the American Chemical Society***, 147 (2025) 9879-9890.
50. Z. Zhang, W. Si, W. Li, H. Zhang, A novel nanoplastic removal method based on solar interface evaporation technology, ***Chemical Engineering Journal***, 496 (2024) 154290.
51. X. Meng, X. Peng, J. Xue, Y. Wei, Y. Sun, Y. Dai, A biomass-derived, all-day-round solar evaporation platform for harvesting clean water from microplastic pollution, ***Journal of Materials Chemistry A***, 9 (2021) 11013-11024.
52. W. Tian, P. Song, H. Zhang, X. Duan, Y. Wei, H. Wang, S. Wang, Microplastic materials in the environment: Problem and strategical solutions, ***Progress in Materials Science***, 132 (2023) 101035.
53. Z. Yu, Y. Li, Y. Zhang, P. Xu, C. Lv, W. Li, B. Maryam, X. Liu, S.C. Tan, Microplastic detection and remediation through efficient interfacial solar evaporation for immaculate water production, ***Nature communications***, 15 (2024) 6081.
54. Y. Li, J. Zhang, L. Xu, R. Li, R. Zhang, M. Li, C. Ran, Z. Rao, X. Wei, M. Chen, L. Wang, Z. Li, Y. Xue, C. Peng, C. Liu, H. Sun, B. Xing, L. Wang, Leaf absorption contributes to accumulation of microplastics in plants, ***Nature***, 641 (2025) 666-673.
55. X. Meng, X. Wang, K. Yin, Y. Jing, L. Gu, Z. Tao, X. Ren, M. Tang, X. Shao, L. Sun, Y. Sun, Y. Dai, Y. Xiong, Integration of photothermal water evaporation with photocatalytic

microplastics upcycling via nanofluidic thermal management, ***Proceedings of the National Academy of Sciences of the United States of America***, 121 (2024) e2317192121.
56. S. Zhang, Q. Liu, T. Chen, Z. Liu, G. Liu, Solar thermoplasmonic convection for sustainable removal of microplastics, ***Solar Energy Materials and Solar Cells***, 282 (2025) 113416.
57. R.C. Thompson, W. Courtene-Jones, J. Boucher, S. Pahl, K. Raubenheimer, A.A. Koelmans, Twenty years of microplastic pollution research-what have we learned?, ***Science***, 386 (2024) eadl2746.
58. X. Cui, Q. Ruan, X. Zhuo, X. Xia, J. Hu, R. Fu, Y. Li, J. Wang, H. Xu, Photothermal Nanomaterials: A Powerful Light-to-Heat Converter, ***Chemical reviews***, 123 (2023) 6891-6952.
59. Z. Huang, J. Wei, Y. Wan, P. Li, J. Yu, J. Dong, S. Wang, S. Li, C.S. Lee, Aligned Millineedle Arrays for Solar Power Seawater Desalination with Site-Specific Salt Formation, ***Small***, 17 (2021) e2101487.
60. H. Liu, B. Chen, Y. Chen, M. Zhou, F. Tian, Y. Li, J. Jiang, W. Zhai, Bioinspired Self-Standing, Self-Floating 3D Solar Evaporators Breaking the Trade-Off between Salt Cycle and Heat Localization for Continuous Seawater Desalination, ***Advanced materials***, 35 (2023) e2301596.
61. M. Wang, J. Hu, M. Li, L. Zhang, M. Salimi, M. Amidpour, Z. Chen, Bioinspired design of photothermal anti-fouling fabrics for solar-driven sustainable seawater desalination, ***Nano Energy***, 136 (2025) 110726.
62. S. Tian, Z. Huang, J. Tan, X. Cui, Y. Xiao, Y. Wan, X. Li, Q. Zhao, S. Li, C.-S. Lee, Manipulating Interfacial Charge-Transfer Absorption of Cocrystal Absorber for Efficient Solar Seawater Desalination and Water Purification, ***ACS Energy Letters***, 5 (2020) 2698-2705.
63. L. Zhou, Y. Tan, J. Wang, W. Xu, Y. Yuan, W. Cai, S. Zhu, J. Zhu, 3D self-assembly of aluminium nanoparticles for plasmon-enhanced solar desalination, ***Nature Photonics***, 10 (2016) 393-398.
64. H. Bai, J. Hu, S.H. Lam, Y. Guo, X.-M. Zhu, Z. Yang, J. Wang, Turning Dielectric MoO3 Nanospheres from White to Black through Doping for Efficient Solar Seawater Desalination, ***ACS Materials Letters***, 4 (2022) 1584-1592.
65. Y. Xia, S. Yuan, Y. Li, L. Gao, X. Zhang, Solar-driven brine desalination and concentration by controlled salt excretion, ***EcoMat***, 3 (2021).
66. G. Ni, S.H. Zandavi, S.M. Javid, S.V. Boriskina, T.A. Cooper, G. Chen, A salt-rejecting floating solar still for low-cost desalination, ***Energy & Environmental Science***, 11 (2018) 1510-1519.
67. G. Li, W.-C. Law, K.C. Chan, Floating, highly efficient, and scalable graphene membranes for seawater desalination using solar energy, ***Green Chemistry***, 20 (2018) 3689-3695.
68. M. Xiong, S. Dong, Y. Xu, C. Liu, Y. Di, F. Xing, S. Wu, J. Zhang, L. Yu, L. Dong, Z. Gan, Highly-Efficient Solar Steam Generation with Real Time Salinity Monitoring for Seawater Desalination, ***Advanced Sustainable Systems***, 6 (2021).
69. K. Xu, C. Wang, Z. Li, S. Wu, J. Wang, Salt Mitigation Strategies of Solar-Driven Interfacial Desalination, ***Advanced Functional Materials***, 31 (2020).
70. L. Song, P. Mu, L. Geng, Q. Wang, J. Li, A Novel Salt-Rejecting Linen Fabric-Based Solar Evaporator for Stable and Efficient Water Desalination under Highly Saline Water, ***ACS Sustainable Chemistry & Engineering***, 8 (2020) 11845-11852.
71. Y. Tian, H. Yang, S. Wu, J. Yan, K. Cen, T. Luo, G. Xiong, Y. Hou, Z. Bo, K. Ostrikov, Beyond lotus: Plasma nanostructuring enables efficient energy and water conversion and use, ***Nano Energy***, 66 (2019) 104125.
72. J. Hu, Y. Sun, Z. Liu, B. Zhu, L. Zhang, N. Xu, M. Zhu, J. Zhu, Z. Chen, Photothermal fabrics for solar-driven seawater desalination, ***Progress in Materials Science***, 150 (2025) 101407.
73. E. Chiavazzo, M. Morciano, F. Viglino, M. Fasano, P. Asinari, Passive solar high-yield seawater desalination by modular and low-cost distillation, ***Nature Sustainability***, 1 (2018) 763-772.

74. M. Wang, Y. Wei, R. Li, X. Wang, C. Wang, N. Ren, S.H. Ho, Sustainable Seawater Desalination and Energy Management: Mechanisms, Strategies, and the Way Forward, ***Research***, 6 (2023) 0290.
75. Y. Chen, L. Shen, Z. Qi, Z. Luo, X. Li, H. Bao, Large-scale implementation of solar interfacial desalination, ***Nature Sustainability***, (2025).
76. X. Wu, Y. Lu, X. Ren, P. Wu, D. Chu, X. Yang, H. Xu, Interfacial Solar Evaporation: From Fundamental Research to Applications, ***Advanced materials***, 36 (2024) e2313090.
77. L. Li, J. Zhang, Water harvesting from desert soil via interfacial solar heating under natural sunlight, ***Journal of colloid and interface science***, 607 (2022) 1986-1992.
78. Q. Fang, T. Li, Z. Chen, H. Lin, P. Wang, F. Liu, Full Biomass-Derived Solar Stills for Robust and Stable Evaporation To Collect Clean Water from Various Water-Bearing Media, ***ACS applied materials & interfaces***, 11 (2019) 10672-10679.
79. P. Xiao, J. He, Y. Liang, C. Zhang, J. Gu, J. Zhang, Y. Huang, S.-W. Kuo, T. Chen, Rationally Programmable Paper-Based Artificial Trees Toward Multipath Solar-Driven Water Extraction from Liquid/Solid Substrates, ***Solar RRL***, 3 (2019).
80. X. Li, G. Zhang, C. Wang, L. He, Y. Xu, R. Ma, W. Yao, Water harvesting from soils by light-to-heat induced evaporation and capillary water migration, ***Applied Thermal Engineering***, 175 (2020) 115417.
81. J. Wang, L. Hua, C. Li, R. Wang, Atmospheric water harvesting: critical metrics and challenges, ***Energy & Environmental Science***, 15 (2022) 4867-4871.
82. J. Lord, A. Thomas, N. Treat, M. Forkin, R. Bain, P. Dulac, C.H. Behroozi, T. Mamutov, J. Fongheiser, N. Kobilansky, S. Washburn, C. Truesdell, C. Lee, P.H. Schmaelzle, Global potential for harvesting drinking water from air using solar energy, ***Nature***, 598 (2021) 611-617.
83. A.K. Rao, A.J. Fix, Y.C. Yang, D.M. Warsinger, Thermodynamic limits of atmospheric water harvesting, ***Energy & Environmental Science***, 15 (2022) 4025-4037.
84. Y. Tu, R. Wang, Y. Zhang, J. Wang, Progress and Expectation of Atmospheric Water Harvesting, ***Joule***, 2 (2018) 1452-1475.
85. X. Zhou, H. Lu, F. Zhao, G. Yu, Atmospheric Water Harvesting: A Review of Material and Structural Designs, ***ACS Materials Letters***, 2 (2020) 671-684.
86. C. Zhang, H. Guo, C. Li, F. Wang, X. Guo, A. Li, S. Gong, H. Zhang, X. Zhang, X. Qin, Atmospheric water extraction – a review from materials to devices, ***Journal of Materials Chemistry A***, 11 (2023) 22041-22057.
87. P. Poredoš, H. Shan, C. Wang, F. Deng, R. Wang, Sustainable water generation: grand challenges in continuous atmospheric water harvesting, ***Energy & Environmental Science***, 15 (2022) 3223-3235.
88. Z. Yu, T. Zhu, J. Zhang, M. Ge, S. Fu, Y. Lai, Fog Harvesting Devices Inspired from Single to Multiple Creatures: Current Progress and Future Perspective, ***Advanced Functional Materials***, 32 (2022).
89. Z. Yu, H. Zhang, J. Huang, S. Li, S. Zhang, Y. Cheng, J. Mao, X. Dong, S. Gao, S. Wang, Z. Chen, Y. Jiang, Y. Lai, Namib desert beetle inspired special patterned fabric with programmable and gradient wettability for efficient fog harvesting, ***Journal of Materials Science & Technology***, 61 (2021) 85-92.
90. Z. Li, X. Xu, X. Sheng, P. Lin, J. Tang, L. Pan, Y.V. Kaneti, T. Yang, Y. Yamauchi, Solar-Powered Sustainable Water Production: State-of-the-Art Technologies for Sunlight-Energy-Water Nexus, ***ACS nano***, 15 (2021) 12535-12566.
91. A. LaPotin, Y. Zhong, L. Zhang, L. Zhao, A. Leroy, H. Kim, S.R. Rao, E.N. Wang, Dual-Stage Atmospheric Water Harvesting Device for Scalable Solar-Driven Water Production, ***Joule***, 5 (2021) 166-182.
92. H. Shan, Z. Ye, J. Yu, R. Wang, Z. Xu, Improving solar water harvesting via airflow restructuring using 3D vapor generator, ***Device***, 1 (2023) 100065.
93. B.A. Kashiwa, C.B. Kashiwa, The solar cyclone: A solar chimney for harvesting atmospheric water, ***Energy***, 33 (2008) 331-339.

94. D.K. Nandakumar, Y. Zhang, S.K. Ravi, N. Guo, C. Zhang, S.C. Tan, Solar Energy Triggered Clean Water Harvesting from Humid Air Existing above Sea Surface Enabled by a Hydrogel with Ultrahigh Hygroscopicity, ***Advanced materials***, 31 (2019) e1806730.
95. H. Shan, P. Poredos, Z. Ye, H. Qu, Y. Zhang, M. Zhou, R. Wang, S.C. Tan, All-Day Multicyclic Atmospheric Water Harvesting Enabled by Polyelectrolyte Hydrogel with Hybrid Desorption Mode, ***Advanced materials***, 35 (2023) e2302038.
96. P. Zhu, Z. Yu, H. Sun, D. Zheng, Y. Zheng, Y. Qian, Y. Wei, J. Lee, S. Srebnik, W. Chen, G. Chen, F. Jiang, 3D Printed Cellulose Nanofiber Aerogel Scaffold with Hierarchical Porous Structures for Fast Solar-Driven Atmospheric Water Harvesting, ***Advanced materials***, 36 (2024) e2306653.
97. S.Y. Hyunho Kim, Sameer R. Rao, Shankar Narayanan,Eugene A. Kapustin, Hiroyasu Furukawa, Ari S. Umans, Omar M. Yaghi, Evelyn N. Wang, Water harvesting from air with metal-organic frameworks powered by natural sunlight, ***Science***, 356 (2017) 430-434.
98. J. Xu, T. Li, J. Chao, S. Wu, T. Yan, W. Li, B. Cao, R. Wang, Efficient Solar-Driven Water Harvesting from Arid Air with Metal-Organic Frameworks Modified by Hygroscopic Salt, ***Angewandte Chemie***, 59 (2020) 5202-5210.
99. H. Kim, S.R. Rao, E.A. Kapustin, L. Zhao, S. Yang, O.M. Yaghi, E.N. Wang, Adsorption-based atmospheric water harvesting device for arid climates, ***Nature communications***, 9 (2018) 1191.
100. A. LaPotin, H. Kim, S.R. Rao, E.N. Wang, Adsorption-Based Atmospheric Water Harvesting: Impact of Material and Component Properties on System-Level Performance, ***Accounts of chemical research***, 52 (2019) 1588-1597.
101. A. Mulchandani, P. Westerhoff, Geospatial Climatic Factors Influence Water Production of Solar Desiccant Driven Atmospheric Water Capture Devices, ***Environmental science & technology***, 54 (2020) 8310-8322.
102. H. Lu, W. Shi, Y. Guo, W. Guan, C. Lei, G. Yu, Materials Engineering for Atmospheric Water Harvesting: Progress and Perspectives, ***Advanced materials***, 34 (2022) e2110079.
103. M.N. Dods, S.C. Weston, J.R. Long, Prospects for Simultaneously Capturing Carbon Dioxide and Harvesting Water from Air, ***Advanced materials***, 34 (2022) e2204277.
104. Y. Zhong, L. Zhang, X. Li, B. El Fil, C.D. Díaz-Marín, A.C. Li, X. Liu, A. LaPotin, E.N. Wang, Bridging materials innovations to sorption-based atmospheric water harvesting devices, ***Nature Reviews Materials***, 9 (2024) 681-698.
105. T. Li, M. Wu, J. Xu, R. Du, T. Yan, P. Wang, Z. Bai, R. Wang, S. Wang, Simultaneous atmospheric water production and 24-hour power generation enabled by moisture-induced energy harvesting, ***Nature communications***, 13 (2022) 6771.
106. X. Li, W. Xie, J. Zhu, Interfacial Solar Steam/Vapor Generation for Heating and Cooling, ***Advanced science***, 9 (2022) e2104181.
107. W. Wang, Y. Shi, C. Zhang, R. Li, M. Wu, S. Zhuo, S. Aleid, P. Wang, Conversion and storage of solar energy for cooling, ***Energy & Environmental Science***, 15 (2022) 136-145.
108. L. Zhu, L. Tian, S. Jiang, L. Han, Y. Liang, Q. Li, S. Chen, Advances in photothermal regulation strategies: from efficient solar heating to daytime passive cooling, ***Chemical Society reviews***, 52 (2023) 7389-7460.
109. J. Tang, Z. Song, X. Lu, N. Li, L. Yang, T. Sun, Y. Wang, Y. Shao, H. Liu, G. Xue, Achieving excellent thermal transfer in highly light absorbing conical aerogel for simultaneous passive cooling and solar steam generation, ***Chemical Engineering Journal***, 429 (2022) 132089.
110. R. Li, W. Wang, Y. Shi, C.T. Wang, P. Wang, Advanced Material Design and Engineering for Water-Based Evaporative Cooling, ***Advanced materials***, 36 (2024) e2209460.
111. Z. Lu, E. Strobach, N. Chen, N. Ferralis, J.C. Grossman, Passive Sub-Ambient Cooling from a Transparent Evaporation-Insulation Bilayer, ***Joule***, 4 (2020) 2693-2701.
112. Z. Mao, Y. Yao, J. Shen, J. Liu, Y. Chen, B. Zhou, Y. Chen, Q. Wang, J. Lu, Passive interfacial cooling-induced sustainable electricity–water cogeneration, ***Nature Water***, 2 (2024) 93-100.

113. X. Li, J. Li, J. Lu, N. Xu, C. Chen, X. Min, B. Zhu, H. Li, L. Zhou, S. Zhu, T. Zhang, J. Zhu, Enhancement of Interfacial Solar Vapor Generation by Environmental Energy, ***Joule***, 2 (2018) 1331-1338.
114. R. Li, Y. Shi, M. Wu, S. Hong, P. Wang, Photovoltaic panel cooling by atmospheric water sorption–evaporation cycle, ***Nature Sustainability***, 3 (2020) 636-643.
115. W. Wang, S. Aleid, Y. Shi, C. Zhang, R. Li, M. Wu, S. Zhuo, P. Wang, Integrated solar-driven PV cooling and seawater desalination with zero liquid discharge, ***Joule***, 5 (2021) 1873-1887.
116. C. Wang, L. Hua, H. Yan, B. Li, Y. Tu, R. Wang, A Thermal Management Strategy for Electronic Devices Based on Moisture Sorption-Desorption Processes, ***Joule***, 4 (2020) 435-447.
117. S. Pu, J. Fu, Y. Liao, L. Ge, Y. Zhou, S. Zhang, S. Zhao, X. Liu, X. Hu, K. Liu, J. Chen, Promoting Energy Efficiency via a Self-Adaptive Evaporative Cooling Hydrogel, ***Advanced materials***, 32 (2020) e1907307.
118. A. Aili, X. Yin, R. Yang, Passive sub-ambient cooling: radiative cooling versus evaporative cooling, ***Applied Thermal Engineering***, 202 (2022) 117909.
119. B. Shao, Y. Song, Z. Song, Y. Wang, Y. Wang, R. Liu, B. Sun, Electricity Generation from Phase Transitions between Liquid and Gaseous Water, ***Advanced Energy Materials***, 13 (2023).
120. L. Li, S. Feng, Y. Bai, X. Yang, M. Liu, M. Hao, S. Wang, Y. Wu, F. Sun, Z. Liu, T. Zhang, Enhancing hydrovoltaic power generation through heat conduction effects, ***Nature communications***, 13 (2022) 1043.
121. K. Ni, B. Xu, Z. Wang, Q. Ren, W. Gu, B. Sun, R. Liu, X. Zhang, Ion-Diode-Like Heterojunction for Improving Electricity Generation from Water Droplets by Capillary Infiltration, ***Advanced materials***, 35 (2023) e2305438.
122. S. Sun, H. Li, Y. Wang, F. Zhang, Q. Tian, P. Yang, H.-Y. Mi, C. Liu, C. Shen, Ion-doped water-induced electricity generator with enhanced electrical output in water-segregated connection mode, ***Nano Energy***, 123 (2024) 109355.
123. J. Wang, X. Cao, X. Cui, H. Wang, H. Zhang, K. Wang, X. Li, Z. Li, Y. Zhou, Recent Advances of Green Electricity Generation: Potential in Solar Interfacial Evaporation System, ***Advanced materials***, 36 (2024) e2311151.
124. Q. Hu, X. Lin, G. Ren, J. Lü, W. Wang, D. Zhang, S. Zhou, Hydrovoltaic electricity generation induced by living leaf transpiration, ***Nature Water***, 2 (2024) 988-998.
125. K. Ni, Q. Ren, S. Liu, B. Sun, Y.-C. Lai, X. Zhang, R. Liu, Advances in asymmetric moist-electric generators with innovative heterogeneous structures, ***Energy & Environmental Science***, 17 (2024) 9406-9424.
126. G. Xue, Y. Xu, T. Ding, J. Li, J. Yin, W. Fei, Y. Cao, J. Yu, L. Yuan, L. Gong, J. Chen, S. Deng, J. Zhou, W. Guo, Water-evaporation-induced electricity with nanostructured carbon materials, ***Nature nanotechnology***, 12 (2017) 317-321.
127. S. Fang, J. Li, Y. Xu, C. Shen, W. Guo, Evaporating potential, ***Joule***, 6 (2022) 690-701.
128. X. Li, P. Bista, A.Z. Stetten, H. Bonart, M.T. Schür, S. Hardt, F. Bodziony, H. Marschall, A. Saal, X. Deng, R. Berger, S.A.L. Weber, H.-J. Butt, Spontaneous charging affects the motion of sliding drops, ***Nature Physics***, 18 (2022) 713-719.
129. Y. Chen, J. He, C. Ye, S. Tang, Achieving Ultrahigh Voltage Over 100 V and Remarkable Freshwater Harvesting Based on Thermodiffusion Enhanced Hydrovoltaic Generator, ***Advanced Energy Materials***, 14 (2024).
130. G. Hu, H. Liu, K. Liu, H. Wang, X. Wen, L. Liu, Y. She, L. Feng, R. Niu, J. Gong, All-In-One Carbon Foam Evaporators for Efficient Co-Generation of Freshwater and Electricity, ***Advanced Functional Materials***, (2025).
131. M. Ebrahimian Mashhadi, M.M. Hassan, R. Yang, Q. Lu, All-in-One Hybrid Solar-Driven Interfacial Evaporators for Cogeneration of Clean Water and Electricity, ***Advanced Functional Materials***, (2024).
132. A.V. Delgado, F. Gonzalez-Caballero, R.J. Hunter, L.K. Koopal, J. Lyklema, P. International Union of, P. Applied Chemistry, I.T.R. Biophysical Chemistry Division, Measurement and

interpretation of electrokinetic phenomena, ***Journal of colloid and interface science***, 309 (2007) 194-224.
133. Z. Zhai, J. Chen, X. Li, Q. Jiang, J. Bao, Y. Wang, Q. Liu, Y. Li, X. Li, PECVD-derived oxygen-doped vertical graphene-skinned carbon cloth toward efficient solar steam and water-evaporation-induced electricity cogeneration, ***Nano Energy***, 134 (2025) 110543.
134. C. Ge, D. Xu, X. Feng, X. Yang, Z. Song, Y. Song, J. Chen, Y. Liu, C. Gao, Y. Du, Z. Sun, W. Xu, J. Fang, Recent Advances in Fibrous Materials for Hydroelectricity Generation, ***Nano-micro letters***, 17 (2024) 29.
135. J. Lan, W. Hong, H. Li, S. Wang, C. Dong, Y. Li, Water-flow-induced high-efficiency solar vapor generation and electricity collection, ***Desalination***, 580 (2024) 117545.
136. F. Gao, J. Tu, J. Qu, J. Ge, Q. Yin, Y. Zang, W. Zhong, Z. Jiao, Dual mechanisms based on synergistic effects of evaporation potential and streaming potential for natural water evaporation, ***Journal of colloid and interface science***, 663 (2024) 251-261.
137. W. Deng, G. Feng, L. Li, X. Wang, H. Lu, X. Li, J. Li, W. Guo, J. Yin, Capillary front broadening for water-evaporation-induced electricity of one kilovolt, ***Energy & Environmental Science***, 16 (2023) 4442-4452.
138. Y. Xu, J. Xu, J. Zhang, X. Li, B. Fu, C. Song, W. Shang, P. Tao, T. Deng, All-in-one polymer sponge composite 3D evaporators for simultaneous high-flux solar-thermal desalination and electricity generation, ***Nano Energy***, 93 (2022) 106882.
139. C. Liu, C. Ye, Y. Wu, Y. Liu, Z. Liu, Z. Chen, R. Ma, N. Sakai, L. Xue, J. Sun, W. Zhang, W. Zhang, X. Wang, T. Sasaki, P. Xiong, J. Zhu, Atomic-scale engineering of cation vacancies in two-dimensional unilamellar metal oxide nanosheets for electricity generation from water evaporation, ***Nano Energy***, 110 (2023) 108348.
140. X. Ma, Z. Li, Z. Deng, D. Chen, X. Wang, X. Wan, Z. Fang, X. Peng, Efficiently cogenerating drinkable water and electricity from seawater via flexible MOF nanorod arrays, ***Journal of Materials Chemistry A***, 9 (2021) 9048-9055.
141. X. Zhang, J. Guo, Z. Zou, B. Luo, N. Li, Z. Yang, S. Xiong, X. Wang, Y. Xu, Y. Li, Bioinspired and 3D-printed solar evaporators for highly efficient freshwater-electricity co-generation, ***Materials horizons***, (2025).
142. H. Xia, W. Zhou, X. Qu, W. Wang, X. Wang, R. Qiao, Y. Zhang, X. Wu, C. Yang, B. Ding, L.Y. Hu, Y. Ran, K. Yu, S. Hu, J.F. Li, H.M. Cheng, H. Qiu, J. Yin, W. Guo, L. Qiu, Electricity generated by upstream proton diffusion in two-dimensional nanochannels, ***Nature nanotechnology***, 19 (2024) 1316-1322.
143. L. Li, Z. Zheng, C. Ge, Y. Wang, H. Dai, L. Li, S. Wang, Q. Gao, M. Liu, F. Sun, T. Zhang, A Flexible Tough Hydrovoltaic Coating for Wearable Sensing Electronics, ***Advanced materials***, 35 (2023) e2304099.
144. X. Lu, C. Mu, Y. Liu, L. Wu, Z. Tong, K. Huang, Recent advances in solar-driven interfacial evaporation coupling systems: Energy conversion, water purification, and seawater resource extraction, ***Nano Energy***, 120 (2024) 109180.
145. L. Cheng, L. Kong, X. Zhang, X. Kong, Form-stable phase change nanocapsules with photo and electric dual responses for multipurpose applications in energy storage and conversion, ***Solar Energy Materials and Solar Cells***, 235 (2022) 111461.
146. N. Xu, P. Zhu, Y. Sheng, L. Zhou, X. Li, H. Tan, S. Zhu, J. Zhu, Synergistic Tandem Solar Electricity-Water Generators, ***Joule***, 4 (2020) 347-358.
147. Q. Ji, N. Li, S. Wang, S. Li, F. Li, L. Yu, P. Murto, X. Xu, Synergistic solar-powered water-electricity generation via rational integration of semitransparent photovoltaics and interfacial steam generators, ***Journal of Materials Chemistry A***, 9 (2021) 21197-21208.
148. N. Zhang, T. Jiang, C. Guo, L. Qiao, Q. Ji, L. Yin, L. Yu, P. Murto, X. Xu, High-performance semitransparent polymer solar cells floating on water: Rational analysis of power generation, water evaporation and algal growth, ***Nano Energy***, 77 (2020) 105111.
149. M. Rastgar, K. Moradi, C. Burroughs, A. Hemmati, E. Hoek, M. Sadrzadeh, Harvesting Blue Energy Based on Salinity and Temperature Gradient: Challenges, Solutions, and Opportunities, ***Chemical reviews***, 123 (2023) 10156-10205.

150. B. Wang, X. Huang, Z. Liu, J. Zhang, C. Wei, B. Cai, P. Xie, A. Zhou, Solar interfacial evaporation systems with multi-field synergies boost water purification and blue energy harvesting technologies, ***Energy & Environmental Science***, 17 (2024) 7600-7626.
151. G. Liu, T. Chen, J. Xu, K. Wang, Blue energy harvesting on nanostructured carbon materials, ***Journal of Materials Chemistry A***, 6 (2018) 18357-18377.
152. Y. Mei, C.Y. Tang, Recent developments and future perspectives of reverse electrodialysis technology: A review, ***Desalination***, 425 (2018) 156-174.
153. C. Klaysom, T.Y. Cath, T. Depuydt, I.F.J. Vankelecom, Forward and pressure retarded osmosis: potential solutions for global challenges in energy and water supply, ***Chemical Society reviews***, 42 (2013) 6959.
154. M. Simoncelli, N. Ganfoud, A. Sene, M. Haefele, B. Daffos, P.-L. Taberna, M. Salanne, P. Simon, B. Rotenberg, Blue Energy and Desalination with Nanoporous Carbon Electrodes: Capacitance from Molecular Simulations to Continuous Models, ***Physical Review X***, 8 (2018).
155. N.Y. Yip, D. Brogioli, H.V.M. Hamelers, K. Nijmeijer, Salinity Gradients for Sustainable Energy: Primer, Progress, and Prospects, ***Environmental science & technology***, 50 (2016) 12072-12094.
156. B.E. Logan, M. Elimelech, Membrane-based processes for sustainable power generation using water, ***Nature***, 488 (2012) 313-319.
157. J. Yang, B. Tu, G. Zhang, P. Liu, K. Hu, J. Wang, Z. Yan, Z. Huang, M. Fang, J. Hou, Q. Fang, X. Qiu, L. Li, Z. Tang, Advancing osmotic power generation by covalent organic framework monolayer, ***Nature nanotechnology***, 17 (2022) 622-628.
158. P. Yang, K. Liu, Q. Chen, J. Li, J. Duan, G. Xue, Z. Xu, W. Xie, J. Zhou, Solar-driven simultaneous steam production and electricity generation from salinity, ***Energy & Environmental Science***, 10 (2017) 1923-1927.
159. M. Wang, Y. Wei, X. Wang, R. Li, S. Zhang, K. Wang, R. Wang, H. Chang, C. Wang, N. Ren, S.-H. Ho, An integrated system with functions of solar desalination, power generation and crop irrigation, ***Nature Water***, 1 (2023) 716-724.
160. G. Liu, T. Chen, J. Xu, G. Li, K. Wang, Solar evaporation for simultaneous steam and power generation, ***Journal of Materials Chemistry A***, 8 (2020) 513-531.
161. L. Zhu, T. Ding, M. Gao, C.K.N. Peh, G.W. Ho, Shape Conformal and Thermal Insulative Organic Solar Absorber Sponge for Photothermal Water Evaporation and Thermoelectric Power Generation, ***Advanced Energy Materials***, 9 (2019).
162. Y. Du, P. Liu, H. Zhang, L. Zou, K. Deng, X. Li, W. Tian, J. Ji, Nature-Inspired Structure-Engineered TiN/TiO2 Nanotubes Array Toward Solar Desalination Synergy with Photothermal-Enhanced Degradation and Thermoelectric Generation, ***Advanced Functional Materials***, 34 (2023).
163. H. Li, H. Cheng, B. Wu, W. Wang, Y. Zhang, J. Han, Solar-driven hydrovoltaic-pyroelectric hybrid generator for efficiently harvesting water transformation energy, ***Chemical Engineering Journal***, 495 (2024) 153497.
164. D. Kraemer, B. Poudel, H.P. Feng, J.C. Caylor, B. Yu, X. Yan, Y. Ma, X. Wang, D. Wang, A. Muto, K. McEnaney, M. Chiesa, Z. Ren, G. Chen, High-performance flat-panel solar thermoelectric generators with high thermal concentration, ***Nature materials***, 10 (2011) 532-538.
165. P. Cao, L. Zhao, Z. Yang, P. Yuan, Y. Zhang, Q. Li, Carbon Nanotube Network-Based Solar-Thermal Water Evaporator and Thermoelectric Module for Electricity Generation, ***ACS Applied Nano Materials***, 4 (2021) 8906-8912.
166. X. Mu, J. Zhou, P. Wang, H. Chen, T. Yang, S. Chen, L. Miao, T. Mori, A robust starch–polyacrylamide hydrogel with scavenging energy harvesting capacity for efficient solar thermoelectricity–freshwater cogeneration, ***Energy & Environmental Science***, 15 (2022) 3388-3399.
167. Z. Fu, D. Zhong, S. Zhou, L. Zhang, W. Long, J. Zhang, X. Wang, J. Xu, J. Qin, J. Gong, L. Li, L. Xia, B. Yu, W. Xu, Scalable Asymmetric Fabric Evaporator for Solar Desalination and Thermoelectricity Generation, ***Advanced science***, 11 (2024) e2406474.

168. Y. Cui, J. Liu, Z. Li, M. Ji, M. Zhao, M. Shen, X. Han, T. Jia, C. Li, Y. Wang, Donor–Acceptor-Type Organic-Small-Molecule-Based Solar-Energy-Absorbing Material for Highly Efficient Water Evaporation and Thermoelectric Power Generation, ***Advanced Functional Materials***, 31 (2021).
169. A.D. Poletayev, I.S. McKay, W.C. Chueh, A. Majumdar, Continuous electrochemical heat engines, ***Energy & Environmental Science***, 11 (2018) 2964-2971.
170. M. Rahimi, A.P. Straub, F. Zhang, X. Zhu, M. Elimelech, C.A. Gorski, B.E. Logan, Emerging electrochemical and membrane-based systems to convert low-grade heat to electricity, ***Energy & Environmental Science***, 11 (2018) 276-285.
171. Y. Wang, Y. Zhang, X. Xin, J. Yang, M. Wang, R. Wang, P. Guo, W. Huang, A.J. Sobrido, B. Wei, X. Li, In situ photocatalytically enhanced thermogalvanic cells for electricity and hydrogen production, ***Science***, 381 (2023) 291-296.
172. J. Duan, G. Feng, B. Yu, J. Li, M. Chen, P. Yang, J. Feng, K. Liu, J. Zhou, Aqueous thermogalvanic cells with a high Seebeck coefficient for low-grade heat harvest, ***Nature communications***, 9 (2018).
173. B.Y. Yu, J.J. Duan, H.J. Cong, W.K. Xie, R. Liu, X.Y. Zhuang, H. Wang, B. Qi, M. Xu, Z.L. Wang, J. Zhou, Thermosensitive crystallization-boosted liquid thermocells for low-grade heat harvesting, ***Science***, 370 (2020) 342-+.
174. J. Zhao, X. Wu, H. Yu, Y. Wang, P. Wu, X. Yang, D. Chu, G. Owens, H. Xu, Regenerable aerogel-based thermogalvanic cells for efficient low-grade heat harvesting from solar radiation and interfacial solar evaporation systems, ***EcoMat***, 5 (2022).
175. D. Zhang, H. Wu, C.R. Bowen, Y. Yang, Recent Advances in Pyroelectric Materials and Applications, ***Small***, 17 (2021).
176. C.R. Bowen, J. Taylor, E. LeBoulbar, D. Zabek, A. Chauhan, R. Vaish, Pyroelectric materials and devices for energy harvesting applications, ***Energy Environ. Sci.***, 7 (2014) 3836-3856.
177. T. Ding, L. Zhu, X.Q. Wang, K.H. Chan, X. Lu, Y. Cheng, G.W. Ho, Hybrid Photothermal Pyroelectric and Thermogalvanic Generator for Multisituation Low Grade Heat Harvesting, ***Advanced Energy Materials***, 8 (2018).
178. M. Jiang, Q. Shen, J. Zhang, S. An, S. Ma, P. Tao, C. Song, B. Fu, J. Wang, T. Deng, W. Shang, Bioinspired Temperature Regulation in Interfacial Evaporation, ***Advanced Functional Materials***, 30 (2020).
179. F.R. Fan, W. Tang, Z.L. Wang, Flexible Nanogenerators for Energy Harvesting and Self-Powered Electronics, ***Advanced materials***, 28 (2016) 4283-4305.
180. C. Qiu, B. Wang, N. Zhang, S. Zhang, J. Liu, D. Walker, Y. Wang, H. Tian, T.R. Shrout, Z. Xu, L.-Q. Chen, F. Li, Transparent ferroelectric crystals with ultrahigh piezoelectricity, ***Nature***, 577 (2020) 350-354.
181. X.Q. Wang, C.F. Tan, K.H. Chan, X. Lu, L. Zhu, S.W. Kim, G.W. Ho, In-built thermo-mechanical cooperative feedback mechanism for self-propelled multimodal locomotion and electricity generation, ***Nature communications***, 9 (2018) 3438.
182. L. Zhu, M. Gao, C.K.N. Peh, X. Wang, G.W. Ho, Self-Contained Monolithic Carbon Sponges for Solar-Driven Interfacial Water Evaporation Distillation and Electricity Generation, ***Advanced Energy Materials***, 8 (2018).
183. S. Meng, C.Y. Tang, J. Jia, J. Yang, M.B. Yang, W. Yang, A Wave-Driven Piezoelectric Solar Evaporator for Water Purification, ***Advanced Energy Materials***, 12 (2022).
184. M. Gao, C.K. Peh, H.T. Phan, L. Zhu, G.W. Ho, Solar Absorber Gel: Localized Macro-Nano Heat Channeling for Efficient Plasmonic Au Nanoflowers Photothermic Vaporization and Triboelectric Generation, ***Advanced Energy Materials***, 8 (2018).
185. J. Luo, Z.L. Wang, Recent advances in triboelectric nanogenerator based self-charging power systems, ***Energy Storage Materials***, 23 (2019) 617-628.
186. G. Chen, N. Li, J. He, L. Qiao, F. Li, S. Wang, L. Yu, P. Murto, X. Li, X. Xu, Design of self-righting steam generators for solar-driven interfacial evaporation and self-powered water wave detection, ***Journal of Materials Chemistry A***, 8 (2020) 24664-24674.

187. F. Huang, P. Yang, Z. Liu, D. Yang, L. Huang, Y. Shi, X. Tao, Y. Chen, H. Li, X. Chen, Z. Bian, A hybrid nanogenerator for collecting both water wave and steam evaporation energy, ***Nano Energy***, 110 (2023) 108346.
188. M. Wang, Y. Wei, R. Li, X. Wang, C. Wang, N. Ren, S.-H. Ho, Sustainable Seawater Desalination and Energy Management: Mechanisms, Strategies, and the Way Forward, ***Research***, 6 (2023).
189. X. Li, Y. Liu, Y. Xu, P. Tao, T. Deng, Solid–Liquid Phase Change Composite Materials for Direct Solar–Thermal Energy Harvesting and Storage, ***Accounts of Materials Research***, 4 (2023) 484-495.
190. J. Lan, W. Hong, C. Dong, H. Zhang, B. Hu, L. Zhang, Y. Li, X. Niu, H. Li, Enhancing the sustainability of interfacial evaporation to mitigate solar intermittency via phase change thermal storage, ***Chemical Engineering Journal***, 500 (2024) 156855.
191. L. Geng, L. Li, H. Zhang, M. Zhong, P. Mu, J. Li, Interfacial solar evaporator synergistic phase change energy storage for all-day steam generation, ***Journal of Materials Chemistry A***, 10 (2022) 15485-15496.
192. Y. Zhang, J. Tang, J. Chen, Y. Zhang, X. Chen, M. Ding, W. Zhou, X. Xu, H. Liu, G. Xue, Accelerating the solar-thermal energy storage via inner-light supplying with optical waveguide, ***Nature communications***, 14 (2023) 3456.
193. R. Niu, J. Ren, J.J. Koh, L. Chen, J. Gong, J. Qu, X. Xu, J. Azadmanjiri, J. Min, Bio-Inspired Sandwich-Structured All-Day-Round Solar Evaporator for Synergistic Clean Water and Electricity Generation, ***Advanced Energy Materials***, 13 (2023).
194. M. Zhang, K. Sun, Z. Zheng, H. Liu, X. Wang, Development of MXene-decorated sodium alginate/SiO2@n-docosane hierarchical phase-change microcapsules for solar-driven sustainable seawater desalination, ***Desalination***, 550 (2023) 116380.
195. W. Li, Z. Zheng, Z. Qian, H. Liu, X. Wang, Phase Change Material Boosting Electricity Output and Freshwater Production through Hierarchical-Structured 3D Solar Evaporator, ***Advanced Functional Materials***, 34 (2024).
196. B. Gong, H. Yang, S. Wu, Y. Tian, J. Yan, K. Cen, Z. Bo, K. Ostrikov, Phase change material enhanced sustained and energy-efficient solar-thermal water desalination, ***Applied Energy***, 301 (2021) 117463.
197. Z. Zheng, H. Liu, D. Wu, X. Wang, Polyimide/MXene hybrid aerogel-based phase-change composites for solar-driven seawater desalination, ***Chemical Engineering Journal***, 440 (2022) 135862.
198. Z.Y. Ho, R. Bahar, C.H. Koo, Passive solar stills coupled with Fresnel lens and phase change material for sustainable solar desalination in the tropics, ***Journal of Cleaner Production***, 334 (2022) 130279.
199. P. Tao, C. Chang, Z. Tong, H. Bao, C. Song, J. Wu, W. Shang, T. Deng, Magnetically-accelerated large-capacity solar-thermal energy storage within high-temperature phase-change materials, ***Energy & Environmental Science***, 12 (2019) 1613-1621.
200. Z. Wang, Z. Tong, Q. Ye, H. Hu, X. Nie, C. Yan, W. Shang, C. Song, J. Wu, J. Wang, H. Bao, P. Tao, T. Deng, Dynamic tuning of optical absorbers for accelerated solar-thermal energy storage, ***Nature communications***, 8 (2017) 1478.
201. Y. Liang, J. Guo, J.J. Li, J. Mao, A.Q. Xie, L. Zhu, S. Chen, Robust and Flexible 3D Photothermal Evaporator with Heat Storage for High-Performance Solar-Driven Evaporation, ***Advanced Sustainable Systems***, 6 (2022).
202. X. Wu, M.E. Robson, J.L. Phelps, J.S. Tan, B. Shao, G. Owens, H. Xu, A flexible photothermal cotton-CuS nanocage-agarose aerogel towards portable solar steam generation, ***Nano Energy***, 56 (2019) 708-715.
203. X. Wang, Q. Liu, S. Wu, B. Xu, H. Xu, Multilayer Polypyrrole Nanosheets with Self-Organized Surface Structures for Flexible and Efficient Solar-Thermal Energy Conversion, ***Advanced materials***, 31 (2019) e1807716.

204. C. Chang, X. Nie, X. Li, P. Tao, B. Fu, Z. Wang, J. Xu, Q. Ye, J. Zhang, C. Song, W. Shang, T. Deng, Bioinspired roll-to-roll solar-thermal energy harvesting within form-stable flexible composite phase change materials, ***Journal of Materials Chemistry A***, 8 (2020) 20970-20978.
205. M.S. Irshad, N. Arshad, J. Zhang, C. Song, N. Mushtaq, M. Alomar, T. Shamim, V.-D. Dao, H. Wang, X. Wang, H. Zhang, Wormlike Perovskite Oxide Coupled with Phase-Change Material for All-Weather Solar Evaporation and Thermal Storage Applications, ***Advanced Energy and Sustainability Research***, 4 (2023).
206. X. Li, J. Zhang, Y. Liu, Y. Xu, K. Cui, Z. Yao, B. Fu, C. Song, W. Shang, P. Tao, T. Deng, Supercooled sugar alcohols stabilized by alkali hydroxides for long-term room-temperature phase change solar-thermal energy storage, ***Chemical Engineering Journal***, 452 (2023) 139328.
207. W. Aftab, A. Usman, J. Shi, K. Yuan, M. Qin, R. Zou, Phase change material-integrated latent heat storage systems for sustainable energy solutions, ***Energy & Environmental Science***, 14 (2021) 4268-4291.
208. L.A. Weinstein, J. Loomis, B. Bhatia, D.M. Bierman, E.N. Wang, G. Chen, Concentrating Solar Power, ***Chemical reviews***, 115 (2015) 12797-12838.
209. X. Li, J. Zhang, B. Fu, C. Song, W. Shang, P. Tao, T. Deng, Erythritol impregnated within surface-roughened hydrophilic metal foam for medium-temperature solar-thermal energy harvesting, ***Energy Conversion and Management***, 222 (2020) 113241.
210. A.L. Cottrill, A.T. Liu, Y. Kunai, V.B. Koman, A. Kaplan, S.G. Mahajan, P. Liu, A.R. Toland, M.S. Strano, Ultra-high thermal effusivity materials for resonant ambient thermal energy harvesting, ***Nature communications***, 9 (2018).
211. F. Tao, M. Green, A.V. Garcia, T. Xiao, A.T. Van Tran, Y. Zhang, Y. Yin, X. Chen, Recent progress of nanostructured interfacial solar vapor generators, ***Applied Materials Today***, 17 (2019) 45-84.
212. J. Deng, J. Li, P. Chen, X. Fang, X. Sun, Y. Jiang, W. Weng, B. Wang, H. Peng, Tunable Photothermal Actuators Based on a Pre-programmed Aligned Nanostructure, ***Journal of the American Chemical Society***, 138 (2015) 225-230.
213. S.R. Mishra, J.B. Tracy, Sequential Actuation of Shape-Memory Polymers through Wavelength-Selective Photothermal Heating of Gold Nanospheres and Nanorods, ***ACS Applied Nano Materials***, 1 (2018) 3063-3067.
214. M.J. Villangca, D. Palima, A.R. Bañas, J. Glückstad, Light-driven micro-tool equipped with a syringe function, ***Light: Science & Applications***, 5 (2016) e16148-e16148.
215. J. Li, R. Zhang, L. Mou, M. Jung de Andrade, X. Hu, K. Yu, J. Sun, T. Jia, Y. Dou, H. Chen, S. Fang, D. Qian, Z. Liu, Photothermal Bimorph Actuators with In-Built Cooler for Light Mills, Frequency Switches, and Soft Robots, ***Advanced Functional Materials***, 29 (2019).
216. H.K. Bisoyi, A.M. Urbas, Q. Li, Soft Materials Driven by Photothermal Effect and Their Applications, ***Advanced Optical Materials***, 6 (2018).
217. Y.-Y. Xiao, Z.-C. Jiang, J.-B. Hou, Y. Zhao, Desynchronized liquid crystalline network actuators with deformation reversal capability, ***Nature communications***, 12 (2021).
218. Z.F. Sun, Y. Yamauchi, F. Araoka, Y.S. Kim, J. Bergueiro, Y. Ishida, Y. Ebina, T. Sasaki, T. Hikima, T. Aida, An Anisotropic Hydrogel Actuator Enabling Earthworm-Like Directed Peristaltic Crawling, ***Angew Chem Int Edit***, 57 (2018) 15772-15776.
219. Z. Chen, J. Li, Y. Zheng, Heat-Mediated Optical Manipulation, ***Chemical reviews***, 122 (2022) 3122-3179.
220. H. Sipova-Jungova, D. Andren, S. Jones, M. Kall, Nanoscale Inorganic Motors Driven by Light: Principles, Realizations, and Opportunities, ***Chemical reviews***, 120 (2020) 269-287.
221. J. Frueh, S. Rutkowski, T. Si, M. Gai, C. Gao, S.I. Tverdokhlebov, G. Qiu, J. Schmitt, Q. He, Y.-X. Ren, J. Wang, Propulsion Mechanisms of Light-Driven Plasmonic Colloidal Micromotors, ***Advanced Photonics Research***, 3 (2021).
222. I. Buttinoni, G. Volpe, F. Kummel, G. Volpe, C. Bechinger, Active Brownian motion tunable by light, ***Journal of physics. Condensed matter : an Institute of Physics journal***, 24 (2012) 284129.

223. M. Liu, T. Zentgraf, Y. Liu, G. Bartal, X. Zhang, Light-driven nanoscale plasmonic motors, ***Nature nanotechnology***, 5 (2010) 570-573.
224. J.G. Gibbs, Y.P. Zhao, Autonomously motile catalytic nanomotors by bubble propulsion, ***Applied Physics Letters***, 94 (2009).
225. W. Qin, T. Peng, Y. Gao, F. Wang, X. Hu, K. Wang, J. Shi, D. Li, J. Ren, C. Fan, Catalysis-Driven Self-Thermophoresis of Janus Plasmonic Nanomotors, ***Angewandte Chemie***, 129 (2016) 530-533.
226. F. Meng, W. Hao, S. Yu, R. Feng, Y. Liu, F. Yu, P. Tao, W. Shang, J. Wu, C. Song, T. Deng, Vapor-Enabled Propulsion for Plasmonic Photothermal Motor at the Liquid/Air Interface, ***Journal of the American Chemical Society***, 139 (2017) 12362-12365.
227. X. Wang, L. Baraban, V.R. Misko, F. Nori, T. Huang, G. Cuniberti, J. Fassbender, D. Makarov, Visible Light Actuated Efficient Exclusion Between Plasmonic Ag/AgCl Micromotors and Passive Beads, ***Small***, 14 (2018) e1802537.
228. M. Franzl, S. Muinos-Landin, V. Holubec, F. Cichos, Fully Steerable Symmetric Thermoplasmonic Microswimmers, ***ACS nano***, 15 (2021) 3434-3440.
229. X. Wang, V. Sridhar, S. Guo, N. Talebi, A. Miguel-López, K. Hahn, P.A. van Aken, S. Sánchez, Fuel-Free Nanocap-Like Motors Actuated Under Visible Light, ***Advanced Functional Materials***, 28 (2018).
230. M. Xuan, Z. Wu, J. Shao, L. Dai, T. Si, Q. He, Near Infrared Light-Powered Janus Mesoporous Silica Nanoparticle Motors, ***Journal of the American Chemical Society***, 138 (2016) 6492-6497.
231. H. Ding, Z. Chen, C. Ponce, Y. Zheng, Optothermal rotation of micro-/nano-objects, ***Chemical communications***, 59 (2023) 2208-2221.
232. L. Shao, M. Käll, Light-Driven Rotation of Plasmonic Nanomotors, ***Advanced Functional Materials***, 28 (2018).
233. L. Shao, Z.J. Yang, D. Andrén, P. Johansson, M. Käll, Gold Nanorod Rotary Motors Driven by Resonant Light Scattering, ***ACS nano***, 9 (2015) 12542-12551.
234. L. Xin, C. Zhou, X. Duan, N. Liu, A rotary plasmonic nanoclock, ***Nature communications***, 10 (2019) 5394.
235. B. Zeng, Y. Wang, M.E. Zaytsev, C. Xia, H.J.W. Zandvliet, D. Lohse, Giant plasmonic bubbles nucleation under different ambient pressures, ***Physical review. E***, 102 (2020) 063109.
236. Y. Wang, M.E. Zaytsev, G. Lajoinie, H.L. The, J.C.T. Eijkel, A. van den Berg, M. Versluis, B.M. Weckhuysen, X. Zhang, H.J.W. Zandvliet, D. Lohse, Giant and explosive plasmonic bubbles by delayed nucleation, ***Proceedings of the National Academy of Sciences of the United States of America***, 115 (2018) 7676-7681.
237. S. Jones, D. Andren, T.J. Antosiewicz, A. Stilgoe, H. Rubinsztein-Dunlop, M. Kall, Strong Transient Flows Generated by Thermoplasmonic Bubble Nucleation, ***ACS nano***, 14 (2020) 17468-17475.
238. B. Han, Y.L. Zhang, Q.D. Chen, H.B. Sun, Carbon-Based Photothermal Actuators, ***Advanced Functional Materials***, 28 (2018).
239. W.Y. Tsai, J.S. Huang, C.B. Huang, Selective trapping or rotation of isotropic dielectric microparticles by optical near field in a plasmonic archimedes spiral, ***Nano letters***, 14 (2014) 547-552.
240. M. Xuan, R. Mestre, C. Gao, C. Zhou, Q. He, S. Sánchez, Noncontinuous Super-Diffusive Dynamics of a Light-Activated Nanobottle Motor, ***Angewandte Chemie***, 130 (2018) 6954-6958.
241. J. Feng, X. Li, T. Xu, X. Zhang, X. Du, Photothermal-driven micro/nanomotors: From structural design to potential applications, ***Acta biomaterialia***, 173 (2024) 1-35.
242. J. Wang, Y. Kong, Z. Liu, H. Wang, Solar-driven interfacial evaporation: Design and application progress of structural evaporators and functional distillers, ***Nano Energy***, 108 (2023) 108115.

243. Y. Zhou, T. Ding, M. Gao, K.H. Chan, Y. Cheng, J. He, G.W. Ho, Controlled heterogeneous water distribution and evaporation towards enhanced photothermal water-electricity-hydrogen production, ***Nano Energy***, 77 (2020) 105102.
244. S. Chu, A. Majumdar, Opportunities and challenges for a sustainable energy future, ***Nature***, 488 (2012) 294-303.
245. H.T. Kim, L. Philip, A. McDonagh, M. Johir, J. Ren, H.K. Shon, L.D. Tijing, Recent Advances in High-Rate Solar-Driven Interfacial Evaporation, ***Advanced science***, 11 (2024) e2401322.
246. M. Gao, L. Zhu, C.K. Peh, G.W. Ho, Solar absorber material and system designs for photothermal water vaporization towards clean water and energy production, ***Energy & Environmental Science***, 12 (2019) 841-864.
247. C. Dang, Y. Cao, H. Nie, W. Lang, J. Zhang, G. Xu, M. Zhu, Structure integration and architecture of solar-driven interfacial desalination from miniaturization designs to industrial applications, ***Nature Water***, 2 (2024) 115-126.
248. H. Li, K. Wahab Ahmed, M.A. Abdelsalam, M. Fowler, X.-Y. Wu, From seawater to hydrogen via direct photocatalytic vapor splitting: A review on device design and system integration, ***Frontiers in Energy***, 18 (2024) 291-307.
249. P. Shi, J. Li, Y. Song, N. Xu, J. Zhu, Cogeneration of Clean Water and Valuable Energy/Resources via Interfacial Solar Evaporation, ***Nano Letters***, 24 (2024) 5673-5682.
250. Y. Lu, H. Zhang, Y. Wang, X. Zhu, W. Xiao, H. Xu, G. Li, Y. Li, D. Fan, H. Zeng, Z. Chen, X. Yang, Solar-Driven Interfacial Evaporation Accelerated Electrocatalytic Water Splitting on 2D Perovskite Oxide/MXene Heterostructure, ***Advanced Functional Materials***, 33 (2023).
251. Y. Lu, H. Zhang, D. Fan, Z. Chen, X. Yang, Coupling solar-driven photothermal effect into photocatalysis for sustainable water treatment, ***Journal of hazardous materials***, 423 (2022) 127128.
252. S. Guo, X. Li, J. Li, B. Wei, Boosting photocatalytic hydrogen production from water by photothermally induced biphase systems, ***Nature communications***, 12 (2021).
253. M. Gao, P.K.N. Connor, G.W. Ho, Plasmonic photothermic directed broadband sunlight harnessing for seawater catalysis and desalination, ***Energy & Environmental Science***, 9 (2016) 3151-3160.
254. M. Gao, C.K. Peh, L. Zhu, G. Yilmaz, G.W. Ho, Photothermal Catalytic Gel Featuring Spectral and Thermal Management for Parallel Freshwater and Hydrogen Production, ***Advanced Energy Materials***, 10 (2020).
255. C. Pornrungroj, A.B. Mohamad Annuar, Q. Wang, M. Rahaman, S. Bhattacharjee, V. Andrei, E. Reisner, Hybrid photothermal–photocatalyst sheets for solar-driven overall water splitting coupled to water purification, ***Nature Water***, 1 (2023) 952-960.
256. W.H. Lee, C.W. Lee, G.D. Cha, B.H. Lee, J.H. Jeong, H. Park, J. Heo, M.S. Bootharaju, S.H. Sunwoo, J.H. Kim, K.H. Ahn, D.H. Kim, T. Hyeon, Floatable photocatalytic hydrogel nanocomposites for large-scale solar hydrogen production, ***Nature nanotechnology***, 18 (2023) 754-762.
257. O.K. Varghese, M. Paulose, T.J. Latempa, C.A. Grimes, High-rate solar photocatalytic conversion of CO2 and water vapor to hydrocarbon fuels, ***Nano letters***, 9 (2009) 731-737.
258. V. Andrei, G.M. Ucoski, C. Pornrungroj, C. Uswachoke, Q. Wang, D.S. Achilleos, H. Kasap, K.P. Sokol, R.A. Jagt, H. Lu, T. Lawson, A. Wagner, S.D. Pike, D.S. Wright, R.L.Z. Hoye, J.L. MacManus-Driscoll, H.J. Joyce, R.H. Friend, E. Reisner, Floating perovskite-BiVO(4) devices for scalable solar fuel production, ***Nature***, 608 (2022) 518-522.
259. O. Neumann, A.D. Neumann, S. Tian, C. Thibodeaux, S. Shubhankar, J. Müller, E. Silva, A. Alabastri, S.W. Bishnoi, P. Nordlander, N.J. Halas, Combining Solar Steam Processing and Solar Distillation for Fully Off-Grid Production of Cellulosic Bioethanol, ***ACS Energy Letters***, 2 (2016) 8-13.
260. Q. Zhu, Y. Xuan, D. Zhao, H. Yu, J. Wang, X. Liu, A lotus-inspired triphase continuous CO2 photo-thermal reduction system with wide applicability and tunable interface environment, ***Materials Today Energy***, 48 (2025) 101777.

261. M. Oelgemöller, Solar Photochemical Synthesis: From the Beginnings of Organic Photochemistry to the Solar Manufacturing of Commodity Chemicals, ***Chemical reviews***, 116 (2016) 9664-9682.
262. G. Chen, G.I.N. Waterhouse, R. Shi, J. Zhao, Z. Li, L.Z. Wu, C.H. Tung, T. Zhang, From Solar Energy to Fuels: Recent Advances in Light-Driven C1 Chemistry, ***Angewandte Chemie International Edition***, 58 (2019) 17528-17551.
263. D. Spasiano, L. del Pilar Prieto Rodriguez, J.C. Olleros, S. Malato, R. Marotta, R. Andreozzi, TiO2/Cu(II) photocatalytic production of benzaldehyde from benzyl alcohol in solar pilot plant reactor, ***Applied Catalysis B: Environmental***, 136-137 (2013) 56-63.
264. S. Piao, P. Ciais, Y. Huang, Z. Shen, S. Peng, J. Li, L. Zhou, H. Liu, Y. Ma, Y. Ding, P. Friedlingstein, C. Liu, K. Tan, Y. Yu, T. Zhang, J. Fang, The impacts of climate change on water resources and agriculture in China, ***Nature***, 467 (2010) 43-51.
265. M.S. Diallo, M.R. Kotte, M. Cho, Mining Critical Metals and Elements from Seawater: Opportunities and Challenges, ***Environmental science & technology***, 49 (2015) 9390-9399.
266. L. Li, X. Sun, J. Miao, H. Wang, Y. Song, D. Tang, Gradient engineering in interfacial evaporation for water, energy, and mineral harvesting, ***Energy & Environmental Science***, (2025).
267. Y. Song, S. Fang, N. Xu, J. Zhu, Solar-driven interfacial evaporation technologies for food, energy and water, ***Nature Reviews Clean Technology***, 1 (2025) 55-74.
268. P. Shi, J. Li, Y. Song, N. Xu, J. Zhu, Cogeneration of Clean Water and Valuable Energy/ Resources via Interfacial Solar Evaporation, ***Nano letters***, 24 (2024) 5673-5682.
269. S. Sun, C. Shi, Y. Kuang, M. Li, S. Li, H. Chan, S. Zhang, G. Chen, A. Nilghaz, R. Cao, J. Tian, 3D-printed solar evaporator with seashell ornamentation-inspired structure for zero liquid discharge desalination, ***Water research***, 226 (2022) 119279.
270. H. Peng, D. Wang, S. Fu, Artificial transpiration with asymmetric photothermal textile for continuous solar-driven evaporation, spatial salt harvesting and electrokinetic power generation, ***Chemical Engineering Journal***, 426 (2021) 131818.
271. L. Wang, Z. Liu, J. Xu, K. Wang, Q. Wang, G. Liu, Conical solar-thermo-radiative evaporator for sustainable desalination and salt recovery, ***Desalination***, 567 (2023) 116993.
272. C. Zhang, Y. Shi, L. Shi, H. Li, R. Li, S. Hong, S. Zhuo, T. Zhang, P. Wang, Designing a next generation solar crystallizer for real seawater brine treatment with zero liquid discharge, ***Nature communications***, 12 (2021) 998.
273. Y. Shi, C. Zhang, R. Li, S. Zhuo, Y. Jin, L. Shi, S. Hong, J. Chang, C. Ong, P. Wang, Solar Evaporator with Controlled Salt Precipitation for Zero Liquid Discharge Desalination, ***Environmental science & technology***, 52 (2018) 11822-11830.
274. Y. Bian, Z. Ye, G. Zhao, K. Tang, Y. Teng, S. Chen, L. Zhao, X. Yuan, S. Zhu, J. Ye, H. Lu, Y. Yang, L. Fu, S. Gu, Enhanced Contactless Salt-Collecting Solar Desalination, ***ACS applied materials & interfaces***, (2022).
275. X.Q. Li, R.X. Lin, G. Ni, N. Xu, X.Z. Hu, B. Zhu, G.X. Lv, J.L. Li, S.N. Zhu, J. Zhu, Three-dimensional artificial transpiration for efficient solar waste-water treatment, ***National science review***, 5 (2018) 70-77.
276. N. Xu, J. Li, Y. Wang, C. Fang, X. Li, Y. Wang, L. Zhou, B. Zhu, Z. Wu, S. Zhu, J. Zhu, A water lily-inspired hierarchical design for stable and efficient solar evaporation of high-salinity brine, ***Science advances***, 5 (2019) eaaw7013.
277. L. Wu, Z. Dong, Z. Cai, T. Ganapathy, N.X. Fang, C. Li, C. Yu, Y. Zhang, Y. Song, Highly efficient three-dimensional solar evaporator for high salinity desalination by localized crystallization, ***Nature communications***, 11 (2020) 521.
278. J. Xu, Z. Wang, C. Chang, B. Fu, P. Tao, C. Song, W. Shang, T. Deng, Solar-driven interfacial desalination for simultaneous freshwater and salt generation, ***Desalination***, 484 (2020) 114423.
279. G. Jiang, L. Wang, Z. Tian, C. Chen, X. Hu, R. Peng, D. Li, H. Zhang, P. Fan, M. Zhong, Boosting water evaporation via continuous formation of a 3D thin film through triple-level super-wicking routes, ***Materials horizons***, 10 (2023) 3523-3535.

280. X. Ma, X. Jia, G. Yao, D. Wen, Double-Sided Suspending Evaporator with Top Water Supply for Concurrent Solar Evaporation and Salt Harvesting, ***ACS Sustain Chem Eng***, 10 (2022) 12843-12851.
281. R. Gu, Z. Yu, Y. Sun, P. Xie, Y. Li, S. Cheng, Enhancing stability of interfacial solar evaporator in high-salinity solutions by managing salt precipitation with Janus-based directional salt transfer structure, ***Desalination***, 524 (2022) 115470.
282. Z. Wang, J. Gao, J. Zhou, J. Gong, L. Shang, H. Ye, F. He, S. Peng, Z. Lin, Y. Li, F. Caruso, Engineering Metal-Phenolic Networks for Solar Desalination with Directional Salt Crystallization, ***Advanced materials***, 35 (2023) e2209015.
283. Y. Zhang, T. Xiong, L. Suresh, H. Qu, X. Zhang, Q. Zhang, J. Yang, S.C. Tan, Guaranteeing Complete Salt Rejection by Channeling Saline Water through Fluidic Photothermal Structure toward Synergistic Zero Energy Clean Water Production and In Situ Energy Generation, ***ACS Energy Letters***, 5 (2020) 3397-3404.
284. Y. Wu, C. Ma, K. Zhu, L. Jin, L. Song, L. Li, Y. Lu, Y. Zheng, Y. Zhang, X. Zheng, S. Wu, Y. Pang, Z. Shen, S.C. Tan, H. Chen, Asymmetric evaporation for efficient and simultaneous extraction of freshwater, salt, and electrical energy from seawater, ***Energy & Environmental Science***, 17 (2024) 9303-9312.
285. L. Li, N. He, B. Jiang, K. Yu, Q. Zhang, H. Zhang, D. Tang, Y. Song, Highly Salt-Resistant 3D Hydrogel Evaporator for Continuous Solar Desalination via Localized Crystallization, ***Advanced Functional Materials***, 31 (2021).
286. R. Gu, Z. Yu, Y. Sun, Y. Su, W. Wu, S. Cheng, Janus 3D solar crystallizer enabling an eco-friendly zero liquid discharge of high-salinity concentrated seawater with antiscalant, ***Desalination***, 537 (2022) 115862.
287. P. Yuan, C. Men, L. Zhao, P. Cao, Z. Yang, Y. Niu, Y. Zhang, Y. Yu, Q. Li, Spontaneous Salt-Preventing Solar-Thermal Water Evaporator with a High Evaporation Efficiency through Dual-Mode Water Transfer, ***ACS applied materials & interfaces***, 14 (2022) 15549-15557.
288. Q. Xiong, D. Wang, B. Shao, H. Yu, X. Wu, Y. Lu, X. Yang, H. Xu, Unlocking Zero Liquid Discharge: A Parallel Water Supply Strategy to Realize Selective Salt Crystallization for Long-Term Interfacial Solar Evaporation, ***Advanced Functional Materials***, (2025).
289. Y. Xia, Q. Hou, H. Jubaer, Y. Li, Y. Kang, S. Yuan, H. Liu, M.W. Woo, L. Zhang, L. Gao, H. Wang, X. Zhang, Spatially isolating salt crystallisation from water evaporation for continuous solar steam generation and salt harvesting, ***Energy & Environmental Science***, 12 (2019) 1840-1847.
290. J. Gao, L. Zhang, J. You, Z. Ye, Y. Zhong, R. Wang, E.N. Wang, Z. Xu, Extreme salt-resisting multistage solar distillation with thermohaline convection, ***Joule***, 7 (2023) 2274-2290.
291. J. Liu, S. Zhang, J. Wang, Q. Lan, Recent progress in solar-driven interfacial evaporation: Evaporators, condensers, applications and prospects, ***Desalination***, 597 (2025) 118356.
292. S. Zhang, X. Wei, X. Cao, M. Peng, M. Wang, L. Jiang, J. Jin, Solar-driven membrane separation for direct lithium extraction from artificial salt-lake brine, ***Nature communications***, 15 (2024) 238.
293. X. Zhang, Z. Li, J. Liu, F. Xu, L. Zheng, S. De Wolf, Z. Lai, X. Lu, Solar-driven ultrafast lithium extraction from low-grade brine using microfluidics-mediated vortex in scalable electrochemical reactors, ***Chemical Engineering Journal***, 454 (2023) 140074.
294. Q. Xia, Z. Deng, S. Sun, W. Zhao, J. Ding, B. Xi, G. Gao, C. Wang, Solar-enhanced lithium extraction with self-sustaining water recycling from salt-lake brines, ***Proceedings of the National Academy of Sciences of the United States of America***, 121 (2024) e2400159121.
295. Z.H. Foo, D. Rehman, A.T. Bouma, S. Monsalvo, J.H. Lienhard, Lithium Concentration from Salt-Lake Brine by Donnan-Enhanced Nanofiltration, ***Environmental science & technology***, 57 (2023) 6320-6330.
296. Z. Li, I.C. Chen, L. Cao, X. Liu, K.W. Huang, Z. Lai, Lithium extraction from brine through a decoupled and membrane-free electrochemical cell design, ***Science***, 385 (2024) 1438-1444.
297. S. Santoro, M. Aquino, C. Rizza, J. Occhiuzzi, D. Mastrippolito, G. D'Olimpio, A.H. Avci, J. De Santis, V. Paolucci, L. Ottaviano, L. Lozzi, A. Ronen, M. Bar-Sadan, D.S. Han,

A. Politano, E. Curcio, Lithium recovery through WS2 nanofillers-promoted solar photothermal membrane crystallization of LiCl, ***Desalination***, 546 (2023) 116186.

298. Z. Li, C. Li, X. Liu, L. Cao, P. Li, R. Wei, X. Li, D. Guo, K.-W. Huang, Z. Lai, Continuous electrical pumping membrane process for seawater lithium mining, ***Energy & Environmental Science***, 14 (2021) 3152-3159.
299. G.C. Sosso, J. Chen, S.J. Cox, M. Fitzner, P. Pedevilla, A. Zen, A. Michaelides, Crystal Nucleation in Liquids: Open Questions and Future Challenges in Molecular Dynamics Simulations, ***Chemical reviews***, 116 (2016) 7078-7116.
300. J. Farahbakhsh, F. Arshadi, Z. Mofidi, M. Mohseni-Dargah, C. Kök, M. Assefi, A. Soozanipour, M. Zargar, M. Asadnia, Y. Boroumand, V. Presser, A. Razmjou, Direct lithium extraction: A new paradigm for lithium production and resource utilization, ***Desalination***, 575 (2024) 117249.
301. S. Yang, F. Zhang, H. Ding, P. He, H. Zhou, Lithium Metal Extraction from Seawater, ***Joule***, 2 (2018) 1648-1651.
302. X. Chen, M. Yang, S. Zheng, F. Temprano-Coleto, Q. Dong, G. Cheng, N. Yao, H.A. Stone, L. Hu, Z.J. Ren, Spatially separated crystallization for selective lithium extraction from saline water, ***Nature Water***, 1 (2023) 808-817.
303. Y. Song, S. Fang, N. Xu, M. Wang, S. Chen, J. Chen, B. Mi, J. Zhu, Solar transpiration-powered lithium extraction and storage, ***Science***, 385 (2024) 1444-1449.
304. M.L. Vera, W.R. Torres, C.I. Galli, A. Chagnes, V. Flexer, Environmental impact of direct lithium extraction from brines, ***Nature Reviews Earth & Environment***, 4 (2023) 149-165.
305. S. Xu, J. Song, Q. Bi, Q. Chen, W.-M. Zhang, Z. Qian, L. Zhang, S. Xu, N. Tang, T. He, Extraction of lithium from Chinese salt-lake brines by membranes: Design and practice, ***Journal of Membrane Science***, 635 (2021) 119441.
306. C. Liu, P.-C. Hsu, J. Xie, J. Zhao, T. Wu, H. Wang, W. Liu, J. Zhang, S. Chu, Y. Cui, A half-wave rectified alternating current electrochemical method for uranium extraction from seawater, ***Nature Energy***, 2 (2017).
307. H. Li, F. Zhai, D. Gui, X. Wang, C. Wu, D. Zhang, X. Dai, H. Deng, X. Su, J. Diwu, Z. Lin, Z. Chai, S. Wang, Powerful uranium extraction strategy with combined ligand complexation and photocatalytic reduction by postsynthetically modified photoactive metal-organic frameworks, ***Applied Catalysis B: Environmental***, 254 (2019) 47-54.
308. Y. Ye, J. Jin, W. Han, S. Miao, Y. Feng, Z. Qin, X. Tang, C. Li, Y. Chen, F. Chen, Y. Wang, Spontaneous electrochemical uranium extraction from wastewater with net electrical energy production, ***Nature Water***, 1 (2023) 887-898.
309. Y. Xie, Z. Liu, Y. Geng, H. Li, N. Wang, Y. Song, X. Wang, J. Chen, J. Wang, S. Ma, G. Ye, Uranium extraction from seawater: material design, emerging technologies and marine engineering, ***Chemical Society reviews***, 52 (2023) 97-162.
310. N. Wang, X. Zhao, J. Wang, B. Yan, S. Wen, J. Zhang, K. Lin, H. Wang, T. Liu, Z. Liu, C. Ma, J. Li, Y. Yuan, Accelerated Chemical Thermodynamics of Uranium Extraction from Seawater by Plant-Mimetic Transpiration, ***Advanced science***, 8 (2021) e2102250.
311. K. Yu, H. Pan, Y. Jiang, T. Zhang, H. Zhang, F. Ma, H. Song, Y. Yuan, J. Pan, Anti-biological contamination strategies for enhanced uranium extraction from seawater, ***Desalination***, 566 (2023) 116893.
312. N. Li, J. Wu, R. Su, N. Zhang, J. Zhao, Z. Wang, Bioinspired green tea waste/graphene aerogel for solar-enhanced uranium extraction from seawater, ***Desalination***, 545 (2023) 116153.
313. Y. Yuan, B. Niu, Q. Yu, X. Guo, Z. Guo, J. Wen, T. Liu, H. Zhang, N. Wang, Photoinduced Multiple Effects to Enhance Uranium Extraction from Natural Seawater by Black Phosphorus Nanosheets, ***Angewandte Chemie***, 59 (2020) 1220-1227.
314. W.R. Cui, C.R. Zhang, W. Jiang, F.F. Li, R.P. Liang, J. Liu, J.D. Qiu, Regenerable and stable sp(2) carbon-conjugated covalent organic frameworks for selective detection and extraction of uranium, ***Nature communications***, 11 (2020) 436.
315. C.W. Abney, R.T. Mayes, T. Saito, S. Dai, Materials for the Recovery of Uranium from Seawater, ***Chemical reviews***, 117 (2017) 13935-14013.

316. Y. Yuan, Q. Yu, M. Cao, L. Feng, S. Feng, T. Liu, T. Feng, B. Yan, Z. Guo, N. Wang, Selective extraction of uranium from seawater with biofouling-resistant polymeric peptide, ***Nature Sustainability***, 4 (2021) 708-714.
317. M. Keener, C. Hunt, T.G. Carroll, V. Kampel, R. Dobrovetsky, T.W. Hayton, G. Menard, Redox-switchable carboranes for uranium capture and release, ***Nature***, 577 (2020) 652-655.
318. T. Li, X. Lin, Z. Zhang, L. Yang, Y. Qian, L. Fu, S. Zhou, W. Chen, Q. Wang, X. Li, X.Y. Kong, H. Xiao, L. Jiang, L. Wen, Photothermal-Enhanced Uranium Extraction from Seawater: A Biomass Solar Thermal Collector with 3D Ion-Transport Networks, ***Advanced Functional Materials***, 33 (2023).
319. Y. Wang, Z. Lin, H. Zhang, Q. Liu, J. Yu, J. Liu, R. Chen, J. Zhu, J. Wang, Anti-bacterial and super-hydrophilic bamboo charcoal with amidoxime modified for efficient and selective uranium extraction from seawater, ***Journal of colloid and interface science***, 598 (2021) 455-463.
320. Y. Pu, T. Qiang, L. Ren, Anti-biofouling bio-adsorbent with ultrahigh uranium extraction capacity: One uranium resource recycling solution, ***Desalination***, 531 (2022) 115721.
321. L. Yang, H. Xiao, Y. Qian, X. Zhao, X.-Y. Kong, P. Liu, W. Xin, L. Fu, L. Jiang, L. Wen, Bioinspired hierarchical porous membrane for efficient uranium extraction from seawater, ***Nature Sustainability***, 5 (2021) 71-80.
322. S. Xu, K. Zhao, Y. Zhou, K. Zheng, Z. Wang, Z. Yu, N. Cao, X. Liu, Ion pump-inspired biomimetic interfacial evaporation platform for simultaneous seawater desalination, uranium extraction, and electricity generation, ***Nano Energy***, 131 (2024) 110232.
323. Z. Wang, Q. Meng, R. Ma, Z. Wang, Y. Yang, H. Sha, X. Ma, X. Ruan, X. Zou, Y. Yuan, G. Zhu, Constructing an Ion Pathway for Uranium Extraction from Seawater, ***Chem***, 6 (2020) 1683-1691.
324. W.-R. Cui, C.-R. Zhang, R.-P. Liang, J.-D. Qiu, Covalent organic framework hydrogels for synergistic seawater desalination and uranium extraction, ***Journal of Materials Chemistry A***, 9 (2021) 25611-25620.
325. M. Fu, L. Han, C. Huang, L. Ma, Y. Yao, K. Chen, P. Sun, H. Shao, M. Wu, B. Zhang, H. Ma, Janus photothermal adsorbent for solar-powered simultaneous uranium extraction and co-production of freshwater and sea salt from seawater, ***Chemical Engineering Journal***, 477 (2023) 147103.
326. K. Mao, Y. Zhang, S.C. Tan, Functionalizing solar-driven steam generation towards water and energy sustainability, ***Nature Water***, (2025).
327. H.C. Godfray, J.R. Beddington, I.R. Crute, L. Haddad, D. Lawrence, J.F. Muir, J. Pretty, S. Robinson, S.M. Thomas, C. Toulmin, Food security: the challenge of feeding 9 billion people, ***Science***, 327 (2010) 812-818.
328. M. Zou, Y. Zhang, Z. Cai, C. Li, Z. Sun, C. Yu, Z. Dong, L. Wu, Y. Song, 3D Printing a Biomimetic Bridge-Arch Solar Evaporator for Eliminating Salt Accumulation with Desalination and Agricultural Applications, ***Advanced materials***, 33 (2021) e2102443.
329. R.R. Gentry, E.O. Ruff, S.E. Lester, Temporal patterns of adoption of mariculture innovation globally, ***Nature Sustainability***, 2 (2019) 949-956.
330. X. Chang, S. Li, N. Li, S. Wang, J. Li, C. Guo, L. Yu, P. Murto, X. Xu, Marine biomass-derived, hygroscopic and temperature-responsive hydrogel beads for atmospheric water harvesting and solar-powered irrigation, ***Journal of Materials Chemistry A***, 10 (2022) 18170-18184.
331. C. Zhang, s. zhuang, X. Chen, L. Dai, Z. Long, Z. He, Biomass-based photothermal fabrics and superhydrophobic aerogel for self-floating solar evaporators with high energy efficiency in fresh water production from seawater, ***Chemical Engineering Journal***, 502 (2024) 157948.
332. S. Guo, Y. Zhang, H. Qu, M. Li, S. Zhang, J. Yang, X. Zhang, S.C. Tan, Repurposing face mask waste to construct floating photothermal evaporator for autonomous solar ocean farming, ***EcoMat***, 4 (2022).

333. P. Wu, X. Wu, H. Yu, J. Zhao, Y. Wang, K. Pi, G. Owens, H. Xu, An interfacial solar evaporation enabled autonomous double-layered vertical floating solar sea farm, ***Chemical Engineering Journal***, 473 (2023) 145452.
334. I. Banerjee, R.C. Pangule, R.S. Kane, Antifouling coatings: recent developments in the design of surfaces that prevent fouling by proteins, bacteria, and marine organisms, ***Advanced materials***, 23 (2011) 690-718.
335. L. Hou, S. Li, Y. Qi, J. Liu, Z. Cui, X. Liu, Y. Zhang, N. Wang, Y. Zhao, Advancing Efficiency in Solar-Driven Interfacial Evaporation: Strategies and Applications, ***ACS nano***, 19 (2025) 9636-9683.
336. W. Li, Y. Xu, K. Liu, L. Zhu, T. Xu, G. Wang, C. Si, Engineered Biomass-Based Solar Evaporators for Diversified and Sustainable Water Management, ***Advanced materials***, (2025) e2503658.
337. H. Zou, X. Yang, J. Zhu, F. Wang, Z. Zeng, C. Xiang, D. Huang, J. Li, R. Wang, Solar-driven scalable hygroscopic gel for recycling water from passive plant transpiration and soil evaporation, ***Nature Water***, 2 (2024) 663-673.
338. H. Zou, F. Wang, Z. Zeng, J. Zhu, L. Zha, D. Huang, J. Li, R. Wang, Next-generation water-saving strategies for greenhouses using a nexus approach with modern technologies, ***Nature communications***, 16 (2025) 2091.
339. L. Zhang, Z. Yuan, X. Fu, S. Shi, X. Chen, P. Chen, D. Ye, Biomass-Derived Gradient and Aligned Structured Aerogel for Sustainable Agricultural Irrigation, ***Nano letters***, 25 (2025) 5383-5390.
340. Y. Guo, J. Bae, Z. Fang, P. Li, F. Zhao, G. Yu, Hydrogels and Hydrogel-Derived Materials for Energy and Water Sustainability, ***Chemical reviews***, 120 (2020) 7642-7707.
341. C. Wang, H. Zou, S. Du, D. Huang, R. Wang, Water and heat recovery for greenhouses in cold climates using a solid sorption system, ***Energy***, 270 (2023) 126919.
342. J. Yang, X. Zhang, H. Qu, Z.G. Yu, Y. Zhang, T.J. Eey, Y.W. Zhang, S.C. Tan, A Moisture-Hungry Copper Complex Harvesting Air Moisture for Potable Water and Autonomous Urban Agriculture, ***Advanced materials***, 32 (2020) e2002936.
343. A. Entezari, O.C. Esan, X. Yan, R. Wang, L. An, Sorption-Based Atmospheric Water Harvesting: Materials, Components, Systems, and Applications, ***Advanced materials***, 35 (2023) e2210957.
344. R. Li, M. Wu, S. Aleid, C. Zhang, W. Wang, P. Wang, An integrated solar-driven system produces electricity with fresh water and crops in arid regions, ***Cell Reports Physical Science***, 3 (2022) 100781.
345. Y. Xu, C. Dang, X.E. Cao, Y. Cao, J. Huang, Y. Xu, M. Shan, R. Liu, P. Li, G. Xu, M. Zhu, Artificial phytoremediation solar interface evaporator for efficient heavy metal salt separation and saline soil remediation, ***Journal of Environmental Chemical Engineering***, 12 (2024) 113114.
346. Y. Chen, J. Yang, L. Zhu, S. Wang, X. Jia, Y. Li, D. Shao, L. Feng, H. Song, Marangoni-driven biomimetic salt secretion evaporator, ***Desalination***, 548 (2023) 116287.
347. P. Wu, X. Wu, H. Xu, G. Owens, Interfacial solar evaporation driven lead removal from a contaminated soil, ***EcoMat***, 3 (2021).
348. P. Wu, X. Wu, Y. Wang, J. Zhao, H. Xu, G. Owens, Boosting extraction of Pb in contaminated soil via interfacial solar evaporation of multifunctional sponge, ***Green Energy & Environment***, 8 (2023) 1459-1468.
349. P. Wu, X. Wu, Y. Wang, H. Xu, G. Owens, A biomimetic interfacial solar evaporator for heavy metal soil remediation, ***Chemical Engineering Journal***, 435 (2022) 134793.
350. K. Liu, Y. Li, P. Wu, Y. Guo, Y. Ye, J. Jiang, J. Hu, X. Yang, Y. Shi, G. Owens, K. Pi, Cotton-based bionic tree-shaped photothermal evaporator for extraction of heavy metals from river sediment, ***Journal of Environmental Chemical Engineering***, 11 (2023) 111063.
351. K. Liu, P. Wu, Y. Ye, J. Jiang, Y. Guo, Y. Li, X. Yang, Y. Shi, G. Owens, K. Pi, Interfacial solar evaporation driven Cd removal from contaminated sediments, ***Journal of Environmental Chemical Engineering***, 12 (2024) 112920.

352. P. Wu, X. Wu, Y. Wang, H. Xu, G. Owens, Towards sustainable saline agriculture: Interfacial solar evaporation for simultaneous seawater desalination and saline soil remediation, ***Water research***, 212 (2022) 118099.
353. P. Wu, D. Zhou, Y. Li, X. Yang, Y. Shi, X. Shi, G. Owens, K. Pi, A bioinspired photothermal evaporator for enhanced salt separation during saline soil remediation, ***Desalination***, 602 (2025) 118647.
354. M. Jelusic, D. Vodnik, I. Macek, D. Lestan, Effect of EDTA washing of metal polluted garden soils. Part II: Can remediated soil be used as a plant substrate?, ***The Science of the total environment***, 475 (2014) 142-152.
355. H.W. Lim, S.H. Park, S.J. Lee, 3D thermoresponsive hydrogel with enhanced water uptake and active evaporation for effective interfacial solar steam generation, ***Desalination***, 550 (2023) 116368.
356. X. Jin, Y. Liang, Y. Liu, H. Ye, R. Yang, Y. Wu, W. Zhao, C. Sonne, C. Xia, Hierarchical Self-Assembly of Fe3O4 and TiO2 on the Surface of Wood Sawdust for Flowthrough Antibiotic Removal, ***Advanced Sustainable Systems***, (2025).

Chapter 7
Summary and Outlook

Solar evaporation has emerged as a versatile solution for addressing the water-energy-resource nexus. Significant progress has been made in the development of photothermal and substrate materials, along with various structural engineering strategies focused on light absorption, water transportation, thermal insulation, and interfacial properties. These advancements are applied in clean water production, sustainable energy generation, and valuable resource recovery [1–5]. In the domain of clean water, technologies such as solar desalination, pollutant degradation, and antibacterial disinfection have demonstrated substantial potential, providing scalable and cost-effective solutions to water scarcity and contamination. For energy sustainability, this approach supports energy storage, electricity generation, and fuel production, thus enriching the renewable energy landscape. Moreover, the extraction of resources like salt and lithium, as well as augmented food production via solar-driven processes, highlights its transformative impact on resource management and agricultural practices. Despite the current exciting achievements, further research and technological advancements are essential to optimize efficiency, reduce costs, and expand the applicability of solar evaporation across diverse global contexts:

A deeper understanding of light transfer, water transport, thermal diffusion behaviors and kinetics, and their corresponding structure-property relationships is critical for optimizing evaporator designs [6, 7]. The influence of pore structure, particularly pore size and tortuosity, and surface properties on water transport behavior remains to be explored. Additionally, a more fundamental understanding of thermal diffusion behaviors, including conduction, radiation, and convection, is needed [8, 9]. The atomistic interaction mechanisms between functional materials and water molecules, such as dynamic adsorption/desorption of water molecules/ions on solid surfaces and their coupling with interfacial charges, are poorly understood. Combining simulation with experimental observations can provide valuable insights into the basic behaviors and kinetics of light, water, vapor, and thermal-related processes [10]. Real-time monitoring through in situ characterization

G. Liu, T. Chen, *Solar Evaporation for Water-Energy-Resource Nexus*,
https://doi.org/10.1007/978-981-95-3815-7_7

technologies is also desirable for gaining fundamental insights into these underlying mechanisms. Developing advanced characterization techniques, such as in situ electrochemical measurements with broader length scales and higher resolution, will further probe the physical and chemical processes at the water/solid interface.

From a materials and systems perspective, challenges include maintaining efficient vapor release and energy generation while utilizing low-cost materials during scaling. Innovations in material reinforcement, surface modifications for antifouling properties, and structural designs that facilitate non-contact configurations and dynamic salt management are essential [11, 12]. For instance, the strategies of nanocomposite reinforcement and surface modification are expected to impart high mechanical properties and antifouling ability. Nanotechnology will continue to play a crucial role by offering diverse and powerful capabilities in material and structural design and engineering. Advancements in smart material and structural design are essential to further enhance this technology. Structural designs that enable photothermal materials with directional or dynamic salt precipitation/exfoliation and noncontact configurations can potentially extend material durability, lifespan, and reduce costs. Hybridization at both material and structural levels will be another strategy to introduce new features, functions, or improved performance. Additionally, system designs that harness other energy sources, such as natural convection, Joule heating, and thermal energy stored in bulk water, offer promising solutions for stabilizing performance amid fluctuating solar conditions. Emulating structures and behaviors found in nature, such as those in plants, animals, or other natural phenomena, represents a promising direction for future solar evaporator designs.

The simulation and modeling approaches for solar evaporation are mainly focused on both macroscopic and microscopic scales [13]. Macroscopic simulations utilize software like COMSOL Multiphysics and ANSYS to model phenomena such as temperature profiles, humidity levels, salt concentration, and water movement during evaporation. These simulations aim to analyze the dynamic equilibrium among water transport, light absorption, and thermal management throughout the evaporation process. Finite element methods employing FDTD are extensively used in areas such as light absorption and radiation, while DFT simulations are prevalent in material computation. MD simulations are widely implemented to study molecular motion. However, the mesoscale methods and multi-scale (coupling of macroscopic- esoscopic-microscopic method) numerical simulation methods and AI techniques are notably underutilized. The adoption of mesoscale techniques like the LBM and the development of multiscale numerical simulations represent significant progress, potentially enhancing the role of numerical simulations in solar evaporation systems. These methodologies enable the analysis of how the behavior and state of microscopic particles influence the properties of macroscopic models, assisting in understanding the variations in physical processes across different scales during evaporation and collection. Further research using AI models on data frameworks is essential for developing smart multifunctional materials and systems.

In clean water production, light energy has harnessed thermal and hydrochemical fields to enhance water evaporation and degradation. This approach has the potential to revolutionize low-energy water production in water-scarce regions and improve decentralized water sanitation and health [14–16]. Integrating multi-stage solar interfacial evaporation with photovoltaics can lead to the development of low-energy desalination or irrigation systems on offshore platforms, offering alternatives to energy-intensive water purification methods like reverse osmosis and addressing freshwater resource shortages. However, this complex integrated system must balance the cost of water productivity and energy recovery equipment. Photothermal catalysis targets water quality, through heat effect and ROS can degrade pollutants in condensate water and sterilization, ensuring water quality for small community-scale. This distributed system will reduce the energy consumption of long-distance water supply networks, and can be adapted to local conditions. However, in complex water quality environments, micron-sized contact areas limit the extent of photodegradation. Systems assisted by oxidizing agents use active particles to initiate a chain reaction of photodegradation below the interface, enhancing degradation efficiency in bulk water. This advancement holds promise for achieving rapid unconventional water purification, such as for municipal sewage and industrial wastewater. Nonetheless, there have been few evaluations of the system factors affecting photodegradation. Anions in water, such as Cl^-, SO_4^{2-} and CO_3^{2-}, can create complex holes, slow down reaction rates, and form halogenated by-products, all of which degrade water quality. Future research should explore the stability of photocatalysts to better understand mechanism synergies.

For electricity generation, optimizing the balance between vapor generation and electrical output through innovative solar absorber designs is necessary [17, 18]. Thus, designing solar absorbers or devices with dynamic configurations is crucial to adjust evaporation and electricity output according to varying water and electricity demands. For electricity generated via salinity gradients, consistent and ample vapor production enhances the salinity gradient, facilitating ion diffusion and boosting electricity output. However, this can result in salt deposition on solar absorbers, diminishing vapor generation rates. Therefore, incorporating antifouling solar absorbers into cogeneration systems is essential to prevent salt crystallization, ensuring high electricity output and consistent vapor generation. In fuel production, integrated systems that combine interfacial solar evaporation and photocatalysis produce at least two gases (water vapor and hydrogen) simultaneously. Efficient separation and purification of specific gases present key challenges, necessitating complex designs for the separation, collection, and purification of clean water and hydrogen.

In resource extraction, improving the selectivity and efficiency of material recovery processes during solar evaporation is imperative. Developing new materials and methods for selective resource recovery and enhancing the microscopic understanding of energy and mass transfer will contribute to better system performance [19–22]. For example, while hydrophobic surfaces can prevent salt deposition, they hinder heat conduction to the water at the evaporation interface due to the gas barrier, thus reducing solar steam generation efficiency. Although initial attempts at selective metal resource recovery have shown promise, related studies remain limited. Consequently, further research on new materials, devices, and methods for selective resource recovery during solar evaporation is needed. Additionally, there is a strong need to understand microscopic energy and mass transport and distribution in various systems, such as salt harvesting and lithium and uranium extraction. The in-depth understanding helps to construct well-contacted transfer interfaces and optimize energy/mass management to enhance the solar energy utilization efficiency. These efforts also contribute to the development of advanced new materials, mechanisms, and strategies for further improvement of these systems.

Integrating new functionalities into solar evaporation to further broaden the applications remains a primary research direction. Promising examples include integrating the systems with cascade solar-driven seawater desalination and green hydrogen production, as well as employing it in the photothermal catalytic transformation of CO_2 [23, 24]. Zero liquid discharge systems are increasingly essential to comply with stricter regulations on brine discharge. Solar-driven zero liquid discharge desalination offers a promising and sustainable solution for cost-effective treatment of brine or hypersaline water [2, 25–27]. Integrating this technology with fields such as nanofluidics, thermoelectricity, rechargeable batteries, catalysis, solar fuel production, and radiative cooling will enhance its potential and applications. With abundant and renewable solar energy and diverse water sources, this interdisciplinary approach is set to expand its impact, establishing it as a key technology for addressing global challenges in the coming decade.

Acknowledgements This research was supported by the Natural Science Foundation of China (52376143).

Declaration of Competing Interest The author declare that they have no known competing financial interests or personal relationships that could have appeared to influence the work reported in this paper.

References

1. X. Wu, Y. Lu, X. Ren, P. Wu, D. Chu, X. Yang, H. Xu, Interfacial Solar Evaporation: From Fundamental Research to Applications, ***Advanced materials***, 36 (2024) e2313090.
2. J. Yao, Q. Zhong, J. Zhang, J. Zhao, Z. Wang, Interfacial solar evaporation for zero liquid discharge desalination, ***Communications Materials***, 5 (2024).
3. J. Hu, Y. Sun, Z. Liu, B. Zhu, L. Zhang, N. Xu, M. Zhu, J. Zhu, Z. Chen, Photothermal fabrics for solar-driven seawater desalination, ***Progress in Materials Science***, 150 (2025) 101407.
4. H.T. Kim, L. Philip, A. McDonagh, M. Johir, J. Ren, H.K. Shon, L.D. Tijing, Recent Advances in High-Rate Solar-Driven Interfacial Evaporation, ***Advanced science***, 11 (2024) e2401322.
5. J. Liu, S. Zhang, J. Wang, Q. Lan, Recent progress in solar-driven interfacial evaporation: Evaporators, condensers, applications and prospects, ***Desalination***, 597 (2025) 118356.
6. T.L. Bergman, Lavine, Adrienne S., Fundamentals of Heat and Mass Transfer, ***John Wiley & Sons Inc***, ISBN 9781119330103 (2017).
7. J.R. Howell, M.P. Mengüç, K. Daun, R. Siegel, Thermal radiation heat transfer, ***CRC Press***, 7th Edition (2020).

8. M. Gao, L. Zhu, C.K. Peh, G.W. Ho, Solar absorber material and system designs for photothermal water vaporization towards clean water and energy production, ***Energy & Environmental Science***, 12 (2019) 841-864.
9. B. Wang, X. Huang, Z. Liu, J. Zhang, C. Wei, B. Cai, P. Xie, A. Zhou, Solar interfacial evaporation systems with multi-field synergies boost water purification and blue energy harvesting technologies, ***Energy & Environmental Science***, 17 (2024) 7600-7626.
10. G. Liu, Thermoplasmonics, ***Springer***, ISBN 978-981-97-8331-1 (2024).
11. X. Yu, S. Fan, B. Zhu, S.I. El-Hout, J. Zhang, C. Chen, Recent progress on photothermal nanomaterials: Design, mechanism, and applications, ***Green Energy & Environment***, (2024).
12. X. Cui, Q. Ruan, X. Zhuo, X. Xia, J. Hu, R. Fu, Y. Li, J. Wang, H. Xu, Photothermal Nanomaterials: A Powerful Light-to-Heat Converter, ***Chemical reviews***, 123 (2023) 6891-6952.
13. Y. Wei, Y. Yang, Q. Zhao, Y. Ma, M. Qiang, L. Fu, Y. Liu, J. Zhang, Z. Qu, W. Que, Numerical Simulation Technologies in Solar-Driven Interfacial Evaporation Processes, ***Small***, 20 (2024) e2312241.
14. P. Zhang, Q. Liao, H. Yao, Y. Huang, H. Cheng, L. Qu, Direct solar steam generation system for clean water production, ***Energy Storage Materials***, 18 (2019) 429-446.
15. J. Wu, G. Yin, J. Liu, Z.Z. Yu, X. Li, Multifunctional solar-driven interfacial evaporation system for simultaneous clean water production and high-value-added ion extraction, ***Materials horizons***, 12 (2025) 2878-2898.
16. F. He, X. Wu, J. Gao, Z. Wang, Solar-driven interfacial evaporation toward clean water production: burgeoning materials, concepts and technologies, ***Journal of Materials Chemistry A***, 9 (2021) 27121-27139.
17. B. Shao, Y. Song, Z. Song, Y. Wang, Y. Wang, R. Liu, B. Sun, Electricity Generation from Phase Transitions between Liquid and Gaseous Water, ***Advanced Energy Materials***, 13 (2023).
18. J. Wang, X. Cao, X. Cui, H. Wang, H. Zhang, K. Wang, X. Li, Z. Li, Y. Zhou, Recent Advances of Green Electricity Generation: Potential in Solar Interfacial Evaporation System, ***Advanced materials***, 36 (2024) e2311151.
19. Y. Song, S. Fang, N. Xu, J. Zhu, Solar-driven interfacial evaporation technologies for food, energy and water, ***Nature Reviews Clean Technology***, 1 (2025) 55-74.
20. S. Yang, F. Zhang, H. Ding, P. He, H. Zhou, Lithium Metal Extraction from Seawater, ***Joule***, 2 (2018) 1648-1651.
21. X. Chen, W. Yu, Y. Zhang, C. Huang, L. Nie, J. Yu, Y. Zhang, C. Zhang, W. Zhai, X. Zhang, Y. Yu, W. Liu, Solar-Driven Lithium Extraction by a Floating Felt, ***Advanced Functional Materials***, 34 (2024).
22. Y. Kong, Y. Gao, Y. Shang, W. Kong, Y. Qi, S. Wang, F. Yin, B. Gao, S. Wang, Q. Yue, Synergistic adjustment of water channels and light absorption pathways to co-generate salt collection and clean water production, ***The Science of the total environment***, 797 (2021) 148912.
23. S.I. In, D.D. Vaughn, 2nd, R.E. Schaak, Hybrid CuO-TiO(2-x)N(x) hollow nanocubes for photocatalytic conversion of CO2 into methane under solar irradiation, ***Angewandte Chemie***, 51 (2012) 3915-3918.
24. Q. Zhu, Y. Xuan, D. Zhao, H. Yu, J. Wang, X. Liu, A lotus-inspired triphase continuous CO2 photo-thermal reduction system with wide applicability and tunable interface environment, ***Materials Today Energy***, 48 (2025) 101777.
25. Y. Li, S. Cheng, Z. Yu, R. Gu, X. He, Stable, zero liquid discharge, and highly efficient solar-driven multistage distillation device based on tree-inspired radial water transfer, ***Journal of Cleaner Production***, 375 (2022) 134025.
26. C. Finnerty, L. Zhang, D.L. Sedlak, K.L. Nelson, B. Mi, Synthetic Graphene Oxide Leaf for Solar Desalination with Zero Liquid Discharge, ***Environmental science & technology***, 51 (2017) 11701-11709.
27. Q. Xiong, D. Wang, B. Shao, H. Yu, X. Wu, Y. Lu, X. Yang, H. Xu, Unlocking Zero Liquid Discharge: A Parallel Water Supply Strategy to Realize Selective Salt Crystallization for Long-Term Interfacial Solar Evaporation, ***Advanced Functional Materials***, (2025).

MIX
Papier aus verantwortungsvollen Quellen
Paper from responsible sources
FSC® C105338

If you have any concerns about our products,
you can contact us on
ProductSafety@springernature.com

In case Publisher is established outside the EU,
the EU authorized representative is:
Springer Nature Customer Service Center GmbH
Europaplatz 3, 69115 Heidelberg, Germany

Printed by Libri Plureos GmbH
in Hamburg, Germany